Protozoa and Human Disease

Protozoa and Human Disease

Mark F. Wiser

LONDON AND NEW YORK

Garland Science
Vice President: Denise Schanck
Editor: Elizabeth Owen
Editorial Assistant: Sarah Holland
Production Editor: Georgina Lucas
Assistant Production Editor: Ioana Moldovan
Illustrator: Oxford Designers & Illustrators
Cover Design: Andrew Magee
Copyeditor: Sally Huish
Typesetting: EJ Publishing Services
Proofreader: Sally Livitt
Indexer: Merrall-Ross International Ltd.

First published 2011 by Garland Science

2 Park Square, Milton Park, Abingdon, Oxfordshire OX14 4RN
52 Vanderbilt Avenue, New York, NY 10017

Routledge is an imprint of the Taylor & Francis Group, an informa business

First issued in paperback 2019

ISBN 978-0-8153-6500-6 (pbk)

Library of Congress Cataloging-in-Publication Data

Wiser, Mark F.
Protozoa and human disease / Mark F. Wiser.
p. ; cm.
Includes bibliographical references.
ISBN 978-0-8153-6500-6 (alk. paper)
1. Protozoan diseases. I. Title.
[DNLM: 1. Protozoan Infections. 2. Eukaryota--pathogenicity.
WC 700]
RC118.7.W57 2011
362.196'936--dc22
2010040729

Front cover images courtesy of Marlene Benchimol, Universidade Santa Ursula, Rio de Janeiro, Brazil (top image, *Giardia* trophozoites); Vince Lovko, Environmental and Aquatic Animal Health, Virginia Institute of Marine Science (middle left image, *Pfiesteria shumwayae*); David Bygott, Tucson, Arizona, USA (bottom right image, blood-feeding tsetse fly).

Mark Wiser is currently on the faculty of the Department of Tropical Medicine at the Tulane University School of Public Health and Tropical Medicine in New Orleans, Louisiana. He is particularly interested in the molecular and cellular biology of protozoan parasites and host–parasite interactions. He earned a B.A. degree in biology from Franklin College in Franklin, Indiana and a Ph.D. degree in cell biology from the University of Minnesota in Minneapolis. Between finishing his formal studies and his current position he carried out post-doctoral research at the Max Planck Institute for Cell Biology in Ladenburg, Germany and had a research faculty position in molecular parasitology at Yale University School of Medicine in New Haven, Connecticut.

Preface

Protozoa and Human Disease is an overview of protozoa which affect human health and the diseases they cause. It is modeled on a course entitled Medical Protozoology which is taught by the author at the Tulane University School of Public Health and Tropical Medicine. The book, as well as the course, strives to pull together the biological aspects of medically important protozoa, the clinical features and pathologies associated with the diseases they cause, and the public health issues involved in their epidemiology and control. To accomplish these goals, life cycles of the various protozoa and pathogeneses of the corresponding diseases are discussed in detail, including molecular and cellular aspects.

This book will be of particular interest to parasitology and microbiology students and can serve as a textbook for medical protozoology courses at an advanced undergraduate, graduate, or professional level. Students in many of the subdisciplines of biology, as well as medical and public health students, will also be interested in the book as a comprehensive reference on medically important protozoa and the diseases they cause. In addition to the basics such as morphological features, life cycles, and the clinical manifestations of the diseases, topics such as the molecular and immunological basis of pathogenesis, metabolic pathways, specialized subcellular structures, ecology of disease transmission, antigenic variation, and molecular epidemiology are discussed for many of the protozoan pathogens. Thus, the book blends classical and medical parasitology with the more modern disciplines.

Each organism is discussed along a taxonomic scheme, and the various pathogens are discussed in reference to the organ systems affected. The life cycle of an organism is clearly fundamental to its biology but also describes host-parasite interactions, which in turn influence pathology. Similarly, embedded in the life cycle is the mechanism of disease transmission which can explain its epidemiology and links to public health. The protozoa are an extremely diverse group and individual protozoa often exhibit some unique biological, pathological, or epidemiological feature which is also included. As the focus of the book is on medically important protozoa pathology, clinical aspects of the diseases they cause and treatment strategies are also covered.

Following an introductory chapter on protozoa is a chapter on the general features of protozoa that inhabit the gastro-intestinal tract and other luminal organs. This is followed by chapters on the individual intestinal protozoa, except for the apicomplexa. The next section consists of a chapter on the general features of kinetoplastids followed by three chapters on the three medically important kinetoplastids. In similar fashion, this is followed by a chapter on apicomplexa in general and individual chapters on the various apicomplexa that infect humans. And finally there is a chapter on free-living protozoa that affect human health.

Many people have contributed directly or indirectly to the medical protozoology course and this book, and I wish I could individually name them all. Of particular note is Antonio D'Alessandro who co-taught Medical Protozoology with me for several years before retiring. In the early days we were referred to as the 'tag team' with me lecturing on the basic biology and more modern aspects and Antonio covering the more clinical aspects. Our distinct personalities and approaches were a nice blend that presented students with an enjoyable learning experience. Hearing his lectures over the years certainly made the transition to teaching the entire course much easier for me and increased my knowledge about medicine. In this same vein H. Norbert Lanners was also an early participant in this course and helped pave the way to include more cell biology. In addition, Norbert has served as an invaluable resource about the cell biology of protozoa over the years. I also acknowledge all of the students over the past two decades who have served as an inspiration for improving the course as well as for writing this book. In particular, the students during 2008 and 2009 read through drafts of the chapters and provided extremely insightful commentary, resulting in a much improved product. Similarly, the expert reviewers quite often corrected errors and pointed to more recent knowledge in the field. I also thank the staff at Garland Science for their courteous help and patience with me throughout this project.

Acknowledgments

The author and publishers would like to thank the following expert reviewers for their suggestions in preparing this edition.

Michael P. Barrett (University of Glasgow, UK); Jeffrey K. Beetham (Iowa State University, USA); Giancarlo Biagini (Liverpool School of Tropical Medicine, UK); Claudia I. Calderon (Universidad de San Carlos de Guatemala, Guatemala); Michael F. Dolan (University of Massachusetts, USA); Rebecca J. Gast (Woods Hole Oceanographic Institution, USA); Jeremy Gray (University College Dublin, Ireland); Linda A. Hufnagel (University of Rhode Island, USA); Naveed Khan (University of Nottingham, UK); Michael Klemba (Virginia Polytechnic Institute and State University, USA); Silvia N.J. Moreno (University of Georgia, USA); William J. Sullivan Jr. (Indiana University School of Medicine, USA); Jonathan Wastling (University of Liverpool, UK); Louis M. Weiss (Albert Einstein College of Medicine, USA); Nigel Yarlett (Pace University, USA).

Contents in Brief

Contents in Detail

Introduction to Medical Protozoology

Many important human and animal diseases are caused by protozoan pathogens (Table 1.1). These diseases impact humans directly in terms of morbidity and mortality as well as having a major impact on agriculture. In addition to being known parasites, some protozoa are considered opportunistic pathogens that primarily affect immunocompromised individuals. In many parts of the world diseases caused by protozoa are on the rise and thus considered as emerging or re-emerging diseases. In addition, these diseases are often neglected because they are found predominantly in developing countries. A wide range of disease manifestations are exhibited by protozoan pathogens reflecting the variety of organs and tissues affected. Most protozoan infections are acquired through the ingestion of specialized infective stages and several protozoan species are transmitted by arthropod vectors.

What Are Protozoa?

Protozoa are not easily defined because they are diverse and are often only distantly related to each other (Figure 1.1). Furthermore, many protozoa exhibit similarities to other eukaryotes and the classification of protozoa is often controversial. Because of the extreme diversity of the protozoa the only feature common to all protozoa is that they are unicellular eukaryotic microorganisms. Thus, protozoa possess typical eukaryotic organelles and in general exhibit the typical features of other eukaryotic cells. For example, a membrane-bound nucleus containing the chromosomes is found

Table 1.1 Major protozoan pathogens infecting humans

Protozoan	Disease	Transmission	Importance
Cryptosporidium species	Acute watery diarrhea	Water-borne outbreaks	Prevalence ranging from 1 to 10%; life threatening in AIDS patients
Entamoeba histolytica	Amebic dysentery	Contaminated food or water	An estimated 50 million infections with 100 000 deaths per year
Giardia lamblia	Diarrhea	Contaminated food or water	Most frequently found protozoa in human stools causing 300 million cases per year
Leishmania species	Cutaneous or visceral disease	Sand flies	1.5–2 million new cases and 70 000 deaths per year
Plasmodium species	Malaria, acute febrile disease	Mosquitoes	40% of world's population at risk; 300–500 million clinical cases with 1–1.5 million deaths per year
Toxoplasma gondii	Birth defects or encephalitis	Ingestion of undercooked meat	Seroprevalence ranging from 6 to 75%
Trichomonas vaginalis	Inflammation of urogenital tract	Sexually transmitted disease (STD)	Most common nonviral STD with 174 million new cases per year
Trypanosoma cruzi	Chagas' disease	Triatomine bug	Major cause of cardiac disease in South and Central America
Trypanosoma gambiense	African sleeping sickness	Tsetse fly	70 000–80 000 new cases with 30 000 deaths per year

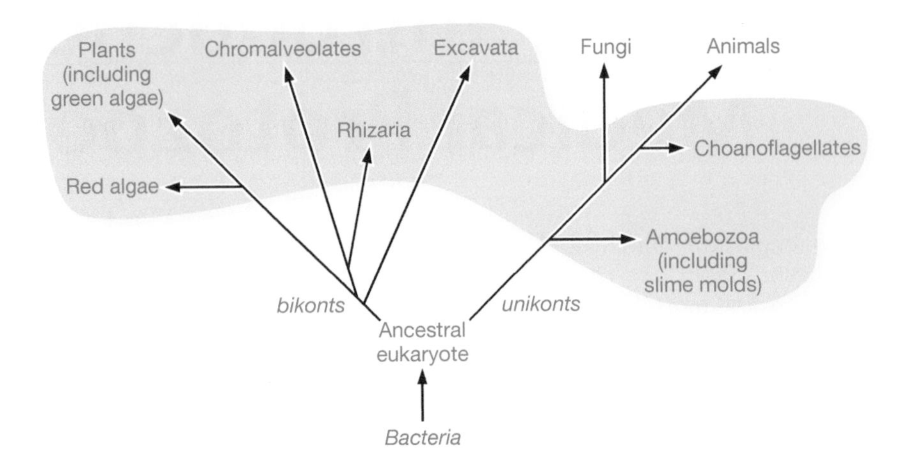

Figure 1.1 Possible evolutionary relationships among organisms. Schematic diagram showing possible evolutionary relationships between living organisms. Shaded area includes groups that contain, have contained, or could contain organisms considered to be protozoa.

in all protozoan species. However, in many protozoan species some of the organelles may be absent, or are morphologically or functionally different from those found in other eukaryotes. In addition, many of the protozoa have organelles that are unique to a particular group of protozoa.

Most protozoa exhibit some form of motility and most are **heterotrophs** in that they must obtain organic material—originating from other living organisms—from the environment as a source of nutrients and energy. This heterotrophic behavior includes absorbing solutes (osmotrophy), pinocytosis (endocytosis of fluid), phagocytosis of particles or other organisms, and predation on bacteria and other protozoa. Historically their motility and heterotrophism led to the protozoa being grouped with animals. In fact the word *protozoan*, first coined by Goldfuss in 1818, literally means first animal and the ancestral eukaryote probably looked somewhat like a protozoan (Box 1.1). However, many protozoa contain plastids (e.g., chloroplasts) and are photosynthetic, or autotrophic, and have been claimed by botanists and phycologists as algae. In some cases the protozoan may still possess a plastid, but it no longer functions in photosynthesis and carries out some other metabolic function. In addition, some protozoan species are mixotrophic in that they are both autotrophic and heterotrophic. Indeed, the separation of unicellular algae from the protozoa in many cases is somewhat superficial in that autotrophic and heterotrophic species can be derived from a common ancestor and are closely related. To deal with these problems, some taxonomy schemes group the traditional protozoa, unicellular algae, diatoms, fungi, and slime molds together into the **protista** (or protoctista). However, protists are probably even more of an artificial taxonomic group than protozoa (see below) and such a classification does not really help to define the protozoa (or the protists) beyond being unicellular eukaryotes.

Protozoa are unicellular, but complex and diverse

Protozoa exhibit an enormous range of morphologies (Figure 1.2) and generally range in size from 10 to 200 μm (microns or 10^{-6} meters). Some protozoa have very well-defined shapes, whereas others, particularly the amebas, can constantly change their shape. Although protozoa are rather small and unicellular they should not be considered as simple organisms. Multicellular animals and plants exhibit complexity through many different types of specialized cells. As a unicellular organism, a protozoan must carry out the functions of an entire organism within the context of a single cell. Thus protozoan cells are often more complex than individual cells from multicellular organisms. This complexity is manifested through specialized organelles and elaborate cytoskeletal elements. In addition, any particular protozoan species may exhibit different morphologies at different times within its life cycle or in response to changing environments. For example

Box 1.1 Origin of eukaryotic cells

The evolutionary jump from prokaryotes to eukaryotes is substantial and there are an estimated 5000 genes found in eukaryotes that are not found in prokaryotes [1]. The first eukaryotic organisms likely resembled what we would now classify as protozoa. However, probably no direct descendants of the early eukaryotes currently exist and the path from prokaryotes to eukaryotes is likely littered with several missing links and extinct branches. Two key events in the evolution of eukaryotes are probably the development of internal cytoskeletal system and phagotrophy (Figure 1A).

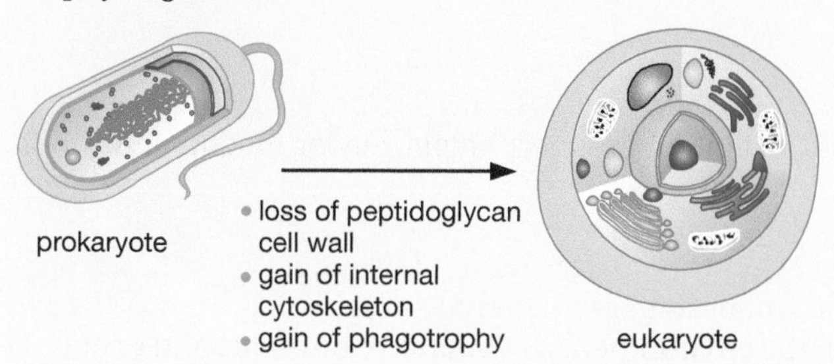

Figure 1A Major events in eukaryote evolution. The major early steps in the evolution of eukaryotes from prokaryotes were likely the loss of the rigid peptidoglycan cell wall and the development of an internal cytoskeletal system and phagocytosis.

A major difference between eukaryotes and prokaryotes is the rigid murein, or peptidoglycan, cell wall of bacteria. Although, many eukaryotes exhibit cell walls, they are fundamentally different than those of bacteria and are likely derived secondarily after the formation of the eukaryotes. For example, cell walls of plant cells are composed primarily of cellulose and fungal cell walls and cyst walls in many protozoa are composed primarily of chitin. The rigidity and strength provided by cell walls protects cells against mechanical stress and confers osmotic stability. Thus, coincident with the loss of the murein cell wall in the early eukaryote was the development of endocytoskeletal systems based on actin and tubulin to compensate for the relative fragility of a naked plasma membrane. In addition, the early eukaryotes likely evolved in a benthic environment and in particular in dense marine microbial mats. The organic solutes excreted by the bacteria and the high salt concentration of such an environment would have had an osmoprotective effect. The dense microbial mats would also provide an environment in which potential prey was abundant and phagocytosis could have evolved without efficient cellular motility.

The ancestral eukaryote probably attached to prey via cell surface N-linked glycoproteins that evolved after the loss of the murein cell wall and then digested them externally by secreting hydrolytic enzymes. To make the process more efficient the primitive predators could have progressively surrounded their prey until the prey were entirely engulfed in host membranes with digestion being carried out internally within phagosomes (Figure 1B). This endocytic mechanism would also require the development of an exocytic mechanism to maintain the surface area of the plasma membrane. Thus, the evolution of phagocytosis not only included an expansion of the function of the actin cytoskeleton but also drove the development of membrane trafficking and subcellular compartments including the nucleus.

The evolution of phagotrophy and endomembranes not only opened up a new way of eating, but also allowed for endosymbiosis and the eventual enslavement of mitochondria. An early eukaryote probably engulfed an α-proteobacterium and formed a symbiotic relationship which evolved into an assimilation of the aerobic metabolism of this organism and the development of the mitochondrion. This enslavement of the mitochondrion was probably another key event in the evolution of the eukaryotes because extant eukaryotes are derived from mitochondria-containing ancestors. Extant anaerobic eukaryotes without mitochondria either have organelles derived from mitochondria, such as mitosomes and hydrogenosomes, or at least have nuclear genes encoding remnant mitochondrial genes. Presumably the amitochondrial eukaryotic ancestor was at a competitive disadvantage and did not survive. Later a cyanobacterium was enslaved by the progenitor of plants and developed into the chloroplasts and similarly these plant predecessors were engulfed and enslaved to form the plastids found in many other algae and protozoa.

The cytoskeleton continued to evolve and expand its functions and in particular began to play a role in cellular motility. Most notably tubulin became a key component in flagella and later myosin evolved to allow for the development of ameboid movement and gliding motility. This motility allowed these early eukaryotes to expand into other niches (e.g., planktonic, terrestrial) and thus this aerobic, phagotrophic and flagellated ancestral eukaryote could radiate into the diverse array of protozoa, and ultimately the other eukaryotes, that currently exist.

1. Cavalier-Smith, T. (2009) Predation and the eukaryote cell origins: a co-evolutionary perspective. *Int. J. Biochem. Cell Biol.* 41: 307–322.

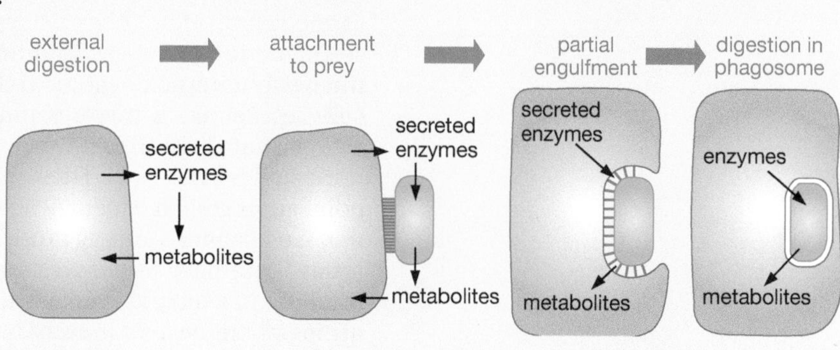

Figure 1B Development of phagotrophy. Phagotrophy may have developed from an original metabolism based on external digestion by attaching to prey or food particles and then gradually engulfing the particles to increase the efficiency of metabolite recovery until the particles were being fully internalized.

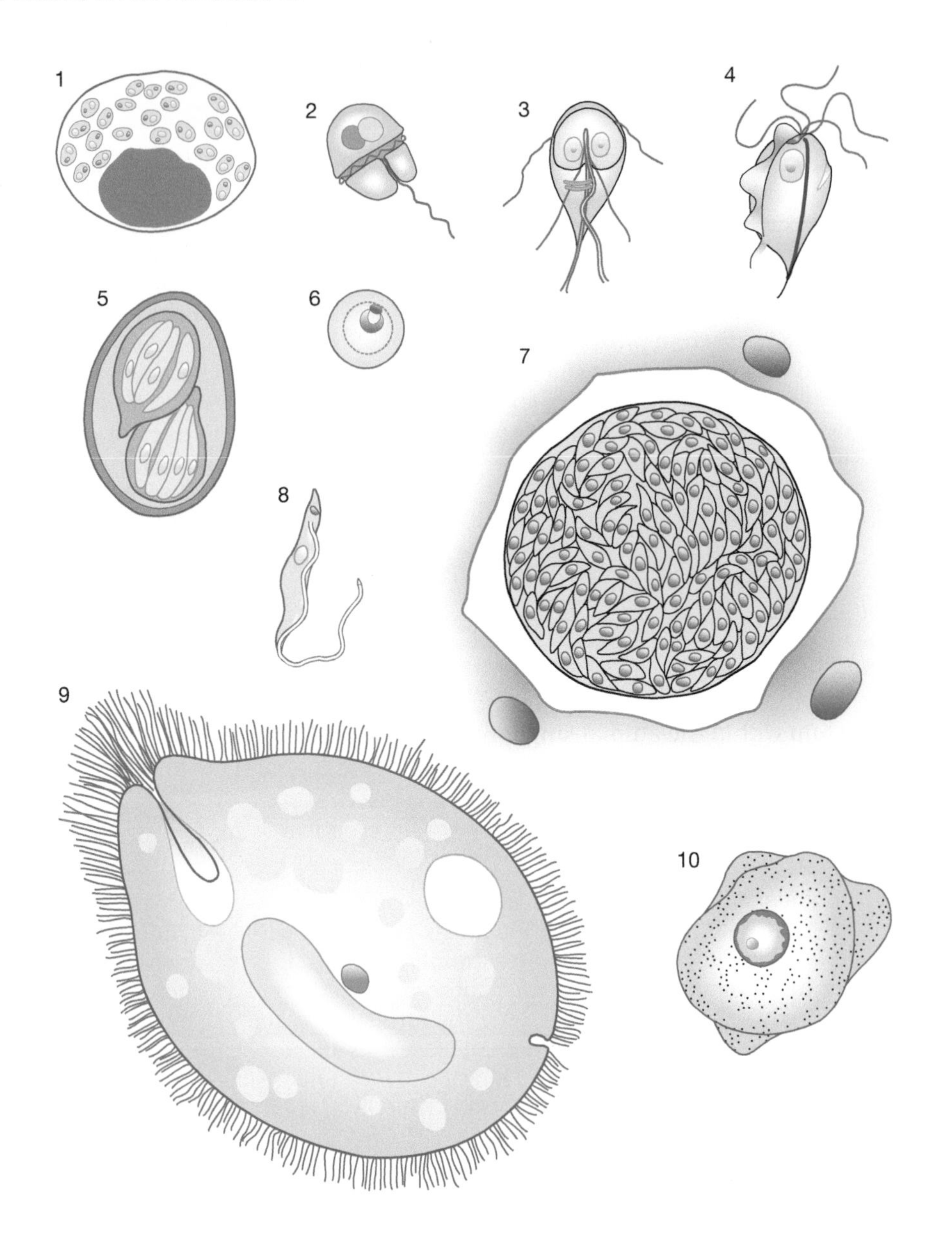

Figure 1.2 Protozoa exhibit a wide variety of shapes and forms. Shown are schematic representations of various protozoa associated with human health. (1) Amastigotes of *Leishmania* within a macrophage. (2) A dinoflagellate associated with shellfish poisoning. (3) *Giardia lamblia* trophozoite. (4) *Trichomonas vaginalis* trophozoite. (5) Oocyst of *Isospora belli*. (6) *Plasmodium* ring stage within an erythrocyte. (7) Tissue cyst of *Toxoplasma gondii*. (8) *Trypanosome* trypomastigote. (9) *Balantidium coli* trophozoite. (10) *Entamoeba coli* trophozoite.

many protozoan species exist in a dynamic feeding and growing form, often called a trophozoite, if conditions are favorable and nutrients are abundant. If conditions become unfavorable the trophozoite form can convert into a dormant cyst form that can survive desiccation and other harsh conditions. Similarly, many parasitic protozoa exhibit complex life cycles with various morphological forms involving multiple hosts and vectors.

Most protozoa generally reproduce asexually whereby the chromosomes are duplicated and distributed between two equal daughter cells as the cell divides. Other modes of mitotic division include budding and segmentation processes. Budding is an unequal division process in which the newly formed daughter cell is substantially smaller than the mother cell. Segmentation involves multiple rounds of nuclear division without cytoplasmic division followed by formation of the progeny from the multinucleated parental cell. These asexual modes of reproduction lead to rapid population growth of genetically identical cells. However, sexual reproduction is widespread among protozoa and often involves complicated life histories. Some have sexual cycles, involving meiosis and the fusion of gametes resulting in a diploid zygote, whereas other protozoa undergo conjugation during which nuclei are exchanged between cells.

Protozoan Taxonomy

Taxonomy, or systematics, is the science of naming and classifying organisms. In addition to assigning hierarchical taxonomic classifications, systematics also attempts to place organisms into groups reflecting evolutionary relationships or phylogenies. For example, a group of organisms all derived from a common ancestor are referred to as being **monophyletic**. However, taxonomic criteria are sometimes arbitrary and not all members of a group may be derived from a common ancestor. In this case the group is referred to as being **polyphyletic**. In addition, taxonomy is always changing to reflect new discoveries and interpretations. Furthermore, systematics of disease-causing organisms is further complicated by the frequent use of utilitarian features—such as type of disease, host range, and geographical distribution—in the naming of pathogenic microorganisms. Even though protozoa were originally defined as being unicellular heterotrophs, it is now recognized that protozoa can utilize multiple nutritional strategies and cannot be regarded simply as either plant-like (autotroph) or animal-like (heterotroph). Thus, the historical use of the term *protozoa* probably does not represent a true taxonomic group and there are some arguments to remove it as a taxonomic category. However, even if the term *protozoa* would no longer be a proper taxonomic name, it is still a useful and functional term.

As discussed previously the protozoa were initially considered as a phylum within the animal kingdom in the original Linnaean two kingdom classification system. Problems with classifying microorganisms as either plant-like or animal-like led to the introduction of a third kingdom (i.e., Protista) to separate microorganisms into their own group (Figure 1.3). This three kingdom classification system evolved into a five kingdom system in which the bacteria (i.e., Monera) and fungi were put into separate groups. In such schemes the protozoa are part of the Protista along with many unicellular algae, slime molds, and other unicellular eukaryotes. However, there has always been dissatisfaction with the protist group. This is due in part to protista being partially defined by negative criterion. In other words, organisms that do not fit into the other four kingdoms are defaulted into the protista. In addition, some protists are phylogenetically more closely related to the other eukaryotic kingdoms than to other protists, and thus the protists are polyphyletic. To solve some of these problems Cavalier-Smith proposes five eukaryotic kingdoms consisting of the basal, and thus paraphyletic, kingdom Protozoa and four kingdoms that are derived from the protozoa. To eliminate the problems of having a paraphyletic basal kingdom there has been a tendency to move away from the concept of kingdoms and divide the eukaryotes into supergroups based primarily on phylogenetic relationships. The currently proposed groups are: Opisthokonta (e.g., animals, fungi, choanoflagellates), Amoebozoa (e.g., lobose ameba, slime molds), Archaiplastida or Plantae (e.g., red algae, green algae, land plants), Chromalveolata (e.g., apicomplexans, ciliates, dinoflagellates, stramenopiles, giant kelps), Rhizaria (e.g., cercomonads, foraminifera), and Excavata (e.g., diplomonads, parabasalids, kinetoplastids). All of these supergroups contain organisms that have been classified as protozoa and most of them contain a mixture of unicellular and multicellular organisms.

Figure 1.3 Examples of hierarchical classification schemes. Some of the prominent schemes used to classify living things into major groups (i.e., kingdoms).

Two Kingdom (Linnaeus, 1735)	Three Kingdom (Haeckel, 1866)	Five Kingdom (Whittaker, 1969)	Six Kingdom (Cavalier-Smith, 1998)	Eukaryotic Supergroups (Simpson and Roger, 2004)
• Animal	• Protista	• Monera	• Prokaryotes	• Opisthokonta
• Plant	• Animal	• Protista	• Protozoa	• Amoebozoa
	• Plant	• Fungi	• Fungi	• Archaeplastida
		• Animal	• Animal	• Chromoalveolata
		• Plant	• Plant	• Rhizaria
			• Chromista	• Excavata

Motility is a defining characteristic of protozoa

The first classification of the protozoa into subgroups was in the late nineteenth century by Bütschli. In this original scheme the phylum Protozoa was divided into four major classes. The primary criterion for separating the protozoa into these various groups was the mechanism of motility because this was easily observed by light microscopy. These four original groups were the amebas, the flagellates, the ciliates, and the sporozoa. Amebas crawl along the substratum utilizing an actin–myosin force generation system. This ameboid movement involves extending out a **pseudopodium**, or false foot, and attaching this extension of its body to the substratum. The remaining body of the ameba is then pulled forward. Another common protozoan locomotory structure is the **flagellum**. Flagella consist of microtubules that can generate a whip-like motion that propel the flagellated protozoa through a liquid medium. **Cilia** have the same subcellular structure as flagella, but are generally more numerous and shorter. Ciliates, in contrast to flagellates which generally have one or a few flagella, are characterized by numerous rows of cilia which beat in unison to propel the organism. The sporozoa were initially defined as having a spore-like stage during their life cycle and exhibiting a gliding motility (Chapter 11). Gliding motility is a substratum-dependent locomotion that does not utilize locomotory organelles (i.e., cilia or flagella) and does not involve cell deformation as seen in ameboid movement.

Ultrastructural and molecular data help define protozoan groups

The application of electron microscopy to the study of protozoa in the 1960s elucidated unique subcellular features which can be used in the classification of the protozoa. For example, Apicomplexa and Microsporidia (now known to be a highly derived fungus) were split out of the sporozoa based upon unique organelles found in these two groups. However, morphological features can sometimes be lost in a particular lineage and placing such descendents can be difficult. More recently, protein and DNA sequences have been used to elucidate evolutionary relationships between the various taxa. The small subunit ribosomal RNA and various conserved proteins can be used to construct phylogenetic trees with the ultimate goal of creating classification schemes compliant with the principles of phylogenetic systematics. But the use of single genes can be an unreliable means of determining evolutionary relationships, especially among distantly related organisms, and molecular data should be interpreted with caution. In addition, sometimes genes are acquired in a horizontal fashion by incorporation of DNA from other species instead of the vertical heredity normally exhibited during reproduction. The availability of large amounts of genomic sequence and expressed sequence tags helps alleviate some of these limitations.

Since the mid-1980s classification of protists has been in a state of flux and afflicted with some philosophical controversies. There is some argument for retaining elements of the Bütschlian scheme due to its familiarity and simplicity. However, these schemes are often in conflict with phylogenetic data. In addition, the hierarchies and ranks of the traditional Linnaean systematics (i.e., phylum, class, order, family, genus, species) does not always fit well with microorganisms, and it is often difficult to decide which hierarchical level is most appropriate for any particular protozoan group. For example, in the model proposed by Cavalier-Smith the protozoa constitute a basal and paraphyletic kingdom and the protozoa are retained as a group. However, some of the protozoan groups are more closely related to the derived kingdoms (Figure 1.4). For example the alveolates, which include many traditional protozoa such as the apicomplexa, ciliates, and dinoflagellates, form a phylogenetic clade with the chromista, but are nonetheless placed in the protozoan kingdom. Thus the chromista are not

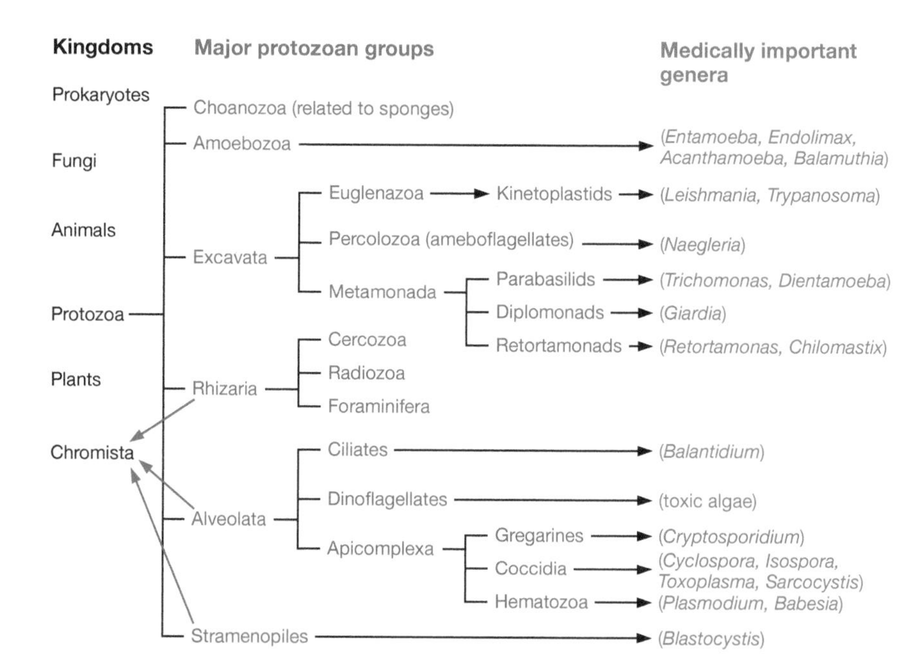

Figure 1.4 Kingdom Protozoa proposed by Cavalier-Smith. A schematic showing the hierarchies of major protozoan groups and the medically important genera within these groups. Gray arrows indicate a closer relationship of the rhizaria, alveolates, and stramenopiles to the chromista than to other protozoa. See also Figure 1.1.

holophyletic in that they do not include all descendants. Moving the alveolates into a new kingdom called Chromalveolata would solve this problem. However, this would probably result in controversy and confusion because many of the alveolates have long been considered to be protozoa. The taxonomy and phylogeny of many other protozoan species is also similarly ambiguous. These taxonomic issues and controversies will probably not be completely resolved anytime in the near future. Thus, it is most convenient to talk about particular groups of the protozoan pathogens combining the traditional classification schemes with the molecular and biochemical phylogenetic data when available and unambiguous.

Protozoan Cell Biology

The overall molecular composition of protozoa is similar to all other organisms and the cell biology of protozoa is similar to other eukaryotic organisms. Eukaryotic organisms were originally defined as those having a true staining nucleus. The nucleus is bound by a double membrane and contains the DNA packaged into chromosomes. Bacteria (i.e., prokaryotes) do not have a membrane-bound nucleus and the DNA and DNA-related functions are in the cytoplasm. Double-stranded DNA functions as the hereditary material in both prokaryotes and eukaryotes and is synthesized in a template-mediated fashion in which each of the strands serves as a template for the synthesis of the complementary strand. The replicated DNA is then segregated into the progeny cells. However, the mechanism of the cell replication in bacteria is simpler and generally involves only a binary fission of the cell. Eukaryotes undergo more complicated replication involving either an asexual replication, called mitosis, or a sexual replication, called meiosis, to insure that the chromosomes are accurately replicated and uniformly segregated into the progeny cells. DNA also encodes genes which are transcribed into RNA and in the case of eukaryotes the RNA is translocated into the cytoplasm where it is translated into proteins. Proteins will either form subcellular structures or play functional roles such as enzymes or receptors.

Other fundamental differences between prokaryotes and eukaryotes include a murein-based cell wall in prokaryotes and an endocytoskeleton in eukaryotes (Table 1.2). The cell wall provides mechanical support and protects the cell from osmotic pressure. These functions are replaced in part by the endocytoskeleton composed of actin microfilaments and microtubules.

Table 1.2 Major distinctions between prokaryotes and eukaryotes

Feature	Prokaryotes	Eukaryotes
Cell wall	Composed of murein	Lacking or composed of cellulose or chitin
Internal cytoskeleton	Lacking	Actin filaments and microtubules
DNA arrangement	Circular and located in cytoplasm	Linear molecules packaged into chromosomes within the nucleus
Reproduction	Binary fission	Mitosis and meiosis
Plasma membrane	No carbohydrates or sterols	Contains sterols and glycoproteins
Organelles	Absent	Present
Size	Typically 0.2–2 μm	Typically 5–100 μm

The development of the phagotrophy and this endocytoskeletal system was probably a key event in the evolution of eukaryotes (Box 1.1). Some protozoa do have cell walls but these are composed of chitin and are chemically distinct from the bacterial cell wall. Also associated with the loss of the murein-based cell wall are changes in the lipid and protein composition of the plasma membrane and an increase in the overall size of the cells. In addition to the nucleus, eukaryotic cells contain other membrane-bound organelles with specialized functions.

Although protozoa have features common to all eukaryotes, many unique features are found in the various protozoa. In some cases this involves an elaboration of subcellular structures found in other eukaryotes or organelles adapting to different or more specialized functions (Table 1.3). In other cases organelles or their functions have been lost, reduced in scope, or significantly modified. These various unique features reflect the unique biology of the various protozoa and the specific niches they occupy. Most of these unique features reflect metabolic specializations adapted to the specific niches occupied. For example, many protozoa inhabiting the gastrointestinal tract and other anaerobic environments no longer need to carry out oxidative phosphorylation and the mitochondrion has lost this function; in some cases it has degenerated into a remnant organelle called the mitosome or adapted to using hydrogen as the terminal electron acceptor. The kinetoplastids contain a microbody-like organelle which contains the enzymes of the glycolytic pathway which presumably increases the efficiency of glycolysis. In addition to metabolic specializations, unique structures are observed in many protozoa. For example, *Giardia* trophozoites have an adhesive disk, which functions in the attachment of the parasite to the intestinal epithelial cells, composed of microtubules, microfilaments and a unique cytoskeletal element called microribbons. Similarly, Apicomplexa contain unique secretory organelles that participate in the invasion of host cells (Chapter 11).

Cytoskeleton and cell motility

Three major cytoskeletal elements are found in eukaryotic cells: microtubules composed of tubulin and associated proteins, microfilaments composed of actin and associated proteins, and intermediate filaments. The microtubules and microfilaments are highly conserved throughout the eukaryotes, whereas intermediate filaments are not found in protozoa and other unicellular eukaryotes. However, some protozoa contain unique arrangements of the microtubules or microfilaments to form unique subcellular structures with specialized functions. In addition, some protozoa have unique cytoskeletal elements not found in other eukaryotes such as the adhesive disk in *Giardia* and the axostyle in trichomonads. In addition to providing mechanical strength and structure to the cell, the cytoskeleton

Table 1.3 Unique subcellular features found in some protozoa

Organelle	Atypical manifestation
Nucleus	Nuclear dimorphism in ciliates (Chapter 6)
Cell surface/ plasma membrane	Multilayered in alveolates and others
	Chitin-based cyst walls in numerous species
	Silica-based shells in diatoms
Mitochondria	Degeneration to mitosome in *Entamoeba* and *Giardia*
	Evolution to hydrogenosome in trichomonads (Chapter 5)
	Tubular cristae or acristate in various species
	Abundance of mini-circle DNA in kinetoplastids (Chapter 7)
Plastids	Loss of photosynthesis and specialized metabolic functions in Apicomplexa
Peroxisomes	Glycosomes in kinetoplastids (Chapter 7)
Lysosomes	Specialized function as food vacuoles in several species
Cytoskeleton/ microtubules	Rows of submembrane microtubules in many species
	Adhesive disk in *Giardia* (Chapter 4)
	Axostyle in trichomonads (Chapter 5)

functions in cellular motility and in this respect the cytoskeleton is more analogous to a cytomusculoskeletal system. Force can be generated through the action of motor proteins—dyneins and kinesis in the case of microtubules and myosins in the case of actin filaments—in conjunction with the other cytoskeletal elements. This force generation is important in chromosome segregation and cytokinesis during mitosis and meiosis as well as in cell motility. In regard to protozoan motility, microtubules and dynein are key components in cilia and flagella and actin and myosin are important in ameboid movement and gliding motility.

Cilia and flagella are similar structures which enable the protozoan to swim in aqueous environments. The primary difference is that cilia tend to be shorter and are generally arranged in rows that are interconnected via a submembrane cytoskeleton that coordinates the ciliary beat. Generally the flagella are longer and usually only one or a few are found per cell. The basic structure of the flagellum (or cilium) consists of rows of microtubules enclosed within a membrane that projects from the body of the cell. This whip-like structure propels the cell via the propagation of wave-like bends along the length of the flagellum. In some species the flagellum folds back to run along and attach to the body of the organism to form an undulating membrane. This membrane moves with the flagellum and increases the motility of the organism. The flagellum originates at a basal body and consists of central pair of microtubules and a ring of nine microtubule doublets (Figure 1.5). The motor protein dynein is attached along the length of one side of the microtubule. The dynein is able to move along the adjacent microtubule doublet coincident with the hydrolysis of ATP. In other words, dynein converts chemical energy into mechanical energy. This movement of dynein along the adjacent microtubule results in the two microtubules sliding past each other. However, links between the microtubules prevent them from sliding and the microtubules bend instead. A wave is propagated along the length of the flagellum through the coordinated release of dynein on one side of the flagellum and dynein movement on the other side. Repetition of this breaking and reforming of dynein links between the microtubules results in a beating of the flagellum and movement of the cell.

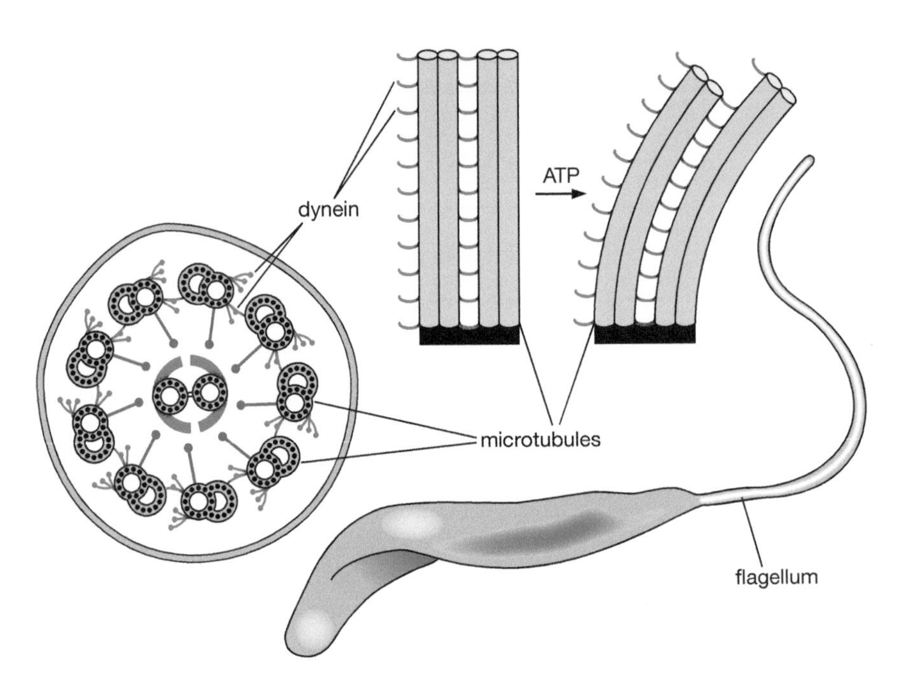

Figure 1.5 Microtubules and flagella. Schematic drawing of a cross section of a microtubule demonstrating the central pair of microtubules, nine microtubule doublets, and dynein interactions with microtubule doublets (left). Schematic drawing demonstrating bending action of microtubules in response to the ATPase-driven dynein motor and cross-linking of the microtubules (upper right). Drawing of flagellated protozoan demonstrating the wave being propagated along the flagellum (lower right).

Many protozoa do not swim, but instead crawl along a substratum. Some protozoa do both in that they can utilize flagella for motility in some life cycle stages as well as crawl along a substratum in other life cycle stages. The two basic means by which protozoa crawl along a substratum are ameboid movement and gliding motility (see Chapter 11). Ameboid movement is based on the temporary extension of part of the cell into one or more processes called pseudopodia, or false feet. Pseudopodia are also often important in feeding as best exemplified by phagocytosis. Distinct morphological types of pseudopodia are recognized: lobopodia (broad and blunt), acanthopodia (pointed or thorn like), filipodia (long and thin), axopodia (long and thin with a central bundle of microtubules), and reticulopodia (thin and branching and forming a network). The morphology of the pseudopodia is often used in classification of the amebas. Lobopodia is the most common type expressed by medically important amebas.

The mechanism of ameboid movement involves protrusion of the pseudopodia, attachment of the pseudopodia to the substratum, and traction, or pulling, of the cell body forward (Figure 1.6). Formation of the pseudopodia requires a reorganization of the cytoplasmic microfilaments. The cytoplasmic microfilaments are cross-linked into a three-dimensional gel which needs to be disassembled. This reorganization of the cytoplasm results in the pseudopodia having a different appearance (Figure 1.6B). This outer clear layer of cytoplasm is called the ectoplasm and the inner more granular layer is the endoplasm. In addition, the protrusion of the pseudopodia requires force involving the actomyosin cytoskeleton. Actin filaments assemble at the leading edge of the pseudopodium and myosin embedded in the plasma membrane forces the membrane forward along the actin filaments. The traction of the cell body forward also requires force involving the actomyosin skeleton. In this case myosin motors may transport the cell body along actin filaments or actin–myosin filaments contract in a mechanism similar to muscle contraction.

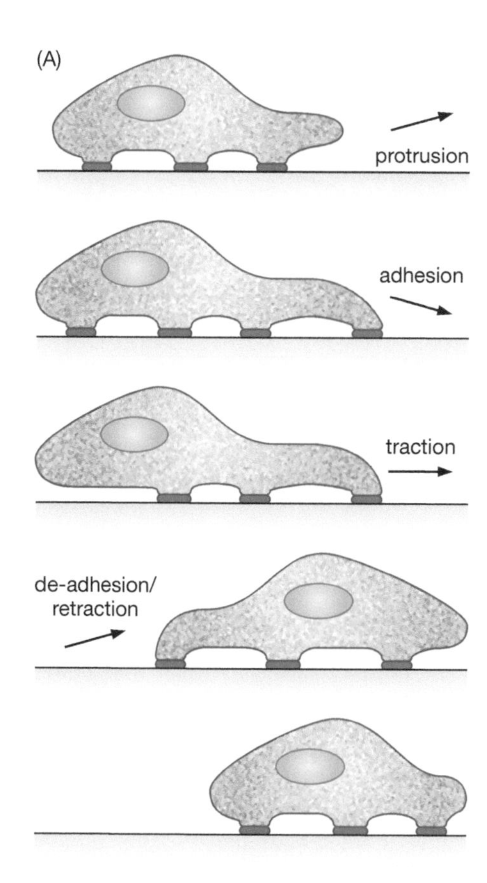

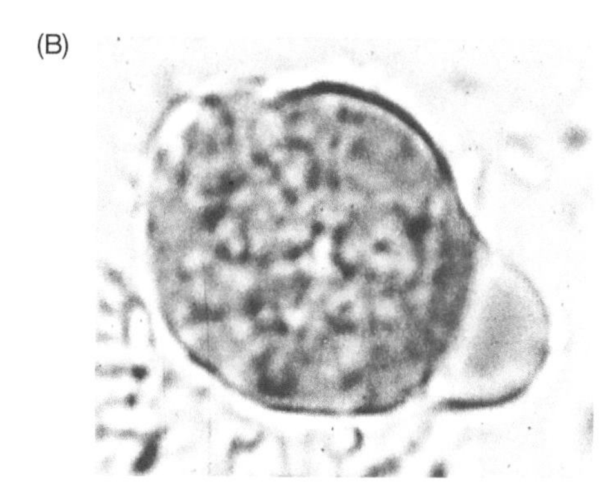

Figure 1.6 Ameboid movement. (A) Schematic of ameboid movement. (B) Light micrograph showing ectoplasm (pseudopodium) and endoplasm (cell body).

Parasitism

Protozoa are ubiquitous and are especially abundant in aquatic or moist environments. In addition, many protozoa exhibit cyst stages that can survive desiccation and other inhospitable environments. Most protozoa are free-living in that they are either autotrophic or heterotrophic and are not dependent upon another specific organism for their survival other than prey in the cases of predatory protozoa. A significant number of protozoa are dependent upon another organism for their survival and are referred to as symbionts. **Symbiosis** refers to a more or less permanent association between organisms from two different species. Different levels of symbiotic relationships can be defined. For example, **mutualism** is a situation in which the two organisms are co-dependent and both derive a benefit from the association. A classic example of mutualism is the lichen which is a union between an alga and a fungus. Another classic example of mutualism is the protozoan *Trichonympha* found in the gut of some termites. *Trichonympha*, with the assistance of a symbiotic bacterium, digests the wood particles (i.e., cellulose) ingested by the termite, resulting in all of the organisms (i.e., bacterium, protozoan, and insect) being metabolically intertwined and thus co-dependent on each other.

Other examples of symbiotic relationships are **commensalism** and **parasitism**. In commensalism one of the organisms benefits from the relationship, whereas the other (i.e., host) is neither benefited nor harmed. Generally this involves a physical proximity and the commensal feeds on substances captured or digested by the host. In parasitism the parasite relies upon the host for its nutrients and a place to live and there is potentially some cost to the host. This cost to the host is generally manifested as damage or pathology (i.e., disease). Organisms causing overt disease are referred to as being pathogenic. In addition, parasites usually reproduce within the host, and in the case of pathogenic protozoa the pathology is usually correlated with the degree of reproduction. The line between parasitism and commensalism is often blurry and the precise definition between the two terms often varies according to author. For example, according to some definitions, commensals, in contrast to parasites, are not physiologically dependent upon the host and can live independently of the host. However, several protozoan species found within the human intestinal tract, that feed upon bacteria and debris and do not cause any human disease, are often referred to as commensals even though they are completely dependent on the host for their survival. In addition, one can also distinguish between obligate symbionts, which cannot live apart from their hosts, and facultative symbionts, which can be either free-living or symbiotic. Similarly, many parasitic protozoa are facultative pathogens in that they do not always cause disease.

The life cycle of many parasitic protozoa can be completed in a single host. These **monoxenous** parasites generally have a life cycle stage outside of the host and transmission almost always involves the ingestion of an infective form. Other parasitic protozoa are **heteroxenous** in that they require multiple host species to complete their life cycle. Quite often one of the hosts is a blood-feeding arthropod referred to as a **vector**. Some parasitic protozoa can only infect a single or limited range of host or vector species; whereas others are more promiscuous and can infect a wide range of hosts. Human diseases caused by parasites normally found in other animals are called **zoonoses** and the animal host is the **reservoir**. Generally zoonotic diseases do not involve a person-to-person transfer.

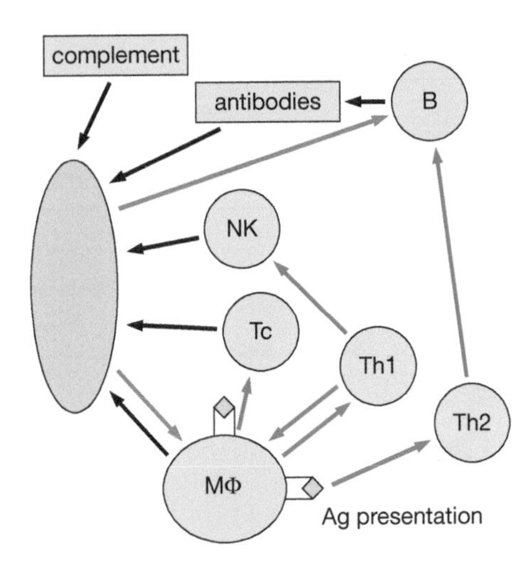

Figure 1.7 Immune effector mechanisms. A schematic drawing illustrating some of the interactions between immune effector mechanisms and pathogens. Black arrows indicate immune components that directly kill or neutralize a pathogen (colored oval) either alone or in conjunction with other effector mechanisms. Gray arrows illustrate pathogen recognition and processing. Blue arrows indicate interactions between immune effector cells via direct contact or cytokines. B = B cell; NK = natural killer cell; Tc = cytotoxic T cell; Th1 = type 1 T-helper cell; Th2 = type 2 T-helper cell; MΦ = macrophage.

Host Immunity

Parasitic protozoa must overcome a rather sophisticated human immune system (Figure 1.7) in order to survive. Generally three levels of the immune system are recognized: (i) physical barriers, (ii) innate nonspecific responses, and (iii) adaptive immune responses. Physical barriers include the skin and epithelial linings of the respiratory, gastrointestinal, and urogenital systems. These internal epithelial layers are bathed in mucus, composed of glycoproteins, proteoglycans and enzymes, which protects against damage and limits microbial infection. Breaching the skin, with its layers of keratinized cells, usually requires some type of physical damage or wound. For example, many vector-transmitted protozoa rely upon a blood feeding arthropod to penetrate or damage the skin and thereby provide access to the internal layers and circulatory system. Other protozoa are internalized via ingestion and live on the mucosal surfaces of the intestines. In some cases they are able to penetrate the intestinal epithelial cells and cause an invasive disease.

Complement and phagocytosis are nonspecific immune effector mechanisms

Pathogens penetrating the skin or epithelial layers will then encounter a second line of host defense. The first step in this second line of immunity, often called the innate immune response, is recognition of a pathogen. Recognition involves soluble proteins or cell-surface receptors binding to either the pathogen, products of the pathogen, or host cells altered by the pathogen. Following recognition, various effector mechanisms can be initiated to destroy and eliminate the pathogen. For example, **complement** proteins bind to many different pathogens. This can mark the pathogen for destruction by effector cells such as macrophages. Alternatively, complement can activate a membrane attack complex which perforates the membrane of the pathogen and results in cell lysis and death. This complement membrane attack complex can also be initiated by antigen–antibody complexes on the surface of the pathogen via the so called classical pathway.

Complement or antibodies on the surface of a pathogen can also lead to recognition and destruction of the pathogen by phagocytic cells such as **macrophages**, neutrophils, also called polymorphonuclear leukocytes, and eosinophils. These phagocytic cells engulf the pathogen, kill it via reactive oxygen intermediates, and break down the pathogen with hydrolytic enzymes within the phagolysosome. Natural killer (NK) cells are another effector cell of the innate immune response which target virus-infected cells. In addition, **cytokines** are produced by various immune effector cells and endothelial cells in response to infection. Cytokines are soluble molecules that recruit other effector cells to the infection site. These cytokines and effector cells also result in inflammation at the infection site. Inflammation is characterized by redness, due to increase blood flow, swelling, fever, and pain and represents a nonspecific immune response against the invading pathogen.

Adaptive immunity is the third line of host defense

Adaptive, or acquired, immunity is specific against a particular pathogen and has little effect on other pathogens. As in innate immunity, receptors on the effector cells of adaptive immunity (i.e., **lymphocytes**) recognize pathogens or pathogen products and are activated to destroy the pathogen. However, these receptors only recognize a specific pathogen. The genes for these receptors can be rearranged resulting in billions of variants which are differentially expressed on various subpopulations of lymphocytes (i.e., B and T cells). During infection those lymphocytes which recognize the pathogen expand in number and mature into effector cells which will destroy the pathogen or activate other immune effector cells. This adaptive response is relatively slow and takes days to weeks after the infection to become functional. However, the lymphocytes that expand during the infection persist in the body and provide long-term immunological memory. Thus, subsequent encounters with the same pathogen produce quicker and stronger responses resulting in elimination of the pathogen with less disease. This is often referred to as acquired immunity because it requires a prior exposure to the pathogen to be fully functional.

Effector cells involved in the adaptive immune response include B cells, cytotoxic T cells, helper T cells, and antigen-presenting cells. Under some situations B cells can respond directly to immunogens produced by the pathogen; whereas, T cells require antigen presentation. Presentation involves processing and binding an immunogen (i.e., antigen) to a surface receptor on the membrane of the antigen-presenting cell. These surface receptors are products of genes called the major histocompatibility complex (**MHC**). Class I MHC proteins are expressed on most nucleated cells and present endogenous proteins produced by intracellular parasites, whereas class II MHC proteins are restricted to professional antigen-presenting cells such as macrophages, dendritic cells and B cells. Endogenous parasite proteins are digested by proteases (e.g., proteasome) in the cytoplasm of the host cell and the resulting peptides are translocated into the lumen of the endoplasmic reticulum. The peptides then bind to the MHC class I proteins and are transported to the cell surface by the secretory pathway. Exogenous proteins or extracellular pathogens are taken up by the antigen-presenting cell via phagocytosis and then degraded within the phagolysosome. The resulting peptides bind to the class II MHC proteins within the lysosome and the complex moves to the cell surface.

The peptide associated with the MHC protein is then recognized by receptors found on the surface of T cells. The gene for the T-cell receptor can rearrange and produce millions of variants of the T-cell receptor. Each of these variants will only recognize a single peptide bound to the MHC molecule. Those T cells recognizing the specific peptide are stimulated to replicate, or undergo "clonal selection," and mature into effector cells. For example, cytotoxic T cells (aka, $CD8^+$ cells) recognize peptides associated with MHC class I proteins and are stimulated to directly kill the cell expressing the MHC–peptide complex. Helper T cells (aka $CD4^+$ cells) recognize peptides associated with MHC class II proteins and are stimulated to secrete cytokines that activate other immune effector cells such as macrophages and B cells. The helper T cells are further distinguished into Th1 and Th2 cells. The Th1 cells secrete inflammatory cytokines that activate macrophages and natural killer kills, or induce a cell-mediated response. Th2 cells induce a humoral response by secreting cytokines which activate B cells and antibody production.

Antibodies and B cells

Antibodies, or immunoglobulins, are proteins secreted by B cells that bind specifically to products of the pathogen. Immunoglobulins are characterized by a constant region and a variable region. The variable region recognizes specific antigenic determinants, or epitopes, and billions of different variants can be generated through gene rearrangements. Immunoglobulins expressed on the surface of B cells function as antigen receptors. In the presence of antigen (i.e., a pathogen), B cells that recognize the specific antigen are stimulated to divide and mature into plasma cells in a manner similar to the clonal selection of T cells. The plasma cells secrete antibodies with the same specificity. The first antibody secreted by mature B cells is immunoglobulin M (IgM). Later with the help of Th1 cells the immunoglobulin type switches to IgG (blood), IgA (mucosal surfaces), or IgE. In addition, continued exposure to the pathogen results in selection for antibodies with higher affinities for the pathogen. The secreted antibodies bind to antigens and either neutralize the pathogen or toxin, promote opsonization of the pathogen, activate antibody-dependent cell-mediated cytotoxicity, or activate the classical complement pathway. Neutralization refers to a situation in which binding of the antibody prevents the replication of the pathogen or prevents the pathogen or toxin from interacting with the host cell. Binding of antibodies to the surface of pathogens also promotes phagocytosis, or opsonization. Macrophages and neutrophils have cell surface receptors that bind to the constant region of IgG and pathogens covered with IgG are more efficiently phagocytosed. Natural killer cells can also recognize the constant region of IgG and are stimulated to release hydrolases which can kill the pathogen.

Protozoa use various mechanisms to evade the host immune system

Protozoa, as well as other pathogens, must adapt to the highly sophisticated immune system of the human host. In many cases transmission of the infection depends on establishing a chronic infection in which the protozoa can survive in the host for relatively long periods of time. Probably the most common means of immune evasion is antigenic variation which is utilized by numerous parasitic protozoa (Table 1.4). In this strategy the pathogen continuously alters the expression of surface molecules in an attempt to stay one step ahead of the adaptive immune response. Another strategy of immune evasion utilized by some protozoa is to exhibit a tropism for a tissue with limited immune responses such as the central nervous system or muscle. Similarly, many different protozoa have adapted to an intracellular existence in part as a means by which to avoid the immune system. In other cases the pathogen directly confronts the immune effector mechanism and overcomes its destructive ability. For example, *Leishmania* species have adapted to life within the phagolysosome of the macrophage and are not killed by the macrophage. Similarly, several protozoa are resistant to complement.

Table 1.4 Common mechanisms of immune evasion utilized by protozoa

Mechanism	Examples
Antigenic variation	African trypanosomes, *Plasmodium, Giardia*
Tissue tropisms	*Toxoplasma*, free-living amebas
Intracellularism	Apicomplexa, *Trypanosoma cruzi*
Survival in macrophages	*Leishmania*
Complement resistance	Trypanosomes, *Entamoeba histolytica*

Genomics and Molecular Analyses

The study of the protozoa began with the invention of the microscope, which remained the primary instrument to study protozoa for several centuries. Thus, protozoa were primarily described in regard to morphological features visible with the light microscope and their means of locomotion. With the invention of the electron microscope the ultrastructural features of protozoa were revealed and used to further describe protozoan cell biology and distinguish the various groups of protozoa. During the past four

Table 1.5 Common methods used in biomedical research

Technique	Description
Gel electrophoresis	Separation of proteins or nucleic acids in an electric field
ELISA	Enzyme-linked immunosorbent assay used to quantify antigens or antibodies
Immunoblotting (aka western blot)	Detection of specific proteins with antibodies following gel electrophoresis
Immunofluorescence	Localization of antigens in tissues or cells using antibody probes
Southern blots	Detection of specific DNA fragments with a complementary probe (e.g., RFLP)
Northern blots	Detection of specific RNA molecules with a complementary probe
Polymerase chain reaction (PCR)	Amplification of specific nucleic acid fragments using complementary primers
DNA sequencing and genomics	Determination and analysis of gene sequences
Mass spectrophotometry and proteomics	Determination of protein sequences and identification of genes

decades the application of biochemical, molecular, and immunological techniques has further refined our understanding of protozoa (Table 1.5). This has been especially true in the case of pathogenic protozoa and these methodologies are often used to better define host–parasite interactions and the pathological mechanisms of disease. In addition, these same methods are increasingly being applied to epidemiological studies (Box 1.2).

Box 1.2 Molecular epidemiology

Epidemiology is the study of disease distribution and determinants in populations. Molecular epidemiology refers to the application of molecular approaches to epidemiological problems. Typically molecular markers, such as DNA, RNA, proteins, or antibodies, are evaluated to define disease distribution or to identify potential disease determinants. Molecular epidemiology is based on general epidemiology and utilizes the same designs (i.e., case-control and cohort studies) as those employed by general epidemiology. The studies can either be descriptive or analytical. Descriptive epidemiology is the study of the distribution of disease in populations and characterizes the burden of disease in a population. Incidence and prevalence rates of disease are determined and often correlated to person, place, and time. Analytical epidemiology is the study of the determinants of disease and characterizes the risk factors for disease in populations. These include causative factors (e.g., infectious agents), host characteristics (e.g., susceptibility genes), and environmental exposures (e.g., lifestyles) that influence the risk of developing the disease. Relative risks and odds ratios are evaluated for potential determinants of the disease.

One advantage of utilizing molecular markers in epidemiological studies is the increased sensitivity often afforded by molecular techniques. In other words subclinical situations can be detected. In addition, molecular markers often reduce biases and improve the validity of studies. More importantly, molecular epidemiology can enhance our understanding of the pathogenesis of disease by identifying specific pathways, molecules, and genes that influence the risk of developing disease. Furthermore, specific polymorphisms or gene alleles that are potentially associated with disease can be evaluated. In infectious diseases this includes both the host (e.g., susceptibility genes) and the pathogen (e.g., virulence genes). For example, various subgroups of pathogens based on serotypes and biotypes can be defined and correlated with such phenotypic traits such as virulence or drug resistance

Electrophoresis

Gel electrophoresis is a widely used technique in biochemistry and molecular biology. Molecules are subjected to an electric field in a medium consisting of cross-linked polymers (i.e., gel). The two most common media for gel electrophoresis are polyacrylamide and agarose. In the presence of an electric field molecules migrate through the gel at a rate corresponding to their charge and inversely proportional to their size. In other words, the migration through the gel is determined largely by their charge/size ratios. In the case of nucleic acids, such as DNA and RNA, the negatively charged phosphate backbone results in a uniform charge to mass ratio and the separation of nucleic acids is determined primarily by their sizes. In other words, the smaller nucleic acids will migrate through the gel faster than the larger molecules. Thus a mixture of nucleic acids can be separated according to size. The end result is a gel in which there is a continuous array of nucleic acids with larger molecules near the top of the gel and smaller molecules near the bottom. Individual nucleic molecules, or "bands," can be detected within the gel by incubating with a dye which fluoresces when bound to nucleic acids.

The overall charge of a protein is determined by the numbers of amino acids with acidic (negative charge) and basic (positive charge) side chains and thus separation of a mixture of proteins will be based on differences in the charge/mass ratios of these proteins. The charge of the protein can also be defined according to its isoelectric point. The isoelectric point is the pH at which the net charge of the proteins is zero. A variation of protein electrophoresis is isoelectric focusing which separates proteins primarily according to their isoelectric points. This is a high-resolution technique that can distinguish polymorphisms due to the change of a single amino acid. Another variation of protein electrophoresis utilizes sodium dodecyl sulfate (SDS). SDS is a negatively charged detergent that binds to the proteins and thus gives them a uniform charge/mass ratio. Separation is then determined primarily by size as is the case in nucleic acid electrophoresis. Individual proteins can be detected by staining the gel with a dye that binds to proteins. Proteins can also be partially sequenced by mass spectrophotometry and other methods.

Polyacrylamide gel electrophoresis and isoelectric focusing are high-resolution techniques widely used in protein analysis. In addition, proteins from different species or isolates can be distinguished by size and charge polymorphisms of particular proteins. One example is the use of isoenzymes as genetic markers for particular isolates or strains of protozoa. Some enzymes are still active after electrophoresis and can be detected. Charge polymorphisms result in the enzymes migrating to different positions within the gel. These variants of the same enzyme are referred to as isoenzymes. Several enzymes can be used in conjunction to define a zymodeme. These zymodemes have been used as markers both for particular isolates of protozoan species and for particular phenotypic characteristics. Isoenzymes are no longer widely used as molecular markers and have been largely replaced by DNA markers or antibodies.

Uses of antibodies

Antibodies are proteins produced by B cells which bind to antigens with a high degree of specificity and with a high affinity. This ability of antibodies to specifically recognize other molecules, in particular proteins, makes them powerful reagents in molecular analyses in addition to their role in immunity. A commonly used method to analyze either antigens or antibodies is the enzyme-linked immunosorbent assay (ELISA). This assay involves absorbing the antigen to 96-well plates and then incubating with sera. If antibody which recognizes the antigen is present within the sera it will bind to the antigen and any antibody not specifically recognizing the antigen of interest is washed away. The plates are then incubated with an enzyme-conjugated second antibody that recognizes antibody from the species being evaluated. Binding of the enzyme-conjugated second antibody will be proportional to the amount of the antigen–antibody complex present in the samples being tested and is evaluated by incubation with chromogenic substrates of the enzyme. Specially designed spectrophotometers, called ELISA readers, can then be used to quantify the amount of antigen–antibody complex.

Another popular assay using antibodies is the immunoblot, or western blot. This assay is similar to the ELISA except that the antigen source is first subjected to gel electrophoresis. The separated proteins are then transferred to a membrane resulting in a replica of the gel. This membrane is then incubated with the primary antibody followed by incubation with the secondary enzyme-conjugated antibody in the same fashion as described for the ELISA. This results in an antigen–antibody complex forming on the membrane at the location of the specific protein(s) recognized by the primary sera or antibody. Substrates which form a colored precipitate in the presence of the enzyme are then used to detect the protein(s) of interest. Immunoblotting allows for the detection and analysis of specific protein subunits.

Immunofluorescence is another widely used technique utilizing antibodies. In this case cells or tissue sections are fixed onto microscope slides and incubated with the primary antibody. The antibody binds to the antigen within the cell and the antigen–antibody complex is then detected with a secondary antibody conjugated to a fluorescent dye. The location of antigen can then be determined by examining the sample under ultraviolet radiation with a fluorescent microscope. Structures containing the antigen of interest will fluoresce. In addition to determining the location of antigens, immunofluorescence is also used to evaluate antibodies. Sera are diluted until the fluorescence is no longer detectable. The reciprocal of the titer at which the fluorescence ceases is proportional to the amount of the specific antibody in the sera.

Blotting techniques and restriction fragment length polymorphisms

Analysis of DNA and RNA molecules involves the use of DNA probes homologous to the target nucleic acids. These probes are relatively short single-stranded DNA molecules which are complementary to the specific nucleic acid molecule of interest (i.e., target) and are labeled in some fashion. Means to label the probes include radioactivity, fluorescence, antigenic determinants, biotin, or enzymes. The probe binds (anneals) to the target due to the complementary base pairing between the two strands and the target is then analyzed or quantified by detection of the label on the probe. Common applications utilizing DNA probes include Southern blots, northern blots, and *in situ* hybridization. Southern blotting, named after the inventor, involves the separation of DNA molecules by gel electrophoresis, transferring the DNA molecules to a membrane to make a replica of the gel, and incubating the membrane with the probe. In northern blotting, RNA molecules are subjected to electrophoresis and analysis, thus providing information about gene expression. Fluorescent probes can also be incubated with tissue sections (i.e., *in situ* hybridization) to determine which cells are expressing particular genes or to detect parasites within tissues.

Genomic DNA consists of very long molecules which need to be cut into smaller fragments before they can be analyzed. Restriction endonucleases are bacterial enzymes which cut DNA at specific sites corresponding to a particular sequence which typically consists of four to eight nucleotides.

Each restriction enzyme recognizes a different sequence and thus can reproducibly cut DNA into fragments of a specific size. Nucleotide polymorphisms between individuals within a species can result in the gain or loss of restriction sites and thus will produce different DNA fragment lengths when analyzed by Southern blotting. These restriction fragment length polymorphisms (RFLP) can serve as a genetic fingerprint for the identification of particular genotypes or in molecular epidemiology studies.

Polymerase chain reaction

RFLP analysis and blotting techniques require a fair amount of material and time to carry out. Therefore, these techniques have been largely replaced by the polymerase chain reaction (PCR). PCR utilizes two oligonucleotide primers which flank the target DNA and are complementary to the opposing DNA strands. These primers are functionally analogous to the probes used in blotting techniques in that they specifically anneal to the target DNA (or RNA). The primers anneal to the template DNA and the complementary strands are synthesized by a DNA polymerase. These newly synthesized strands can then serve as a template in another round of primer annealing and DNA synthesis. Using a thermocycler and a heat-stable DNA polymerase, the region of DNA between the primers is amplified up to a million fold. The resulting DNA fragment is then usually detected by gel electrophoresis. Real time PCR combines the primers with fluorescent DNA-binding dyes or fluorescent probes and utilizes a special thermocycler that measures fluorescence during the reaction. This allows detection of the target DNA without the electrophoresis step and provides a means to quantify the amount of starting template nucleic acid.

PCR is a highly sensitive technique in that a few copies of DNA can be detected. Furthermore, extremely specific primers can be designed so that the target DNA (e.g., pathogen) can be detected against a high background of contaminating DNA (e.g., host). Thus PCR is often used in the diagnosis of infectious diseases in situations where there are few organisms. In addition, PCR can be utilized to detect polymorphisms and particular genotypes and thus is useful in molecular epidemiology. Polymorphisms can be detected through primer design, restriction endonucleases, DNA probes, or sequencing.

DNA sequencing and genomics

DNA can be defined by the order of its nucleotides or sequence. This nucleotide sequence can be translated into the protein sequence. DNA and protein sequences contain information and can be used to predict protein structure and function. In addition, comparing sequences can be used to determine genetic and evolutionary relationships between organisms. This comparison of sequences, or molecular phylogeny, has been instrumental in revealing phylogenetic relationships among many protozoan groups. Sequence differences can also be used as markers for particular subgroups, or biotypes, within a species. Phenotypic differences, such a drug resistance, can also be revealed through sequence differences.

The technology involved in sequencing DNA has advanced tremendously in the past two decades and the complete DNA sequence, or genome, is known for many organisms. Several protozoan species have been completely sequenced or nearly so (Table 1.6). In addition to providing insight into the biology of protozoa, genomics will also contribute to developing new therapeutic strategies against pathogenic protozoa. Because both the human genome and the genomes of some disease vectors are known, evaluations of host–parasite–vector interactions can be carried out at the genetic level.

Table 1.6 Completed protozoan genomes

Organism	Disease or comment	Size	Genes	Year
Babesia bovis	Cattle pathogen	8.2	3671	2007
Cryptosporidium hominis	Human intestinal pathogen	10.4	3994	2004
Cryptosporidium parvum	Human and cattle pathogen	16.5	3807	2004
Dictyostelium discoideum	Free-living social ameba	34.0	12 500	2005
Entamoeba histolytica	Cause of amebic dysentery	23.8	9938	2005
Giardia lamblia	Human intestinal pathogen	11.7	6470	2007
Leishmania major	Human pathogen	32.8	8272	2005
Paramecium tetraurelia	Free-living ciliate	72.0	39 642	2006
Plasmodium falciparum	Human malaria parasite	22.9	5268	2002
Plasmodium knowlesi	Simian malaria parasite	23.5	5188	2008
Plasmodium vivax	Human malaria parasite	26.8	5433	2008
Plasmodium yoelii	Rodent malaria parasite	23.1	5878	2002
Tetrahymena thermophila	Free-living ciliate	104.0	27 000	2006
Theileria annulata	Cattle pathogen	8.3	3792	2005
Theileria parva	Cattle pathogen	8.3	4035	2005
Trichomonas vaginalis	Human pathogen	160.0	59 681	2007
Trypanosoma brucei	African sleeping sickness	26.0	9068	2005
Trypanosoma cruzi	Chagas' disease	34.0	22 570	2005

Shown are protozoa whose genomes have been completely sequenced including the genome size in mega (= million) bases (Mb), the predicted number of genes, and the year in which the sequencing was completed.

Summary and Key Concepts

- Protozoa are an extremely diverse group of unicellular eukaryotic micro-organisms. They are found in a wide range of ecological niches and accordingly exhibit a wide range of metabolisms.
- Protozoan cells can be quite complex and many of the protozoan groups are defined by unique subcellular structures not found in multicellular organisms.
- Most protozoa are motile and motility is either microtubule based (i.e., cilia and flagella) or due to actin–myosin cytoskeletal elements.
- Some protozoa are parasitic and several species cause human disease.
- Successful protozoan parasites must overcome a sophisticated host immune system consisting of humoral and cell-mediated elements.
- Molecular techniques have increased our understanding of pathogenic protozoa and the diseases they cause.

Further Reading

Cavalier-Smith, T. (2003) Protist phylogeny and the high-level classification of Protozoa. *Eur. J. Protistol.* 39: 338–348.

Khan, N.A. (2008) *Emerging Protozoan Pathogens.* Taylor and Francis Group, New York.

Parham, P. (2005) *The Immune System,* 2nd Edn. Garland Science, New York.

Patterson, D.J. (2000) Changing views of protistan systematics: the taxonomy of protozoa. In *An Illustrated Guide to the Protozoa,* 2nd Edn (eds J.J. Lee, G.F. Leedale and P. Bradbury). Society of Protozoologists, Lawrence, KS, pp. 2–9.

Simpson, A.G.B. and Roger, A.J. (2004) The real "kingdoms" of eukaryotes. *Curr. Biol.* 14: R693–R696.

Overview of Intestinal Protozoa

A wide variety of protozoa inhabit the gastrointestinal tract of humans and other luminal organs such as the oral cavity and urogenital tract. These organisms include representatives from most of the major protozoan groups (Table 2.1). The amebas and flagellates are not monophyletic groups and the various species are not necessarily related, whereas the apicomplexans, ciliates, and stramenopiles are monophyletic groups. This wide diversity of protozoan types indicates that adaptation to the intestinal tract has occurred numerous times in the evolutionary history of the protozoa. Most of these lumen-dwelling protozoa are nonpathogenic commensals, or only cause mild disease. In regard to the number of cases and potential for disease, *Entamoeba histolytica* and *Giardia lamblia* are the most important among the intestinal protozoa. *Cryptosporidium* is another common

Table 2.1 Luminal protozoa

Group	Organism	Primary location
Amebas	***Entamoeba histolytica***	Large intestine
	Entamoeba dispar	Large intestine
	Entamoeba coli	Intestines
	Entamoeba hartmanni	Large intestine
	Entamoeba gingivalis	Oral cavity
	Endolimax nana	Intestines
	Iodamoeba bütschlii	Large intestine
Flagellates	***Giardia lamblia***	Small intestine
	Dientamoeba fragilis	Large intestine
	Chilomastix mesnili	Intestines
	Enteromonas hominis	Intestines
	Retortamonas intestinalis	Intestines
	Pentatrichomonas hominis	Large intestine
	Trichomonas tenax	Oral cavity
	Trichomonas vaginalis	Urogenital tract
Apicomplexans	***Cryptosporidium hominis***	Intestines
	Cryptosporidium parvum	Intestines
	Cyclospora cayetanensis	Small intestine
	Isospora belli	Small intestine
Ciliates	*Balantidium coli*	Large intestine
Stramenopiles	*Blastocystis hominis*	Intestines

The species in bold are the most important in terms of the numbers of people affected and the severity of the disease.

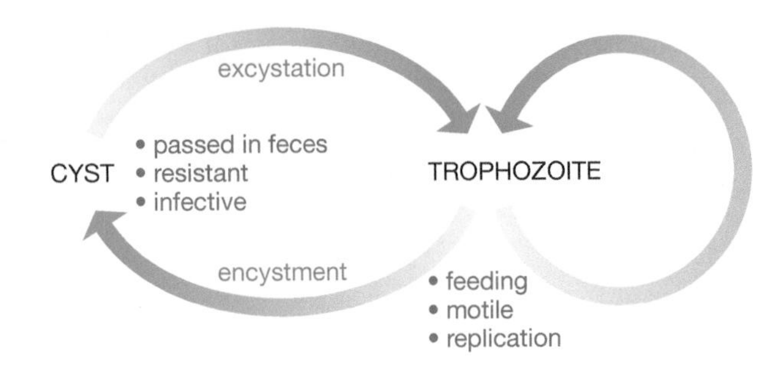

Figure 2.1 Typical fecal–oral life cycle. Infections are acquired by ingestion of the cyst stage. After ingestion the cyst develops into a trophozoite which establishes an infection within the intestinal tract. Trophozoites replicate primarily by binary fission and some trophozoites will develop into cysts which are passed in the feces and serve as the source for spreading the infection from person to person.

pathogen found in the intestinal tract. Although the transmission and pathology associated with *Cryptosporidium* is similar to other intestinal protozoa, it and the other intestinal coccidia will be discussed later with the apicomplexans. *Trichomonas vaginalis* is not found in the intestines, but is a common sexually transmitted disease infecting the urogenital tract of both males and females.

Fecal–Oral Transmission

Intestinal protozoa are transmitted by the **fecal–oral** route and tend to exhibit similar life cycles and modes of transmission (Figure 2.1). The stage or form inhabiting the intestinal tract of the host is generally called a **trophozoite**. The exceptions to this are the apicomplexans which exhibit a more complex life cycle involving intracellular stages. The name trophozoite literally means feeding stage and most of the trophozoites from intestinal protozoa feed by taking up solutes or ingesting debris or bacteria found in the intestinal lumen. Trophozoites are also generally motile and move via pseudopodia, flagella, or cilia. Most importantly in regard to the life cycle, trophozoites are replicating forms which generally reproduce asexually, with binary fission being the most common mode of replication among the intestinal protozoa. In addition, the trophozoite stage is usually responsible for the pathogenesis in disease-causing species.

Some of the trophozoites will develop into **cysts**, as an alternative to asexual replication, and be excreted with the feces. The encystment process usually involves a slowing of the metabolic activity of the trophozoite and a secretion of a cyst wall. The cyst stage represents a dormant or resting stage of the parasite and the cyst wall protects the organism from desiccation and other environmental factors. This allows the parasite to survive in the environment until the cyst is ingested by another appropriate host. After being ingested by the appropriate host the parasite will excyst and develop into a trophozoite, which will establish the infection in a new host and thus complete the life cycle. In general, trophozoites are not infectious and only the cyst stage can establish the infection. However, there are a few intestinal flagellates in which no cyst stage has been identified and the trophozoite is presumed to be the infectious stage as well as the replicative form.

Generally fecal–oral transmission involves the ingestion of food or water contaminated with cysts. Obviously situations leading to food or water being contaminated by fecal matter will promote the transmission of these infections (Table 2.2). In general, the most common mode of transmission involves close personal contact combined with poor personal hygiene. For example, giardiasis outbreaks in day-care centers occur with some frequency and *Entamoeba* infections in institutions such as prisons or mental hospitals are not uncommon. Although intestinal protozoan infections are found throughout the world, the prevalence tends to be higher in developing countries. This is likely due to systemic problems such as insufficient water treatment, lack of indoor plumbing, and a high level of endemicity. However, waterborne outbreaks of *Giardia* and *Cryptosporidium* have also

Table 2.2 Fecal–oral transmission factors	
Personal hygiene issues	Children (e.g., day-care centers)
	Institutions (e.g., prisons, mental hospitals, orphanages)
	Food handlers
	Oral–anal contact
Developing countries	Poor sanitation
	Lack of indoor plumbing
	Endemic presence of species
	Travelers' diarrhea
Waterborne epidemics	Water treatment failures
Possible zoonoses	*Entamoeba* = no
	Cryptosporidium = yes
	Giardia = controversial

occurred in developed countries due to water treatment failures. Generally the intestinal protozoa are relatively host specific and zoonotic transmission (i.e., acquisition from animals) is generally not a problem. However, zoonotic transmission of *Cryptosporidium* does occur and the zoonotic transmission of *Giardia* can occur. Control activities focus on avoiding the ingestion of material contaminated with feces. Health promotion and education aimed at improving personal hygiene (e.g., hand-washing, food handling, etc.) are effective at reducing person-to-person transmission. Better infrastructure and improved socioeconomic status can help lower the level of endemicity.

The Gastrointestinal Tract

The gastrointestinal tract, or alimentary canal, essentially consists of a long tube running from the mouth to the anus. This includes the esophagus, stomach, small intestine, large intestine, and rectum. The small intestine is divided into three sections: the duodenum (first or proximal portion after the stomach); the jejunum (the middle portion); and the ileum (the distal or last portion before the large intestine). Digestive enzymes from the pancreas and bile from the gall bladder are secreted into the duodenum. Most of the digestion of the proteins, carbohydrates, and lipids occurs in this region and the jejunum and ileum function primarily in absorption of nutrients and water. Continued absorption of nutrients and water, and concentration of the feces, is carried out by the large intestines. The pouch-like region where the small intestine connects to the large intestine is called the cecum. The appendix extends from the cecum. The colon is divided into four sections: ascending colon, transverse colon, descending colon, and sigmoid colon. The sigmoid region is followed by the rectum which terminates with the anus. Generally the intestinal protozoa undergo excystation after passing through the stomach and the trophozoites of many species will colonize a specific portion of either the small or large intestine, whereas others are capable of infecting both the small and large intestines.

The intestinal tract consists of four basic layers: the **mucosa**, the submucosa, the muscle layer, and the serosa (Figure 2.2). The mucosa is the innermost layer and consists of epithelial cells, or enterocytes, and a layer of

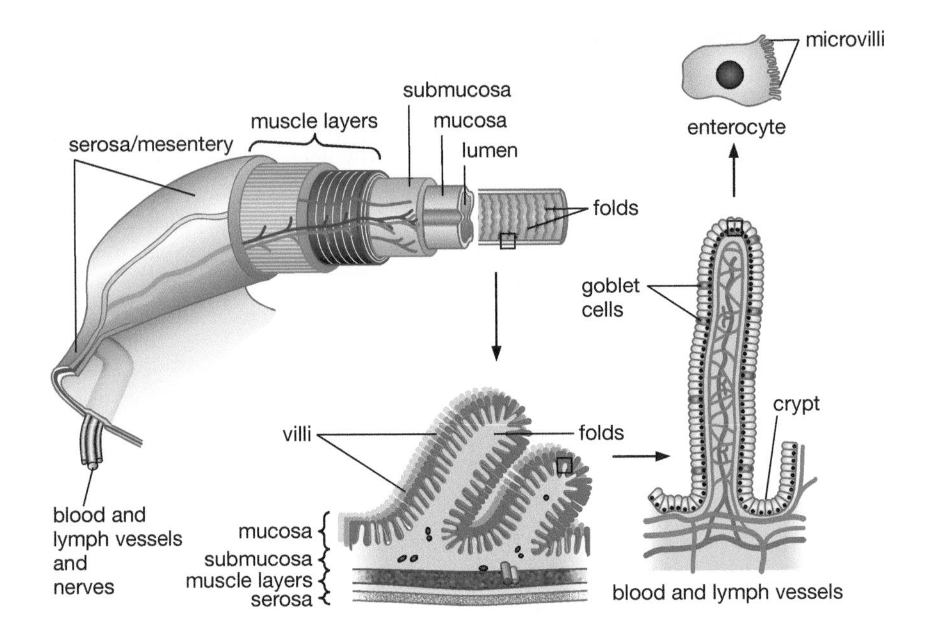

Figure 2.2 Intestinal structure. The intestines are a long tube consisting of multiple layers. The innermost layer, or mucosa, is heavily folded and villi project from the surface. The epithelial layer consists of enterocytes and mucus-secreting goblet cells. The cells found in the crypts of the villi function as stem cells to continuously replace the epithelial layer.

connective tissue called the lamina propria. The epithelial cells are responsible for forming a barrier between the contents of the intestinal lumen and the underlying layers. In addition, the epithelial cells are also responsible for the uptake of nutrients. Thus the mucosa must be able to efficiently take up nutrients and at the same time exclude toxins and microorganisms. The surface area of the intestinal mucosa—and thus the efficiency of uptake—is increased by numerous folds with projections known as villi. In addition, the epithelial cells making up these villi are covered with projections known as microvilli. The cells of the intestinal epithelium exhibit a high turnover rate and are replaced every few days. Stem cells replenishing these cells are found in the crypts (i.e., between the villi) of the small and large intestines. As the progeny of stem cells migrate toward the tips of the villi they differentiate into absorptive enterocytes or mucus-secreting goblet cells.

Below the mucosa is the submucosa, made up of connective tissue, blood vessels, nerves, and immune effector cells. This is followed by two layers of smooth muscle which are responsible for intestinal motility. The inner layer of muscle is oriented in a circular fashion around the lumen of the intestines, whereas the outer layer is orientated lengthwise (i.e., longitudinal). The outermost layer of the intestines consists of a thin connective tissue sheath known as the serosa. The serosa also forms the mesentery, which comprises flat sheets of connective tissue containing nerves and blood vessels, between the folds of the intestines.

The barrier formed by the mucosal layer has two components: an intrinsic barrier and an extrinsic barrier. The intrinsic barrier is formed by the epithelial cells which are connected to each other by tight junctions between the cells. This forms a continuous sheet lining the inner surface of the intestines. The extrinsic barrier consists of secretions, such as mucus, and other factors that affect the epithelial cells and promote the barrier function. Mucus is composed of glycoproteins, known as mucins, and inorganic salts suspended in water. Mucus serves as both a diffusion barrier protecting epithelial cells from noxious substances and a lubricant to minimize shear stresses. Pathogens that manage to cross these intrinsic and extrinsic barriers will encounter lymphocytes and other immune effector cells which are found in the lamina propria and submucosal layer. In addition, specialized lymphoid tissue, often referred to as gut-associated lymphoid tissue (GALT), is found throughout the gastrointestinal tract (Box 2.1).

Box 2.1 Gut-associated lymphoid tissue

The gastrointestinal tract, in addition to its role in nutrient uptake, represents a large surface area (200 times greater than the skin) that functions as an interface between the human host and an extremely high number and diverse array of bacteria and other microorganisms. The immune system of the gastrointestinal tract must continuously monitor for potentially harmful agents (e.g., pathogens) and discriminate these from harmless food and nonpathogenic antigens. The digestive tract is heavily laden with lymphocytes, macrophages, and other immune effector cells, as well as specialized lymphoid tissues collectively referred to as the gut-associated lymphoid tissue (GALT). The GALT consists of Peyer's patches, mesenteric lymph nodes, and other lymphoid aggregates in the large intestine, as well as lymphocytes (primarily IgA-secreting B cells) throughout the lamina propria.

Peyer's patches are macroscopic secondary lymphoid organs found in the submucosa of the small intestine and are similar to other lymph nodes. Smaller lymphoid nodules are also found throughout the intestinal tract. Mature Peyer's patches consist of large B-cell areas, or follicles, interspersed with T-cell areas. In the epithelium above the Peyer's patch are specialized enterocytes known as M (microfold) cells, which lack microvilli and the normal thick layer of mucus. These M cells take up antigens and pathogens from the lumen of the gut and deliver them to the Peyer's patch by a transcytosis process. Dendritic cells and macrophages within the Peyer's patch then take up and process these antigens and present them to lymphocytes (i.e., T and B cells). These primed lymphocytes then drain from the Peyer's patches into the mesenteric lymph nodes for further differentiation before migrating back to the mucosa. Both helper T cells (i.e., $CD4^+$) and cytotoxic T cells (i.e., $CD8^+$) are primed in the Peyer's patch and subsequently distributed to the mucosa. The exact role of the T cells in gastrointestinal immunity is not known. Presumably the helper T cells play some type of regulatory role and in particular may help local B cells produce IgA. The cytotoxic T cells may participate in the direct killing of pathogens. Primed B cells undergo a class switching from IgM to IgA and mature into IgA-secreting plasma cells in the lamina propria. The IgA is transported across the epithelial cells into the lumen of the digestive system and, like other antibodies (Chapter 1), function to neutralize potential pathogens.

Gastrointestinal Tract Pathophysiology and Diarrhea

Symptoms associated with infections of gastrointestinal tract often include nausea, vomiting, diarrhea, and abdominal distress and can be referred to as gastroenteritis. **Diarrhea** is an increase in the frequency of stools and fluid in the stools. Generally it reflects a disease in the small intestine and involves increased fluid and electrolyte loss due to an excess of secretion compared with absorption. Dysentery is associated with blood and mucus in the stools and is often accompanied by pain, fever, and abdominal cramps. Generally dysentery results from a disease of the large intestine. Enterocolitis refers to inflammation of the mucosa of the small and large intestines and colitis is an inflammation of the colon or large intestine.

Diarrhea can be described as being osmotic, inflammatory, or secretory (Table 2.3). Osmotic diarrhea can result from an excessive amount of solutes in the intestinal lumen which prevents the small intestines from effectively absorbing water due to the increased osmolarity of the intestinal lumen. For example, ingestion of poorly absorbed substrates can cause osmotic diarrhea. Some carbohydrates, such as manitol or sorbitol, and some divalent cations are not absorbed and ingestion of large quantities of these can result in osmotic diarrhea. Conditions leading to malabsorption of normally absorbed substrates will also lead to osmotic diarrhea. One common example is lactose intolerance which is caused by a deficiency in the enzyme lactase on the surface of intestinal epithelial cells. Lactase cleaves lactose, which is not taken up by epithelial cells, into glucose and galactose, both of which are taken up by epithelial cells. Therefore, ingestion of dairy products by a lactose-intolerant person will lead to high levels of lactose which will result in decreased absorption of water and consequently diarrhea. Incidentally, infections with *Giardia lamblia* often cause lactose intolerance (Chapter 4).

Table 2.3 Types of diarrhea

Osmotic	Enterocyte malfunction impaired absorption enhanced secretion
	Excessive solutes
Inflammatory	Mucosal invasion
	Leukocytes in stools
Secretory	Toxin associated
	Excessively watery

Damage to intestinal epithelial cells can also contribute to osmotic diarrhea in that absorption of water is impaired and excretion of water is enhanced (Figure 2.3). Most of the absorption is carried out by enterocytes toward the distal ends or tips of the intestinal villi. Coupled with the absorption of solutes is a cotransport of water. When enterocytes are damaged or killed the crypt cells, functioning as stem cells, replicate at higher rates to replace the damaged epithelium. The crypt cells normally secrete large volumes of water associated with the excretion of chloride ions. This water is normally reabsorbed in conjunction with the uptake of sodium ions by mature enterocytes before reaching the large intestine. Several intestinal protozoa damage enterocytes and cause villus atrophy, or blunting, and therefore result in both an enhanced secretion, due to crypt cell hyperplasia, and an impaired absorption due to enterocyte damage.

Disruption of the epithelial cell barrier by pathogens can also lead to ulcers or an inflammatory diarrhea (Figure 2.3). An inflammatory immune response directed at pathogens crossing the epithelial barrier, and the resulting inflammatory cytokines, can stimulate secretion and increase the permeability between the intestinal epithelial cells. Inflammatory diarrhea is often associated with leukocytes in the feces. In addition, disruption of the epithelium can lead to the exudation of serum and blood into the intestinal lumen and cause dysentery. Among the protozoa, only *Entamoeba histolytica* routinely causes dysentery. The types of diarrhea are not mutually exclusive and various mechanisms can simultaneously contribute to the diarrhea. For example, the pathogen and/or reactive oxygen intermediates produced by the inflammatory cells damage or kill the epithelial cells. As discussed above, the immature epithelial cells replacing the damaged epithelium have less absorptive capacity and thus an osmotic component is added to the inflammatory component.

Secretory diarrhea is usually associated with toxins produced by some pathogenic bacteria. No toxins have been definitively identified in intestinal protozoa and thus secretory diarrhea is probably not associated with protozoan infections. The bacterial toxins bind to receptors on the host cell and activate adenylate cyclase. The resulting cyclic AMP and subsequent cascade of the second messenger pathway ultimately activate the chloride channels in the crypt cells leading to a prolonged opening and excessive secretion of water. Massive secretory diarrhea can be lethal and requires aggressive rehydration therapy.

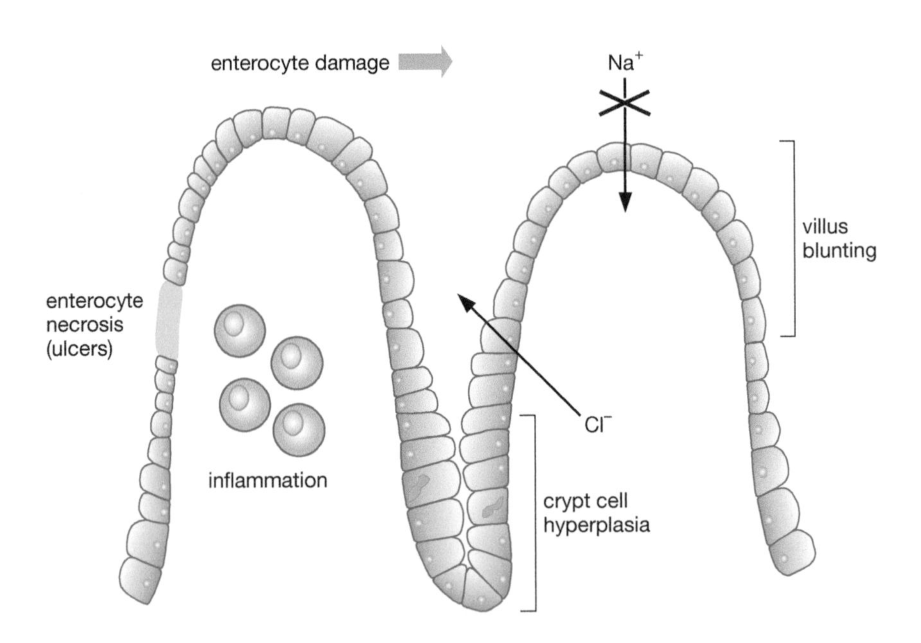

Figure 2.3 Intestinal pathogenesis. Damage of the intestinal epithelial cells (i.e., enterocytes) results in a loss of sodium absorption and crypt cell hyperplasia which results in increased chloride secretion. Both the decreased absorption and increased secretion result in enhanced excretion of water because water is cotransported with the ions. Disruption of the epithelial cell layer will result in inflammation and increased permeability and can produce dysentery.

Intestinal Amebas and Flagellates of Lesser Importance

Many of the protozoa that inhabit the gastrointestinal tract of humans can cause disease, ranging from mild to severe. These protozoa and the diseases they cause will be discussed in subsequent chapters. However, some of the intestinal protozoa exhibit little or no overt pathology and some are considered to be nonpathogenic **commensals**. Commensalism is generally defined as an association between two organisms from which the commensal benefits, but the host neither benefits nor suffers. Among the amebas, *Entamoeba coli, Entamoeba hartmanni, Endolimax nana,* and *Iodamoeba bütschlii* are generally considered to be nonpathogenic commensals. Likewise, *Chilomastix mesnilli, Pentatrichomonas hominis, Retortamonas intestinalis,* and *Enteromonas hominis* are often considered to be nonpathogenic flagellates. Because of their nonpathogenicity little research has been carried out on these species and much less is known about them. For many of these nonpathogenic species our knowledge is primarily limited to morphological descriptions of the cyst and trophozoite stages.

Even though these nonpathogenic amebas and flagellates probably cause no disease, diagnosis of infection should be of some concern because such infections are evidence of ingestion of food or water contaminated with fecal matter. This indicates a risk of more serious infections by organisms such as *Giardia* or *E. histolytica* and some behavior modification may be needed. These nonpathogenic species are also important in that they can potentially complicate diagnosis and treatment. For example, confusion of nonpathogenic species with the potentially pathogenic species can result in unnecessary drug treatment. In addition, such a misdiagnosis is problematic in that the true cause of the symptoms may be missed and the appropriate treatment delayed.

Subtle Differences Help Distinguish Intestinal Protozoa

The small size and somewhat similar morphologies of the intestinal amebas and flagellates can make them difficult to distinguish from one another. Characteristics such as shape, size, nuclear morphology, number of nuclei, and unique structures are used to distinguish the various intestinal amebas and flagellates (Tables 2.4 and 2.5). For example, trophozoites of amebas tend to have an amorphous shape due to the extension of pseudopodia and exhibit ameboid movement (Chapter 1), whereas flagellates tend to be somewhat spindle shaped with a broad blunt end and a tapered end and have flagella (Figure 2.4). However, this is not always true as exemplified by *D. fragilis* which exhibits a typical ameba-like morphology and motility and lacks flagella even though it is more closely related to *Trichomonas* than to the amebas (Chapter 5). Furthermore, quite often intestinal protozoa will round up during collection and storage of samples resulting in

Table 2.4 Distinguishing features of some intestinal amebas*

	Endolimax nana	*Iodoamoeba bütschlii*	*Dientamoeba fragilis*
Trophozoite size	8–10 μm	12–15 μm	8–10 μm
Cyst structure	6–8 μm	10–12 μm	No cyst stage
	4 nuclei	1 nucleus	
Unique feature		Glycogen vacuole	Often binucleated

**Entamoeba* species are discussed in Chapter 3 and *D. fragilis* is actually a flagellate (Chapter 5).

Table 2.5 Distinguishing features of some intestinal flagellates

	Chilomastix mesnili	*Pentatrichomonas hominis*	*Retortamonas intestinalis*	*Enteromonas hominis*
Trophozoites	10–20 μm	7–15 μm	5–7 μm	7–9 μm
	4 flagella	5 flagella	2 flagella	4 flagella
Cysts	6–10 μm	No cyst stage	4–7 μm	6–8 μm
	1 nucleus		1 nucleus	1–4 nuclei
Unique features	Cytostome	Axostyle		
		Undulating membrane		

all trophozoites exhibiting a similar shape. In addition, flagella are not well preserved in stained fecal smears. Cyst stages tend to be round and/or oval shaped in all of the intestinal species. Size can also be used as a distinguishing characteristic in that each species exhibits a defined size range. However, many of the species exhibit overlapping size ranges and this characteristic by itself is not definitive.

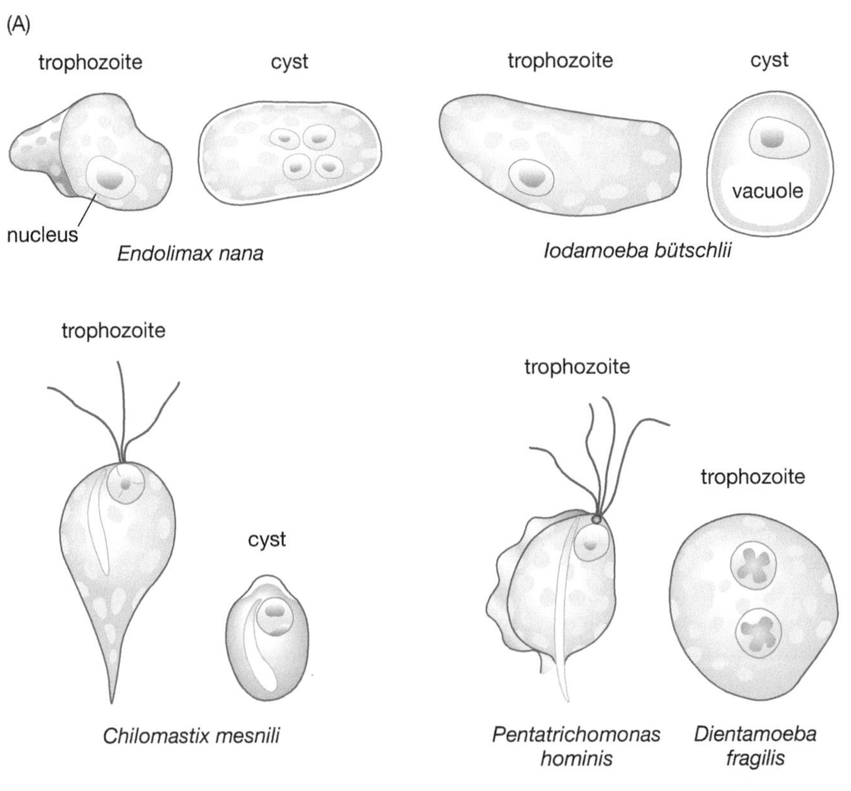

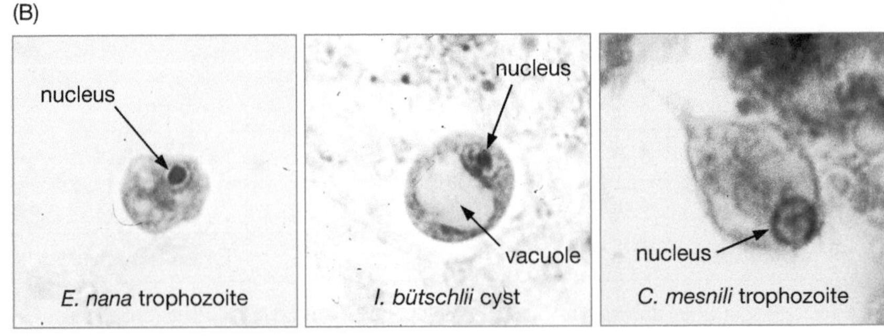

Figure 2.4 Examples of nonpathogenic intestinal protozoa. (A) Schematic representation of selective amebas and flagellates found in human feces. (Adapted from figures by the World Health Organization.) (B) Micrographs of a *Endolimax nana* trophozoite (left), *Iodamoeba bütschlii* cyst (middle), and *Chilomastix mesnili* trophozoite (right) from stained fecal preparations. Arrows denote nuclei and the glycogen vacuole. (Pictures of *I. bütschlii* and *C. mesnili* kindly provided by Thomas Orihel, Tulane University.)

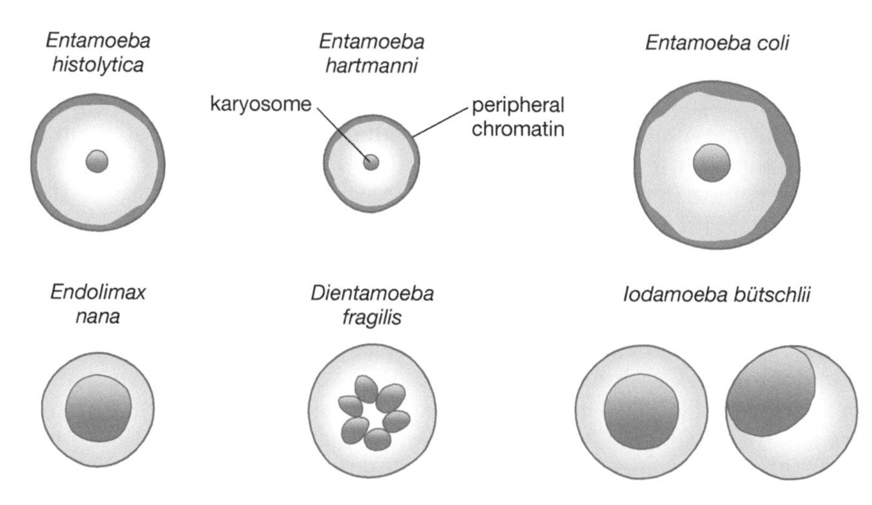

Figure 2.5 Nuclear morphologies of intestinal protozoa. Nuclei of *Entamoeba* species are characterized by a thick ring of peripheral chromatin and a small karyosome (or nucleolus) in contrast to nuclei with a large karyosome and no peripheral chromatin.

This difficulty in distinguishing intestinal protozoan species is especially true in the case of *E. nana*, *I. bütschlii* and *D. fragilis* (Table 2.4). The trophozoite stages of all three species exhibit a similar size range and morphology and the clearest distinguishing feature is the tendency for *Dientamoeba* to have two nuclei. The cysts, on the other hand, are easier to distinguish in that *E. nana* has four nuclei, *I. bütschlii* has a single nucleus and a prominent glycogen vacuole, and *D. fragilis* completely lacks a cyst stage. The glycogen vacuole stains an intense reddish-brown with iodine and appears as a large open area in other types of staining (Figure 2.4).

Nuclear morphology is another criterion that is useful in distinguishing the various amebas found in feces. In general there are two major types of nuclear morphologies (Figure 2.5). In one type the outer edge of the nucleus stains and within this circle is a small dot. The staining of the outer edge is referred to as peripheral chromatin and the small central dot is called a karyosome or nucleolus. The other type of nuclear morphology consists of a large karyosome and no peripheral chromatin. The lack of peripheral chromatin often results in a halo-like appearance around the karyosome.

Among the intestinal flagellates *Giardia* is clearly the most important in terms of causing disease and the number of people infected. The trophozoites and cysts of *Giardia* exhibit several unique features that make them relatively easy to identify (Chapter 4). The other intestinal flagellates, with the exception of *D. fragilis*, are generally considered to be nonpathogenic commensals and exhibit somewhat similar morphologies in that the trophozoites tend to be elongated with a tapered end. Features such as the number of flagella in the trophozoite stages and the number of nuclei in the cyst stages, as well as the presence of unique structures such as a cytostome, axostyle, or undulating membrane, can be used to distinguish these species (Table 2.5).

Diagnosis

Microscopic detection of parasitic protozoa in feces is generally the most common means of diagnosing intestinal infections. It is possible to directly observe mobile trophozoites (i.e., flagellates and amebas) in freshly collected samples and distinguish the organisms to a limited degree by the type of motility. However, direct observation can be a rather insensitive method. Therefore, the most widespread diagnostic methods usually involve microscopy of permanently stained specimens. As part of preparing stained specimens the samples are generally subjected to a fixation step. This serves to preserve the organisms in the sample. This is especially important for the trophozoite stages which tend to be quite labile and do not survive long outside of the host. Several different chemical fixatives are available. The ability to detect subcellular structures for the identification of protozoa is greatly enhanced with biological stains. The two most common stains used for the detection of intestinal protozoa are iron hematoxylin and trichrome. Iron hematoxylin is an older method and is currently not as widely used as the trichrome method. Acid fast is the preferred stain for the detection of *Cryptosporidium* and intestinal coccidia (Chapters 12 and 13).

Some of the intestinal protozoa can be cultured *in vitro*. However, this is not widely used for clinical diagnosis because it can be time consuming and generally is not more sensitive or specific than microscopy. Recently, antigen and DNA detection has been applied to the diagnosis of intestinal protozoa. Some antigen detection kits based on ELISA and immunochromatographic methods are available. These require fresh or frozen samples and do not work with fixed specimens. Similarly, DNA can be detected by PCR with specific primers. The complexity of feces and the presence of substances that inhibit the PCR complicate such analyses and the optimization of fecal DNA extraction procedures is critical for success.

Summary and Key Concepts

- A variety of pathogenic and nonpathogenic protozoa inhabit the human gastrointestinal tract.
- The majority of intestinal protozoan infections are transmitted via a fecal–oral mechanism.
- The mucosal layer of the intestines consists of epithelial cells, functions as a physical barrier, and contains extensive components of the immune system to prevent infection by pathogens.
- Gastroenteritis associated with pathogenic protozoan infections is typically manifested as nausea, vomiting, diarrhea, and abdominal distress.
- Diarrhea is a frequent symptom of gastrointestinal infections and in the case of protozoan infections is generally the result of damage to the epithelium.

Further Reading

Mowat, A.M. (2003) Anatomical basis of tolerance and immunity to intestinal antigens. *Nat. Rev. Immunol.* 3: 331–341.

Entamoeba and Amebiasis

Disease(s)	Amebiasis, amebic dysentery
Etiological agent(s)	*Entamoeba histolytica*
Major organ(s) affected	Colon, liver
Transmission mode or vector	Fecal–oral
Geographical distribution	Worldwide, but more prevalent in tropical and developing countries
Morbidity and mortality	Generally asymptomatic or mild symptoms; severe symptoms include dysentery and spread to other organs that can be fatal
Diagnosis	Detection of parasites in feces
Treatment	Metronidazole
Control and prevention	Avoid fecal contamination of food or water

Several members of the genus *Entamoeba* infect humans. Among these only *E. histolytica* is considered pathogenic and the disease it causes is called amebiasis or amebic dysentery. Humans are the only host of *E. histolytica* and there are no zoonotic reservoirs. *E. dispar* is morphologically identical to *E. histolytica* and the two were previously considered to be the same species. However, genetic and biochemical data clearly indicate that the nonpathogenic *E. dispar* is a distinct species. The two species are found throughout the world, but like many other intestinal protozoa, they are more common in tropical countries or other areas with poor sanitary conditions. High rates of amebiasis occur in the Indian subcontinent, the Far East, western and southern Africa, and parts of South and Central America. In the United States and Europe amebiasis is found primarily in immigrants from endemic areas. It is estimated that up to 10% of the world's population may be infected with either *E. histolytica* or *E. dispar* (or both) and in many tropical countries the prevalence may approach 50%. There are an estimated 50 million clinical cases of amebiasis per year with up to 100 000 deaths.

Life Cycle and Morphology

E. histolytica exhibits a typical fecal–oral life cycle consisting of infectious cysts passed in the feces and trophozoites which replicate within the large intestine. The infection is acquired through the ingestion of cysts and the risk factors are similar to other diseases transmitted by the fecal–oral route (see Chapter 2). Contaminated food and water are probably the primary sources of infection. The higher prevalence in areas of lower socioeconomic status is likely due to poor sanitation and a lack of indoor plumbing. However, *E. histolytica* is rarely the cause of travelers' diarrhea and is usually associated with long-term (>1 month) stays in an endemic area. A higher prevalence of *E. histolytica* infection is also observed in institutions, such as mental hospitals, orphanages, and prisons, where crowding and problems with fecal contamination are contributing factors. A high prevalence among male homosexuals has also been noted in several studies.

Upon ingestion the cysts pass through the stomach and excyst in the lower portion of the small intestine. Excystation involves a disruption of the cyst wall and the quadrinucleated ameba emerges through the opening. The ameba undergoes a round of nuclear division followed by three successive rounds of cytokinesis (i.e., cell division) to produce eight small uninucleated trophozoites, sometimes called amebula. These trophozoites colonize the large intestine, especially the cecal and sigmoidorectal regions, where they feed on bacteria and cellular debris and undergo repeated rounds of binary fission. Like many other intestinal protozoa, *Entamoeba* trophozoites are obligate fermenters and lack enzymes of the tricarboxylic acid cycle and proteins of the electron transport chain. In keeping with this anaerobic metabolism the parasites also lack mitochondria and only have a mitochondrial remnant called a mitosome. Interestingly, *E. histolytica* appears to have obtained many of its metabolic enzymes through lateral gene transfer from bacteria.

E. histolytica trophozoites have an amorphous shape and are generally 15–30 μm in diameter. The trophozoites move by extending a pseudopodium and pulling the rest of the body forward (called ameboid movement). Pseudopodia of *Entamoeba* tend to be the broad blunt type called lobopodia (Chapter 1). The pseudopodia, and sometimes the outer edge of the trophozoite, have a clear refractile appearance which is referred to as the ectoplasm. The rest of the cytoplasm has a granular appearance and is called the endoplasm. Occasionally a large glycogen vacuole is evident. Nuclear morphology in stained specimens is characterized by a granular ring of peripheral chromatin and a centrally located karyosome (i.e., nucleolus).

As an alternative to asexual replication trophozoites can also encyst. The factors responsible for the induction of encystation are not known. However, it has been suggested that aggregation of trophozoites in the mucin layer may trigger encystation. Encystation begins with the trophozoites becoming more spherical and the appearance of **chromatoid bodies** in the cytoplasm. Chromatoid bodies are stained elongated structures with round ends and represent the aggregation of ribosomes. The cyst wall is composed of chitin and has a smooth refractile appearance. Cyst maturation involves two rounds of nuclear replication without cell division and cysts with 1–4 nuclei are found in feces (Figure 3.1). The nuclear morphology of the cyst is similar to that of the trophozoite except that the nuclei become progressively smaller following each division. The chromatoid

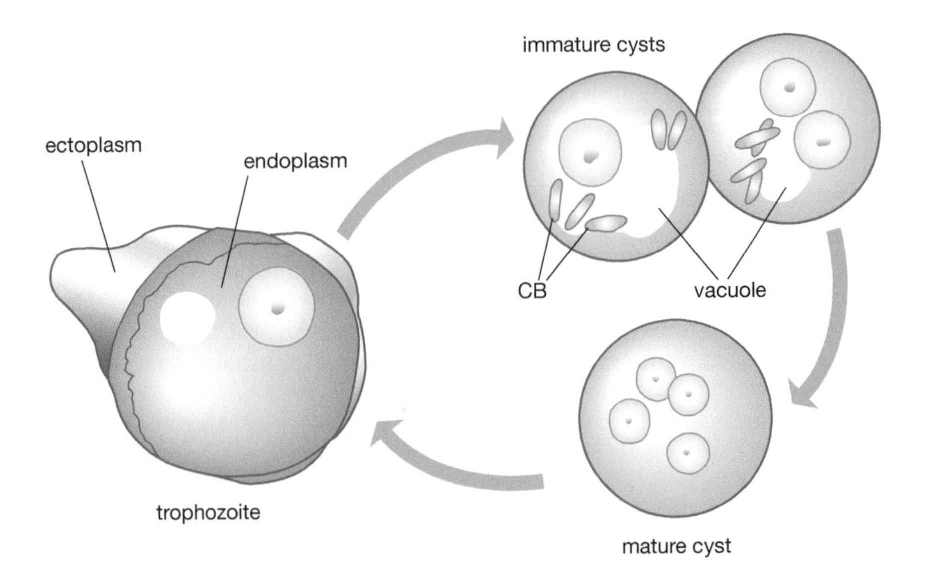

Figure 3.1 Life cycle. Ameboid trophozoites inhabit the colon. Immature cysts, or precysts, with one or two nuclei, as well as the mature quadrinucleated cysts, are commonly found in the feces. Chromatoid bodies (CB) are also found in the precysts and sometimes large glycogen vacuoles are evident.

Table 3.1 Disease manifestations of amebiasis

Noninvasive disease	Invasive disease	Extraintestinal amebiasis
Ameba colony on mucosa surface	Necrosis of mucosa → ulcer	Metastasis via blood stream or direct extension
Asymptomatic cyst passer	Amebic colitis (i.e., dysentery)	Primarily liver abscess
Nondysenteric diarrhea	Ulcer enlargement → fulminating colitis or peritonitis	Other sites less frequent
Other abdominal symptoms	Occasional ameboma	Ameba-free stools common

bodies tend to disappear as the cyst matures. The cysts are generally 12–15 μm in diameter. Cysts are immediately infective upon excretion with the feces and will be viable for weeks to months depending on environmental conditions.

Pathogenesis

The pathogenesis associated with *E. histolytica* infection can range from a noninvasive intestinal disease to an invasive disease which can also include an extraintestinal disease (Table 3.1). The noninvasive disease is often asymptomatic, but can cause diarrhea or other gastrointestinal symptoms such as abdominal pain or cramps. Most infections will exhibit no overt clinical manifestations and self-resolve in a few months. The noninvasive infection can also persist as a chronic noninvasive disease or progress to an invasive disease in which trophozoites penetrate the intestinal mucosa. This invasive disease can become progressively worse and lead to a more serious disease. The amebas can also metastasize to other organs and produce an extraintestinal amebiasis. In other words, *E. histolytica* is a facultative pathogen that exhibits a wide range of virulence. (See Box 3.1 for discussion of difference between pathogenicity and virulence.)

In the invasive disease, trophozoites kill epithelial cells and invade the colonic epithelium. The early lesion is a small area of necrosis, or ulcer, characterized by raised edges and virtually no inflammation between lesions (Figure 3.2). The clinical syndrome associated with this stage of the disease is an amebic **colitis** or **dysentery**. Dysentery is characterized by frequent stools containing blood and mucus. The lesions start off as a small ulcer of the mucosal layer. The ameba will spread laterally and downward in the submucosa (beneath the epithelium) and kill host cells as they progress. This results in the classic "flask-shaped" ulcer with a small opening and a wide base. These invasive amebas kill and ingest host cells as they are expanding through the submucosa. Thus trophozoites with ingested erythrocytes are often evident in the lesions and these **hematophagous** trophozoites are sometimes found in the dysenteric feces. The trophozoites also replicate at a high rate in the host tissues. However, cyst production decreases during the invasive stage of the infection and cysts are never found in the tissue lesions.

Expansion of ulcers results in a fulminating necrotic colitis

The ulcers can exhibit a wide range of sizes. The larger ulcers are characterized by a central area of necrosis due to the destruction of the host cells and tissue by the amebas (Figure 3.2B). Few amebas are found in these

Box 3.1 Pathogenicity vs. virulence

Pathogenicity refers to the ability of an organism to cause disease (i.e., harm the host). This ability represents a genetic component of the pathogen and the overt damage done to the host is a property of the host–pathogen interaction. Commensals and opportunistic pathogens lack this inherent ability to cause disease. However, disease is not an inevitable outcome of the host–pathogen interaction and pathogens can express a wide range of virulence. Virulence refers to the degree of pathology caused by the organism. The extent of the virulence is usually correlated with the ability of the pathogen to multiply within the host and may be affected by other factors (i.e., conditional). In summary, an organism is defined as being pathogenic (or not) and, depending upon conditions, may exhibit different levels of virulence.

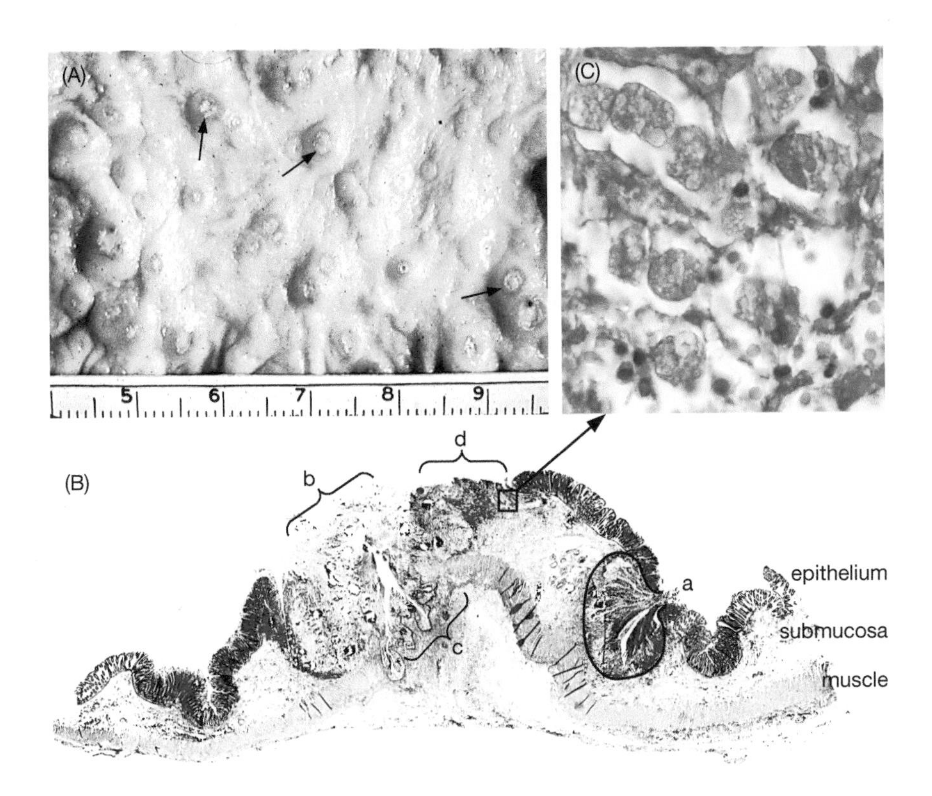

Figure 3.2 Invasion of the intestinal mucosa by *E. histolytica*. (A) The luminal side of the colon from amebiasis case showing several ulcers (a few of which are denoted with arrows). Note raised edges and the lack of inflammation between the lesions. (B) Section of human colon showing two lesions. On the right side is an early lesion (a) with a small opening in the mucosal layer and a comparatively large area of inflammation in the submucosal layer (outlined). A massive lesion with extensive necrosis is found on the left side (bracket b) and has extended completely through the submucosa and penetrated the muscle and serous layers (bracket c). An area of the epithelium has sloughed off (bracket d) and the exposed submucosa is highly inflamed. (C) High power magnification of submucosal region showing numerous *E. histolytica* trophozoites. Note also the lack of intact host cells except for erythrocytes in the lower right corner. (Figure A kindly provided by Lawrence R. Ash, UCLA, School of Public Health.)

areas of necrosis and the trophozoites are most numerous at the boundary between the healthy tissue and the necrotic tissue. The ulcerative process may continue to expand laterally and downward. If large numbers of ulcers are present, they may coalesce leading to extensive mucosal necrosis and even possibly can lead to a localized sloughing off of the intestinal wall. Generally the expansion of the ulcer is limited by the muscle layer. However, on occasion the muscle and serous layers can be penetrated resulting in a perforation of the intestinal wall. This perforation is usually a rather dramatic event and is accompanied by a generalized peritonitis. In addition, erosion of blood vessels can produce hemorrhaging.

Amebic ulcers, especially larger ulcers, can also become secondarily infected with bacteria. Bacterial infections promote further inflammation and the formation of abscesses. In addition to confusing the clinical situation, secondary bacterial infections can intensify the destructive process. *E. histolytica* infection can also occasionally lead to the formation of an amebic granuloma, also called an **ameboma**. The ameboma is an inflammatory thickening of the intestinal wall around the ulcer which can be confused with a tumor.

Extraintestinal amebiasis

Amebiasis can also progress to an extraintestinal infection. Dissemination from the primary intestinal lesion is predominantly via the blood stream. Trophozoites entering capillaries in the large intestine can be carried to other organs. The liver is the most commonly affected organ and this is probably due to the direct transport of trophozoites from the large intestine to the liver via the mesenteric blood vessels feeding into the hepatic portal vein (Figure 3.3). This provides a more or less direct connection between the large intestine and the liver in that the portal vein drains most of the blood from the cecum and ascending colon. Initially the liver lesions are small foci of necrosis which tend to coalesce into larger abscesses as they expand. These hepatic abscesses will continue to enlarge as the trophozoites progressively destroy and ingest host cells. The center of the abscess, consisting of lysed hepatocytes, erythrocytes, bile, and fat, may liquefy and

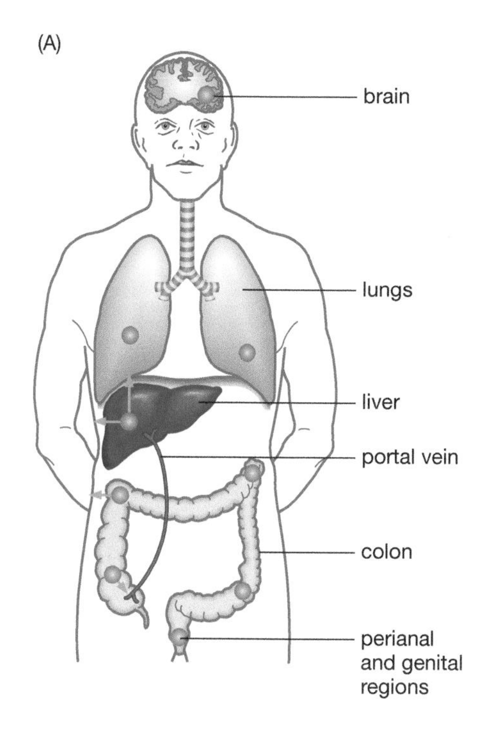

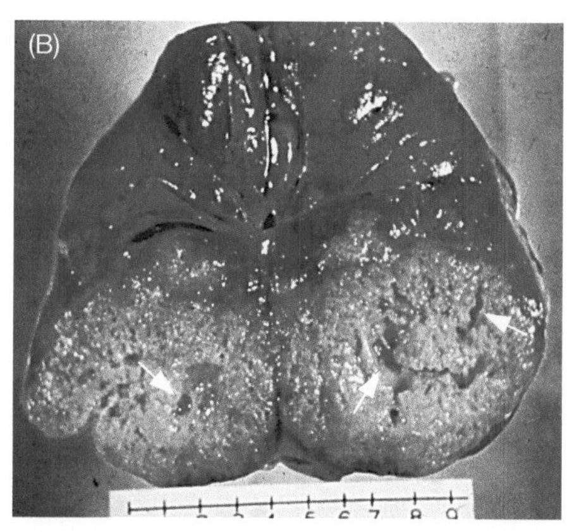

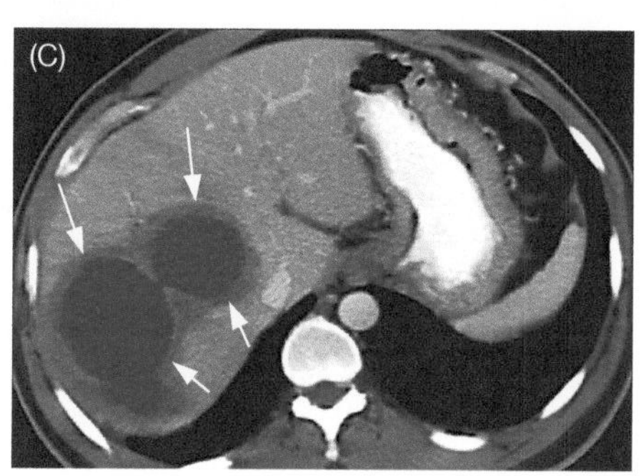

Figure 3.3 Extraintestinal amebiasis. (A) Diagram showing spread of *E. histolytica* from the colon to other organs via a hematogenous route primarily involving the portal vein and the liver. The ameba can also spread via a direct expansion causing pulmonary infections, cutaneous lesions or perianal ulcers. (B) Section of liver showing fluid-filled abscesses (arrows denote a few of them). (C) Computerized tomography (CT scan) showing hepatic lesion (arrows). (Figure B kindly provided by Antonio D'Alessandro, Tulane University, Department of Tropical Medicine. Figure C reprinted from http://imaging.consult.com/case/Liver%20Amebiasis/S1933-0332(07)72777-5 with permission from Elsevier Ltd.)

this necrotic material (sometimes incorrectly called pus) will range in color from yellowish to reddish brown. Secondary bacterial infections in the liver abscess are not common (~2% of the cases).

Hematogenous spread of trophozoites to other sites, such as the lungs or brain, is rare, but does occur. The second most common extraintestinal site after the liver is the lungs. Pulmonary infections generally result from a direct extension of the hepatic lesion across the diaphragm and into the pleura and lungs. Cutaneous lesions formed as a result of hepatic or intestinal fistula can also occur, although are relatively rare. Other cutaneous lesions include perianal ulcers and involvement of the genitalia. These manifestations are likely due to the skin or mucous membranes coming in contact with fluids containing invasive trophozoites. Fistulas forming between the rectum and vaginal have also been reported.

Possible Mechanisms of Pathogenesis

As discussed above, *E. histolytica* is pathogenic that exhibits a wide spectrum of virulence, ranging from an avirulent commensal to a highly invasive and destructive organism. In a historical sense, some of this difference in virulence is explained by the existence of the morphologically identical, but avirulent, *E. dispar*. Diagnosis by microscopy alone cannot distinguish *E. dispar* infections from *E. histolytica* infections and thus in the absence of antigen- or molecular-based tests many infections diagnosed as *E. histolytica* were likely *E. dispar*. The ability to distinguish the two species has shed some light on the issue of pathogenesis. For example, it appears that *E. dispar* is only capable of causing superficial erosions of the colonic mucosa and has not been associated with symptomatic invasive disease and infection does not elicit serum antibodies. However, infection with *E. histolytica* does not always lead to invasive disease and less than 10% of the infected individuals will develop symptomatic invasive amebiasis. In addition, confirmed *E. histolytica* infections are prevalent in asymptomatic individuals from endemic areas and anti-ameba humoral responses are observed in both asymptomatic and symptomatic *E. histolytica* infections. This suggests

Table 3.2 Possible virulence factors

Host factors	Ineffective innate immunity
	Inflammatory response
Parasite factors	Resistance to host response
	Adherence properties
	Cytolytic properties
	Ability to break down tissues
Environmental factors	Bacterial flora

that even in asymptomatic infections there is a limited amount of invasion across the intestinal epithelium in contrast to the situation with *E. dispar*.

The exact factors responsible for the pathogenesis of *E. histolytica* are not well understood. Pathology results from host–parasite interactions, and therefore host factors, parasite factors, environmental factors, or combinations of factors likely contribute to the disease state (Table 3.2). A key feature of the pathogenesis is the ability of *E. histolytica* trophozoites to penetrate the epithelial cell layer, thereby breeching the first line of host defense. Parasite factors that promote cytoadherence, cytotoxicity, and the breakdown of the tissues may contribute to the ability of the parasite to cross the epithelial cell barrier. The bacterial flora has also been speculated to influence the phenotype of *E. histolytica* and to affect the mucus layer. In regard to host factors, the development of invasive disease could be due to quantitative or qualitative aspects of the host immune response. In addition to an ineffective immune response which does not impede trophozoite invasion, an inappropriate immune response could contribute to pathogenesis. For example, recruitment of neutrophils and intense inflammation are noted in the early phases of amebic invasion. This inflammation could accelerate the tissue damage associated with amebic ulcers. However, inflammation surrounding established ulcers and abscesses is often minimal in consideration of the degree of tissue damage. Thus, inflammation may only be important during the initial stages of pathogenesis.

Immunity and parasite resistance to host immune responses

The nature of protective immune responses against *E. histolytica* is not clear. Innate or nonspecific immunity, as well as acquired immunity, are probably both important for the prevention of invasive disease. In regard to innate responses, the mucous layer covering the epithelial cells can prevent contact between trophozoites and host cells. In regard to acquired immunity, mucosal IgA responses occur as a result of infection and fecal IgA directed against a trophozoite surface protein, called the **Gal/GalNAc lectin** (Box 3.2), is correlated with a lower incidence of new *E. histolytica* infections. High titers of serum antibodies also develop in patients with liver abscesses. However, since the invasive disease is often progressive and unremitting, the role of these anti-ameba antibodies is in question. Cell-mediated responses appear to play a larger role in limiting the extent of invasive amebiasis and protecting the host from recurrence following successful treatment.

Resistance to the host immune response is another possible virulence factor which could contribute to the development and exacerbation of invasive disease. For example, one phenotypic difference between *E. dispar* and *E. histolytica* is the resistance of the latter to complement-mediated lysis. In addition, *E. histolytica* rapidly degrades secretory IgA and possibly suppresses T-cell responses to amebic antigens. *E. histolytica* is also able to kill cells, including immune effector cells, in a contact-dependent manner. Lysis of neutrophils and other granulocytes could also release toxic products which contribute to the destruction of host tissue. However, the role of these various immunological phenomena in pathogenesis is not known.

Invasion of intestinal mucosa is mediated by the parasite

Normally trophozoites adhere to the mucous layer covering the epithelial cells and ingest bacteria and debris (Figure 3.4). This adherence *per se* probably does not contribute to pathogenesis and is simply a mechanism for

Box 3.2 Amebic Gal/GalNAc lectin and adherence

Adherence of *E. histolytica* to host cells and colonic mucins is mediated by a lectin activity expressed on the surface of the trophozoites. This lectin binds galactose (Gal) or *N*-acetyl-D-galactosamine (GalNAc) with a high affinity. The contact-dependent killing of target cells is inhibited by galactose and target cells lacking terminal galactose residues on their surface glycoproteins are resistant to trophozoite adherence and cytotoxicity. This suggests that the Gal/GalNAc lectin is an important virulence factor [1]. In addition, the Gal/GalNAc lectin may be involved in resistance to complement-mediated lysis due to its ability to bind components of complement. Because of its potential role in adherence and virulence and since fecal anti-lectin IgA protects against amebic colitis, the Gal/GalNAc lectin is a vaccine candidate [2].

The Gal/GalNAc lectin is a heterodimer consisting of a 170 kDa heavy chain and a 31 or 35 kDa light chain joined by disulfide bonds (Figure 3A). An intermediate subunit of 150 kDa is noncovalently associated with the heterodimer. The heavy chain has a transmembrane domain and a carbohydrate binding domain. All of the subunits are encoded by multigene families. There are five members of the heavy chain family, six to seven members of the light chain family, and 30 members of the intermediate chain family. The members of the heavy chain gene family exhibit 89–95% sequence identity at the amino acid level whereas the light chain family members are less conserved sharing only 79–85% sequence identity.

E. dispar also expresses a galactose-inhibitable lectin on its surface. Both *E. dispar* and *E. histolytica* adhere to the mucous layer which is mediated by the Gal/GalNAc lectin. Mucus is composed of glycoproteins called mucins and the predominant mucin found on the intestinal mucosa is MUC2 which is extensively glycosylated with O-linked GalNAc residues. The sequences of the light and heavy chain genes from *E. dispar* are homologous, but not identical, to those of *E. histolytica*. Antigenic differences between the Gal/GalNAc lectin of *E. dispar* and *E. histolytica* have also been described in that only two epitopes out of six are shared between the two species (see section on diagnosis). It is not known whether these sequence differences can account for the differences in virulence between *E. dispar* and *E. histolytica*. Adherence is obviously important for both species, but it is possible that the adherence is qualitatively or quantitatively different between the two species and that this accounts for the higher level of virulence in *E. histolytica*.

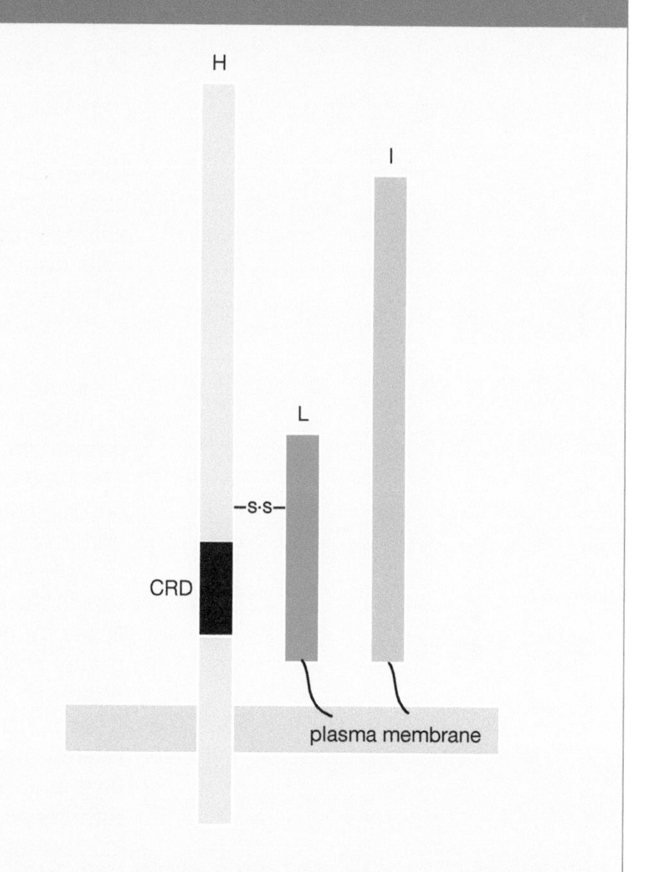

Figure 3A Proposed structure of Gal/GalNAc lectin. The Gal/GalNAc lectin is a heterotrimeric molecule composed of heavy (H), light (L), and intermediate (I) sized protein subunits. The H and L subunits are joined by disulfide bonds and the I subunit is noncovalently associated with the H–L dimer. The H chain has a predicted transmembrane domain and the L and I subunits may have GPI anchors. A carbohydrate recognition domain (CRD) is located in the H subunit. (Modified from Petri et al., 2002, *Annu. Rev. Microbiol.* 56: 39–64.)

1. Petri, W.A., Jr., Haque, R. and Mann, B.J. (2002) The bitter-sweet interaction of parasite and host: lectin–carbohydrate interactions during human invasion by the parasite *Entamoeba histolytica*. *Annu. Rev. Microbiol.* 56: 39–64.

2. Petri, W.A., Jr., Chaudhry, O., Haque, R. and Houpt, E. (2006) Adherence-blocking vaccine for amebiasis. *Arch. Med. Res.* 37: 288–291.

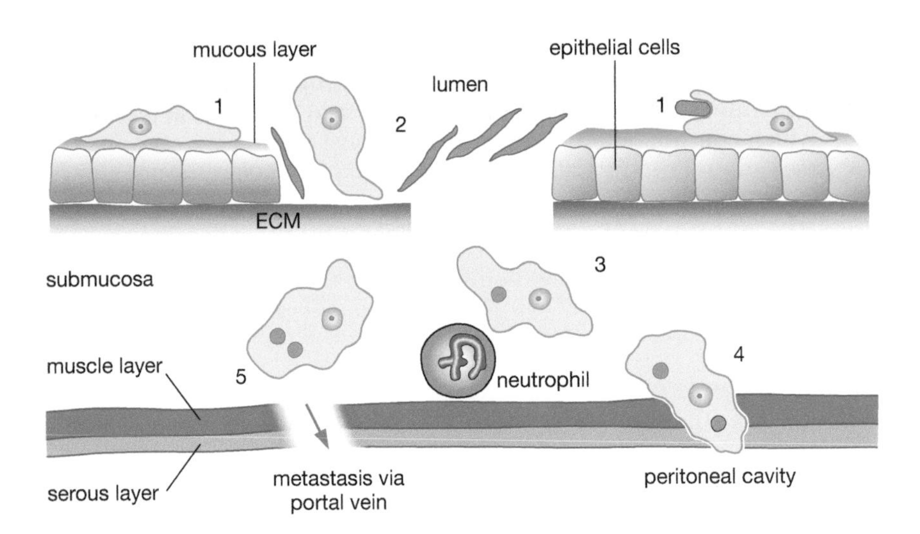

Figure 3.4 Schematic representation of *E. histolytica* pathogenesis. Trophozoites normally crawl along the mucous layer ingesting bacteria and debris (1). Erosion of the mucous layer allows for a contact-dependent killing of enterocytes and access to the lamina propria and submucosal layers (2). Continued host cell killing, including neutrophils and other immune effector cells, and a breakdown of the extracellular matrix (ECM) ensues (3). Perforation of the muscle and serous layers by the trophozoites can lead to peritonitis (4) and access to the circulatory system can result in the spread of the infection to other organs and in particular the liver (5).

the ameba to crawl along the substratum. Depletion of the mucous barrier allows for the trophozoite to come in contact with epithelial cells which are killed by the trophozoites in a contact-dependent manner. Killing epithelial cells leads to a disruption of the intestinal mucosa and gives the trophozoites access to the submucosa. A breakdown of the extracellular matrix is also noted during trophozoite invasion and provides more access to the submucosa. The trophozoites will continue to kill host cells in the submucosa and further disrupt the tissue as they spread laterally and downward. Neutrophils and other immune effector cells are also killed in a contact-dependent manner allowing for continued replication of the trophozoites. The destruction of the tissue also provides access to the circulatory system and metastasis to other organs and can lead to perforation of the colon wall and invasion of the peritoneal cavity.

Adherence, cytotoxicity, and disruption of the tissues are important factors in the pathogenesis of *E. histolytica.* Presumably parasite proteins play a role in these processes and some candidate proteins include: proteases, the Gal/GalNAc lectin, and pore-forming proteins. In addition, a possible approach to understanding the pathogenesis is to compare these factors from *E. histolytica* and *E. dispar.* These two species are somewhat closely related and primarily differ in their capacity to cross the epithelial cell layer and establish an active infection within the submucosa and beyond. Adherence, cytolytic activity, and proteolytic activity are inherent biological features of both species and these activities do not necessarily lead to pathology. However, there are qualitative and quantitative differences between *E. histolytica* and *E. dispar* which may account for the differences in virulence (Table 3.3).

Proteases are enzymes that degrade other proteins and could also contribute to the pathogenesis cause by *E. histolytica.* For example, proteases have been shown to disrupt the polymerization of MUC2, the major component of colonic mucus. This degraded mucin is less efficient at preventing contact between trophozoites and epithelial cells. Similarly, destruction of extracellular matrix proteins may also facilitate trophozoite invasion. Inhibitors of cysteine proteases decrease liver abscess size in experimental models, thus providing evidence for a role of proteases in pathogenesis. In addition, *E. histolytica* expresses and secretes higher levels of cysteine proteases than *E. dispar* (Box 3.3).

E. histolytica can kill cells within minutes of adhering to them in the presence of extracellular calcium. This killing is mediated by the Gal/GalNAc lectin (Box 3.2) in that galactose or antibodies against the protein can inhibit adherence and killing. However, the purified Gal/GalNAc lectin is not directly cytotoxic suggesting that the protein is involved in signaling

Table 3.3 Summary of proteins implicated in pathogenesis

Factor	Possible natural function	Proposed role in pathogenesis	Differences between Eh and Ed*
Cysteine proteases	Various and unknown	Breakdown of mucus and extracellular matrix	More activity in Eh
	Degrade IgA and IgG		2 genes (CP1 and CP2) missing in Ed
Surface lectins	Adherence to mucous layer	Adherence to host cells	Sequence differences
		Contact-dependent killing (apoptosis)	Antigenic differences
Amebapore	Killing bacteria in food vacuole	Lysis of host cells (necrosis)	Glu vs. Pro
			Less activity in Ed

*Eh = *E. histolytica*; Ed = *E. dispar*

the cytotoxic event. Evidence for programmed cell death, or **apoptosis**, has been observed, as well as a direct lysis of host cells (i.e., necrosis). The relative contributions of apoptotic and necrotic cell death to the pathogenesis observed during amebic colitis are unclear. Pore-forming peptides capable of lysing bacteria and eukaryotic cells have been identified (Box 3.4). In theory, pore-forming peptides could poke holes in the plasma membranes of the host cells leading to an osmotic lysis and cell death, and could thereby account for a necrotic type of cell death. **Amebapore** A is the best characterized among these peptides and is found in the food vacuole where its primary function is to kill ingested bacteria. Some studies do suggest a role for amebapore in cytotoxicity, but no clear evidence for the secretion of the amebapore has been demonstrated. Thus the precise role of amebapore is not known.

In summary, the pathogenesis associated with *E. histolytica* infection is primarily due to its ability to invade tissues and kill host cells. Several potential virulence factors have been identified (Figure 3.5). However, it is not clear the exact role these various virulence factors play in the development of invasive disease. The differences between *E. histolytica* and *E. dispar* imply that pathogenesis is, at least in part, an inherent feature of the parasite. However, definitive parasite virulence factors have not yet been identified. Pathogenesis is probably due to the combined effects of several environmental, host, and parasite factors, and the virulence may represent the degree to which the host can control trophozoite invasion and replication.

Box 3.3 Amebic cysteine proteases

Cysteine proteases are a particular type of protease. At least 20 different cysteine protease (CP) genes have been identified in *E. histolytica* [1]. Orthologs of two of the *E. histolytica* cysteine protease genes are not found in *E. dispar*. One of these, designated CP5, is expressed at high levels on the trophozoite surface. Mutants expressing lower levels of CP5 have a reduced ability to generate liver abscesses in a hamster amebiasis model. However, these mutants also have a reduced growth rate and lower erythrophagocytic activity, thus it is not clear whether CP5 directly participates in the invasiveness of *E. histolytica*. Furthermore inhibition of 90% of CP5 activity did not affect the ability of *E. histolytica* trophozoites to destroy cell monolayers *in vitro*. CP1, CP2, and CP5 are the most abundantly expressed cysteine proteases in *E. histolytica*, whereas CP3 is the most abundant in *E. dispar*. Interestingly, overexpression of CP2 in *E. dispar* increased the ability of trophozoites to destroy cell monolayers *in vitro*. However, the overexpression of CP2 did not lead to the ability of *E. dispar* to form liver abscesses in a gerbil model system. Therefore, it is not clear the precise roles proteases may play in pathogenesis.

1. Bruchhaus, I., Loftus, B.J., Hall, N. and Tannich, E. (2003) The intestinal protozoan parasite *Entamoeba histolytica* contains 20 cysteine protease genes, of which only a small subset is expressed during in vitro culture. *Eukaryot. Cell* 2: 501–509.

Box 3.4 Amebapores

A family of pore-forming polypeptides has been identified in *E. histolytica* and *E. dispar*. The three family members are designated as amebapore A, B, and C with amebapore A being expressed at the highest levels. The mature polypeptide is 77 amino acids long and forms dimers at low pH [1]. Three of these dimers then assemble into a hollow ring-shaped structure. This hexamer then can intercalate into membranes and introduce 2 nm pores (i.e., holes) which results in cell death. The pore-forming activity is dependent on this assembly process beginning with the dimerization. Amebapore A is 95% identical (i.e., four residues are different) between *E. histolytica* and *E. dispar*. Despite this high level of amino acid identity, the *E. dispar* amebapore has approximately half the pore-forming activity of *E. histolytica* amebapore. This difference in pore-forming activity has been attributed to a glutamate residue at position 2 in the *E. histolytica* amebapore, as compared with a proline residue in the *E. dispar* amebapore. This particular amino acid residue is important for the formation of the dimers and it is believed that the dimers of *E. dispar* amebapore are less stable [2].

Amebapore is localized to vacuolar compartments (e.g., food vacuoles) within the trophozoite and is most active at acidic pH suggesting that the major function of amebapore is to lyse ingested bacteria. Nonetheless, amebapore is implicated as a virulence factor in that genetic manipulation of *E. histolytica* resulting in decreased expression of amebapore leads to a reduction in pathogenicity (ability to form liver abscesses) as well as a reduction in bactericidal activity. Similarly, modified *E. histolytica* completely devoid of amebapore production are unable to form liver abscesses in model systems. However, these modified amebas are still able to cause inflammation and tissue damage in models for amebic colitis.

1. Leippe, M. and Herbst, R. (2004) Ancient weapons for attack and defense: the pore-forming polypeptides of pathogenic enteric and free-living amoeboid protozoa. *J. Eukaryot. Microbiol.* 51: 516–521.

2. Leippe, M., Bruhn, H., Hecht, O. and Grötzinger, J. (2005) Ancient weapons: the three-dimensional structure of amoebapore A. *Tr. Parasitol.* 21: 5–7.

Clinical Presentation

Amebiasis presents a wide range of clinical syndromes (Table 3.4) which reflect the potential for *E. histolytica* to become invasive and cause a progressive disease. The incubation period can range from a few days to months or years, with 2–4 weeks being the most common for development of symptomatic nondysenteric disease. Transitions from one type of intestinal syndrome to another can occur and intestinal infections can give rise to extraintestinal infections. Many individuals who are diagnosed with *E. histolytica* (or *E. dispar*) infections exhibit no symptoms or have vague and nonspecific abdominal symptoms. This state can persist or progress to a symptomatic infection. Symptomatic nondysenteric infections exhibit variable symptoms ranging from mild and transient to intense and long lasting. Typical symptoms include: diarrhea, cramps, flatulence, nausea, and anorexia. The diarrhea frequently alternates with periods of constipation or soft stools. Stools sometimes contain mucus, but no visible blood.

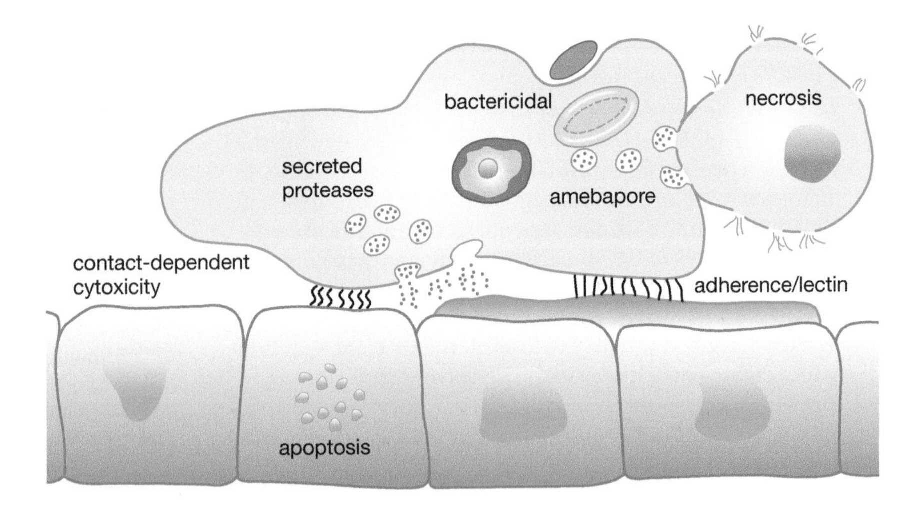

Figure 3.5 Schematic representation of virulence factors and mechanisms of pathogenesis. Adherence to the mucous layer and contact-dependent cytotoxicity is mediated by the Gal/GalNAc lectin. Secreted protease may mediate the breakdown of the mucous layer and subsequently tissue destruction following invasion of the epithelial layer. Amebapore can potentially lyse cells.

Table 3.4 Clinical syndromes associated with amebiasis	
Intestinal disease	Asymptomatic cyst passer
	Symptomatic nondysenteric disease
	Amebic dysentery (acute)
	Fulminant colitis
	Colon perforation (peritonitis)
	Ameboma (amebic granuloma)
	Perianal ulceration
Extraintestinal disease	Liver abscess
	Pleuropulmonary amebiasis
	Brain and other organs
	Cutaneous and genital diseases

Amebic colitis usually starts slowly over several days with abdominal cramps, tenesmus, and occasional loose stools, but progresses to diarrhea with blood and mucus. Blood, mucus, and pieces of necrotic tissue become more evident as the number of stools increases (10–20 or more per day) and stools will often contain little fecal material. A few patients may develop fever, vomiting, abdominal tenderness, or dehydration (especially children) as the severity of the disease increases. Acute necrotizing colitis is a rare but extremely severe form of intestinal amebiasis which can result in death. Such patients present with severe bloody diarrhea, fever, and diffuse abdominal tenderness. Most of the mucosa is involved and mortality exceeds 50%. Peritonitis resulting from perforation of the intestinal wall can also be fatal. A chronic amebiasis, characterized by recurrent attacks of dysentery with intervening periods of mild or moderate gastrointestinal symptoms, can also occur.

Amebomas present as painful abdominal masses which occur most frequently in the cecum and ascending colon. Obstructive symptoms or hemorrhages may also be associated with an ameboma. Amebomas are infrequent and can be confused with carcinomas or tumors. Perianal ulcers are a form of cutaneous amebiasis that results from trophozoites emerging from the rectum and invading the skin around the anus.

Extraintestinal amebiasis

The clinical symptoms associated with extraintestinal amebiasis will depend on the affected organ. Amebic liver abscesses are the most common form of extraintestinal amebiasis. This form of the disease can occur months to years after the intestinal stage of the infection. The onset of hepatic symptoms can be rapid or gradual. Hepatic infections are characterized by fever, hepatomegaly, liver tenderness, pain in the upper right quadrant, and anorexia. Fever sometimes occurs on a daily basis in the afternoon or evening. Liver function tests are usually normal or slightly abnormal and jaundice is unusual. Liver abscesses will occasionally rupture into the peritoneum resulting in peritonitis.

Pulmonary amebiasis generally results from the direct extension of the liver abscess through the diaphragm. Clinical symptoms most often include cough, chest pain, dyspnea (difficult breathing), and fever. The sputum may be purulent or blood-stained and contain trophozoites. A profuse expectoration (i.e., vomica) of purulent material can also occur. Primary metastasis to the lungs is rare, but does occur. Similarly, infection of other organs (e.g., brain, spleen, pericardium) is also rare. Clinical symptoms are related to the affected organ.

Cutaneous amebiasis is the result of skin or mucous membranes being bathed in fluids containing trophozoites. This contact can be the result of fistula (intestinal, hepatic, perineal) or an invasion of the genitalia. Cutaneous lesions have a wet, granular, necrotic surface with prominent borders and can be highly destructive. Clinical diagnosis is difficult and is usually considered with epidemiological risk factors such as living in an endemic area.

Diagnosis, Treatment, and Control

Diagnosis of amebiasis requires the demonstration of *E. histolytica* cysts or trophozoites in feces or tissues (Table 3.5). In the case of intestinal disease the most common method is to microscopically examine stools. Stool specimens should be preserved and stained and microscopically examined. Cysts will tend to predominate in formed stools and trophozoites in diarrheic stools. Fresh stools can also be immediately examined for motile trophozoites which exhibit a progressive motility. According to the World Health Organization, diagnosis by light microscopy alone should be reported as *E. histolytica*/*E. dispar*. Hematophagous trophozoites in feces or trophozoites in tissues correlate with *E. histolytica*. Sigmoidoscopy may reveal the characteristic ulcers, especially in more severe disease. Aspirates or biopsies can also be examined microscopically for trophozoites. Several antigen detection kits are currently available and protocols for extracting fecal DNA and carrying out PCR are available. Antigen and DNA detection methods can be used to distinguish *E. histolytica* from *E. dispar*.

Serology is especially useful for the diagnosis of extraintestinal amebiasis. Seventy to eighty percent of patients with acute invasive colitis or liver abscesses, and greater than 90% of the convalescence patients, exhibit serum antibodies against *E. histolytica*. However, these antibodies can persist for years and distinguishing past and current infections may pose problems in endemic areas. Noninvasive imaging techniques (e.g., ultrasound,

Table 3.5 Diagnosis

Intestinal disease	Stool examination
	cysts and/or trophozoites
	antigen or DNA detection (distinguish *E. histolytica*/*E. dispar*)
	Sigmoidoscopy
	detect lesions
	examine aspirate or biopsy
Extraintestinal (hepatic) disease	Serology
	current or past infection?
	Imaging (CT, MRI, ultrasound)
	Abscess aspiration
	only select cases
	reddish brown liquid
	trophozoites at abscess wall

computerized tomography, magnetic resonance imaging) can be used to detect hepatic abscesses (Figure 3.3). The detection of a space-occupying lesion in the liver combined with positive serology provides a high level of sensitivity and specificity. It is also possible to aspirate hepatic abscesses. However, this is rarely done and only indicated in selected cases (e.g., serology and imaging not available, therapeutic purposes). The aspirate is usually a thick reddish brown liquid that rarely contains trophozoites. Trophozoites are most likely to be found at the abscess wall and not in the necrotic debris at the abscess center.

Multiple drugs are available to treat amebiasis

Several drugs are available for the treatment of amebiasis and the choice of drug(s) depends on the clinical stage (i.e., noninvasive or invasive) of the infection (Table 3.6). Noninvasive or asymptomatic infections are treated with luminal amebicides such as paromomycin, diloxanide furoate, or iodoquinol. These luminal agents are not well absorbed and therefore not effective against the tissues stages. In cases where *E. histolytica* is confirmed or the species (i.e., *dispar* or *histolytica*) is unknown, asymptomatic cyst passers should be treated to prevent the progression to severe disease and to control the spread of the disease. However, in many endemic areas, where the rates of reinfection are high and treatment is expensive, the standard practice is to only treat symptomatic cases.

Metronidazole or tinidazole (if available) is recommended for symptomatic invasive disease. These drugs are absorbed well but do not reach high enough concentrations within the lumen of the intestines. Therefore, treatment with tissue amebicides will not efficiently clear the luminal ameba and should be followed by treatment with a luminal agent to completely cure the infection. The prognosis following treatment is generally good in uncomplicated cases. In addition, these drugs are generally well tolerated by most people and exhibit few side effects.

In the cases of fulminating amebic colitis or perforation of the intestinal wall a broad-spectrum antibiotic can also be used to treat intestinal bacteria in the peritoneum. Necrotic colitis requires urgent hospitalization to restore fluid and electrolyte balance. In addition, emetine or dehydroemetine are sometimes co-administered with the nitroimidazole. This is only done in the most severe cases due to the toxicity of these drugs. Surgery may also be needed to close perforations or a partial colostomy. Abscess drainage of hepatic lesions (i.e., needle aspiration or surgical drainage) is now rarely performed for therapeutic purposes and is only indicated in cases of large abscesses with a high probability of rupture.

Prevention and control

Prevention and control measures are similar to other diseases transmitted by the fecal–oral route (Chapter 2). The major difference is that humans

Table 3.6 Treatment of amebiasis

Drugs	Uses
Iodoquinol, paromomycin, or diloxanide furoate	Luminal agents to treat asymptomatic cases and as a follow up treatment after a nitroimidazole
Metronidazole or tinidazole	Treatment of nondysenteric colitis, dysentery, and extraintestinal infections
Dehydroemetine or emetine	Treatment of severe disease such as necrotic colitis, perforation of intestinal wall, rupture of liver abscess

are the only host for *E. histolytica* and there is no possibility of zoonotic transmission. Control is based on avoiding the contamination of food or water with fecal material. Health education in regard to improving personal hygiene, sanitary disposal of feces, and hand washing are particularly effective. Although waterborne transmission of *Entamoeba* is lower than other intestinal protozoa, protecting water supplies will lower endemicity and epidemics. Like *Giardia*, *Entamoeba* cysts are resistant to standard chlorine treatment, but are killed by iodine or boiling. Sedimentation and filtration processes are quite effective at removing *Entamoeba* cysts. Chemoprophylaxis is not recommended.

Differences Between *E. histolytica* and *E. dispar*

As discussed above, *E. dispar* is morphologically identical to *E. histolytica* and the two species can only be distinguished by molecular, biochemical, and antigenic differences. Historically the two were considered a single species, *E. histolytica*. The first to consider that there may be two morphologically identical species was Brumpt, who in 1925 proposed the existence of a pathogenic species which he called *E. dysenteriae* and a nonpathogenic species he called *E. dispar*. However, this hypothesis did not gain favor and without a means to distinguish the two species was of little practical value. In 1960s investigators started to recognize phenotypic and genotypic differences in *E. histolytica* isolates from invasive cases and noninvasive cases (Table 3.7). Some of the first noted differences were the *in vitro* growth characteristics, agglutination with concanavalin A, and resistance to complement lysis. Pathogenic isolates have the ability to grow in axenic cultures (without bacteria) whereas the nonpathogenic strains required bacteria for *in vitro* growth.

Isoenzyme analysis revealed different zymodemes for the pathogenic and nonpathogenic isolates. Similarly, numerous antigenic differences between pathogenic and nonpathogenic isolates have been described. A well-characterized antigenic difference is in a surface Gal/GalNAc lectin. A panel of six monoclonal antibodies against the Gal/GalNAc lectin can distinguish pathogenic from nonpathogenic isolates. Two of the monoclonal antibodies recognize epitopes shared between isolates (i.e., 1 and 2), whereas the other monoclonal antibodies recognize epitopes only found on pathogenic isolates (i.e., 4–6). These antibodies have been adapted for the differentiation of the two species in antigen detection kits used for diagnosis. The differences in the Gal/GalNAc lectin between the species may also explain the differences in agglutination with concanavalin A and resistance to complement lysis.

Table 3.7 Noninvasive and invasive isolates of *Entamoeba histolytica*

Criteria	Noninvasive	Invasive
In vitro culture	Xenic	Axenic
Concanavalin A agglutination	–	+
Complement resistance	–	+
Zymodemes (i.e., isoenzymes)	I & III	II
Numerous antigenic differences (e.g., Gal/GalNAc lectin epitopes)	1–2	1–6
Numerous DNA sequence differences (e.g., rRNA)	2.2% sequence dissimilarity	

Analysis of DNA and sequencing of several genes also revealed genotypic differences between the pathogenic and nonpathogenic isolates. On average *E. histolytica* and *E. dispar* exhibit approximately 95% sequence identity in coding regions and approximately 85% sequence identity in noncoding regions. The most striking variation is the 2.2% difference between the ribosomal RNA gene sequences of pathogenic and nonpathogenic isolates. Unlike some of the other genes that exhibit sequence differences, rRNA would not potentially contribute to virulence. Furthermore, rRNA sequences of humans and mice differ by less than 2.2%. These biochemical, antigenic, and molecular differences led to the identification of a new species in 1993; it was called *E. dispar* as originally proposed by Brumpt 68 years earlier.

Other *Entamoeba* Species Infecting Humans

Several other *Entamoeba* species in addition to *E. histolytica* and *E. dispar* infect humans. Among the species found in human feces, *E. hartmanni* and *E. coli* are two relatively common commensals and *E. moshkovskii* and *E. polecki* are generally considered to be rare infections. *E. histolytica/dispar*, *E. coli*, and *E. hartmanni* can be distinguished by size and minor morphological differences (Table 3.8). *E. coli* is the largest and is best distinguished by eight nuclei in the mature cyst. The trophozoites of *E. coli* can be difficult to distinguish from *E. histolytica/dispar* since there is some overlap in the size ranges. *E. hartmanni* is quite similar to *E. histolytica* and was previously considered a "small race" of *E. histolytica*. Generally 10 μm is chosen as the boundary between *E. histolytica* and *E. hartmanni*.

E. moshkovskii is generally considered to be a free-living ameba found in environments ranging from clean riverine sediments to brackish coastal pools, as well as sewers. It is indistinguishable in its cyst and trophozoite forms from *E. histolytica* and *E. dispar* and phylogenetic analysis indicates that the three species form a clade and thus suggesting a common ancestor. *E. moshkovskii* can be distinguished by its ability to grow at ambient temperatures during *in vitro* culture, whereas *E. histolytica* and *E. dispar* need to be cultivated at 37°C. *E. moshkovskii* has been isolated from human fecal samples on rare occasions and in some of these cases the patients have exhibited symptoms. However, *E. moshkovskii* infections may be more prevalent than realized since few studies have directly looked for its prevalence in human samples using antigenic or molecular markers.

E. polecki is usually associated with pigs and monkeys, but human cases have been occasionally documented. It appears to be geographically restricted to particular areas, such as Papua New Guinea, and is often

Table 3.8 Intestinal Entamoeba species

	E. histolytica/dispar	*E. coli*	*E. hartmanni*
Trophozoites	15–20 μm (invasive Eh can be > 20 mm)	20–25 μm	8–10 μm
	Extended pseudopodia	Broad blunt pseudopodia	Less progressive than Eh/Ed
	Progressive movement	Sluggish, nondirectional movement	
Cysts	12–15 μm	15–25 μm	6–8 μm
	4 nuclei	8 nuclei	4 nuclei
	Blunt chromatoid bodies	Pointed chromatoid bodies	Blunt chromatoid bodies Chromatoid bodies persist in mature cysts

Eh = *E. histolytica*; Ed = *E. dispar*

associated with contact with pigs. The trophozoites are similar to *E. coli*, except a little smaller, and the cysts are similar to *E. histolytica* except that the mature cyst has a single nucleus. *E. polecki* appears to be nonpathogenic.

E. gingivalis can be recovered from the soft tartar between teeth and exhibits a similar morphology to *E. histolytica* except that it has no cyst stage. *E. gingivalis* can also multiply in bronchial mucus, and thus can appear in the sputum. In such cases it could be confused with *E. histolytica* from a pulmonary abscess. *E. gingivalis* trophozoites will often contain ingested leukocytes which can be used to differentiate it from *E. histolytica*. The trophozoites are most often recovered from patients with periodontal disease, but an etiology between the organism and disease has not been established and *E. gingivalis* is considered to be nonpathogenic.

Summary and Key Concepts

- *E. histolytica* exhibits a typical fecal–oral life cycle consisting of an ameboid trophozoite stage and a cyst containing four nuclei.
- *E. histolytica*, in contrast to the morphologically identical *E. dispar*, is capable of invading the intestinal mucosa and causing serious disease.
- Trophozoites of *E. histolytica* can kill intestinal epithelial cells and produce colonic ulcers leading to amebic colitis (i.e., dysentery).
- Extensive damage of the submucosa by the trophozoites can lead to a fulminating necrotic colitis or perforation of the intestinal wall.
- Trophozoites can metastasize to the other organs, typically the liver, and produce an extraintestinal amebiasis.
- The basis of pathogenesis is not well understood but possible virulence factors, including surface lectins, pore-forming peptides, and cysteine proteases, have been identified.
- The drugs for the treatment of amebiasis are generally effective with minimal toxicity and the prognosis for recovery is generally good if the complications are not severe.
- Molecular- or antibody-based methods are needed to distinguish *E. histolytica* from the morphologically identical, but nonpathogenic, *E. dispar*.

Further Reading

Diamond, L.S. and Clark, C.G. (1993) A redescription of *Entamoeba histolytica* Schaudinn, 1903 (Emended Walker, 1911) separating it from *Entamoeba dispar* Brumpt, 1925. *J. Euk. Microbiol.* 40: 340–344.

Fotedar, R., Stark, D., Beebe, N., Marriott, D., Ellis, J. and Harkness, J. (2007) Laboratory diagnostic techniques for *Entamoeba* species. *Clin. Microbiol. Rev.* 20: 511–532.

Haque, R., Huston, C.D., Hughes, M., Houpt, E. and Petri, W.A., Jr. (2003) Amebiasis. *N. Engl. J. Med.* 348: 1565.

Huston, C.D. (2004) Parasite and host contributions to the pathogenesis of amebic colitis. *Tr. Parasitol.* 20: 23–26.

Loftus, B. et al. (2005) The genome of the protist parasite *Entamoeba histolytica*. *Nature* 433: 865–868.

Ravdin, J.I. (1995) Amebiasis. *Clin. Infect. Dis.* 20: 1453–1566.

Shahran, S.M. and Petri, W.A., Jr. (2008) Intestinal invasion by *Entamoeba histolytica*. *Subcell. Biochem.* 47: 221–232.

Stanley, S.L., Jr. (2003) Amoebiasis. *Lancet* 361: 1025–1034.

Giardiasis

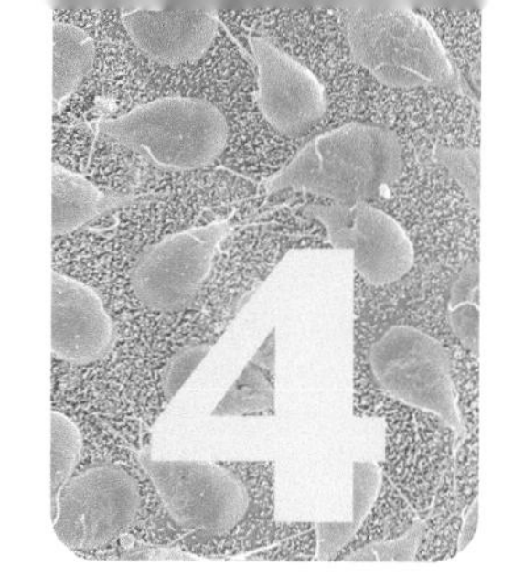

Disease(s)	Giardiasis
Etiological agent(s)	*Giardia duodenalis* (also known as *G. lamblia* and *G. intestinalis*)
Major organ(s) affected	Small intestine
Transmission mode or vector	Fecal–oral
Geographical distribution	Worldwide, but more prevalent in tropical and developing countries
Morbidity and mortality	Ranges from asymptomatic to acute intestinal symptoms, including diarrhea, to chronic wasting syndromes. Never fatal
Diagnosis	Detection of parasites in feces
Treatment	Metronidazole
Control and prevention	Avoid fecal contamination of food or water

Giardia has a worldwide distribution and is the most common protozoan isolated from human stools. In fact it is quite likely that van Leeuwenhoek, one of the early pioneers of light microscopy, first described *Giardia* in 1681 in his own diarrheic stools. This is based upon his description of its characteristic swimming movement. However, van Leeuwenhoek never submitted drawings of the organisms and Lambl, a physician from Prague who described the trophozoites from stools of pediatric patients in 1859, is usually given credit for the discovery of *Giardia* and thus the species infecting humans has historically been referred to as *G. lamblia.* Other common species names include *G. duodenalis* or *G. intestinalis* and the three designations should be considered as synonyms. The more recent literature trend has been to use *G. duodenalis.* However, *G. lamblia* is still widely used in the medical literature.

As the name *G. duodenalis* implies, *Giardia* is a protozoan parasite that colonizes the upper portions of the small intestine. *Giardia* virtually never penetrates the intestinal epithelium and often results in asymptomatic infections. Symptomatic giardiasis is characterized by acute or chronic diarrhea and/or other gastrointestinal manifestations. The worldwide prevalence of symptomatic giardiasis is believed to be more than 200 million cases with 500 000 new cases reported per year.

Life Cycle and Morphology

Giardia exhibits a typical fecal–oral life cycle consisting of infectious cyst stages passed in the feces and replicating trophozoites found in the small intestine (Figure 4.1). The infection is acquired through the ingestion of food or water contaminated with cysts. Factors leading to contamination of food or water with fecal material are correlated with transmission (Chapter 2). For example, in the United States giardiasis is especially prevalent in

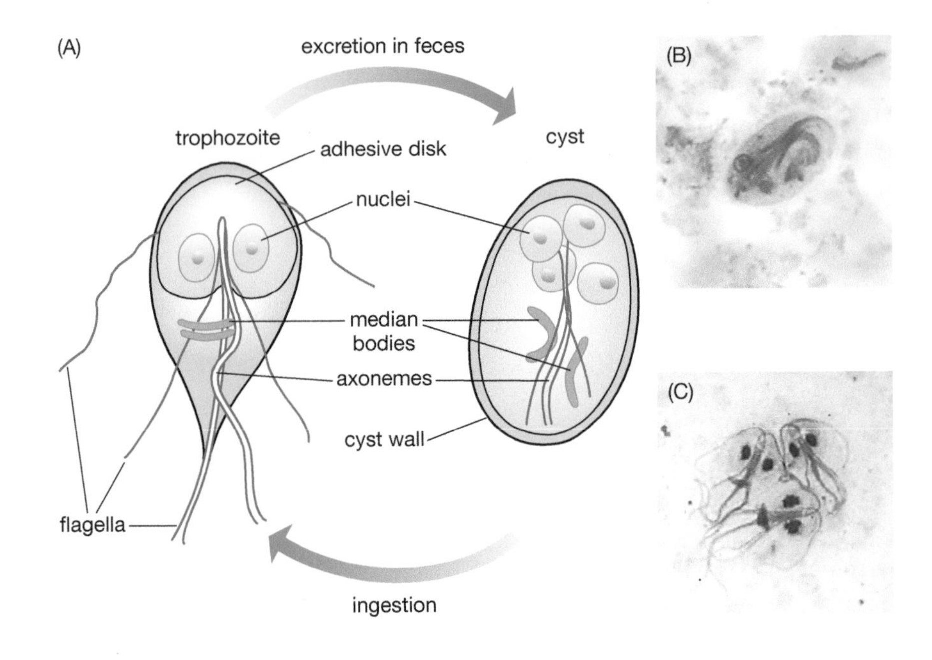

Figure 4.1 Life cycle and morphology of *Giardia*. (A) Schematic drawing showing typical fecal–oral life cycle of *Giardia* and major morphological features of trophozoites and cysts. (B) Micrograph of stained cyst from feces. (C) Micrograph of stained trophozoite from feces.(Figures B and C from the Parasite Image Library, CDC's Division of Parasitic Diseases (DPD) with permission from the Center for Disease Control (CDC).)

children and particularly those children in institutions or day-care centers. In developing countries, poor sanitation contributes to the higher levels of giardiasis and waterborne outbreaks due to inadequate water treatment have also been documented.

Zoonotic transmission potential

Giardia duodenalis infects a wide range of mammals and it has long been considered that *Giardia* infections could be potentially acquired from animals (i.e., zoonotic transmission). For example, in parts of Canada and the northern United States giardiasis has been referred to as "beaver fever" following an outbreak at Banff National Park in Canada that was associated with lakes and streams inhabited by beavers. It has also been called "hiker's diarrhea" in reference to the belief that hikers and campers often acquire *Giardia* infections by drinking from streams contaminated with animal feces. However, the zoonotic transmission of *Giardia* is controversial and has not been unambiguously demonstrated.

A major problem in determining the extent of zoonotic transmission is that *Giardia* isolated from mammals, with the exception of *G. muris* and *G. microti* from rodents, exhibit the same morphologies. Therefore, in the absence of molecular or biochemical markers it was never clear whether *Giardia duodenalis* represented a single species capable of infecting a wide range of animals, or whether each host had their own subtype of *Giardia*. More recent molecular and biochemical evidence, however, suggests that there are distinct subtypes of *Giardia*, some of which exhibit narrow host ranges while others exhibit broader host ranges (Box 4.1). In particular, seven distinct genotypic groups called clades A–G have been described. The two clades that infect humans (A and B) are found in other host species indicating that there is a potential public health risk of obtaining the infection from domestic dogs, cats, or livestock. However, molecular epidemiology studies indicate that *Giardia* transmission between dogs and humans is quite rare in spite of the close contact between the two. The accumulating molecular data to date tend not to support widespread zoonotic transmission of *Giardia* and instead suggest that zoonotic transmission is relatively rare. Regardless of whether zoonotic transmission is possible, person-to-person transmission is probably the most prevalent mode of transmission and major risk factors are close human contact combined with unhygienic conditions.

Box 4.1 Molecular epidemiology of *Giardia*

Most *Giardia* isolates from mammals exhibit similar morphologies. This has resulted in a long history of uncertainty and controversy concerning the classification of *Giardia* species. Biochemical and molecular markers have been used in an attempt to resolve taxonomic issues and questions about the transmission of *Giardia* as well provide some insight into virulence and pathogenesis. Molecular and isoenzyme analyses reveal a substantial level of genetic diversity between *Giardia* isolates [1]. Seven well-defined lineages, or clades, designated A–G have been identified (Figure 4A). Some of these clades are quite specific for a particular host or class of hosts. For example, clades C and D have only been isolated from dogs, clade F only from cats, clade G only from rats, and clade E only from hoofed livestock. Clades A and B on the other hand exhibit a relatively broad host range.

Human isolates fall into either clade A or clade B. However, within both of these clades the human isolates tend to fall into distinct clades that are different from the isolates obtained from other host animals. These observations suggest that most of the *Giardia* subtypes are adapted to specific host species. Indeed, most molecular epidemiology studies indicate that zoonotic transmission of *Giardia* is not a major source of human infection. There have been a few isolated reports clearly showing a particular genotype circulating between humans and dogs. However, most studies indicated that there is little exchange of *Giardia* between humans and dogs despite their close contact. Furthermore, in cases where human genotypes are isolated from animals, it is possible that the primary reservoir hosts are humans and that the infection is being spread from humans to animals. Although zoonotic transmission of *Giardia* is possible, it appears to be rather infrequent.

DNA sequencing and fingerprinting methods confirm that the clades, or biotypes, identified by isoenzyme analysis represent distinct evolutionary lineages. However, the phylogenetic relationships of these biotypes do not reflect the phylogenetic relationships among the mammalian hosts. Thus the different biotypes are probably not the result of coevolution of parasite and host, but rather reflect host switching and host adaptation. Nonetheless, the magnitude of the sequence differences between the biotypes is of a similar scale as differences between species. Thus, a redesignation of *Giardia* species has been proposed based on these genetic and host range differences between these groups [1]. An overall acceptance of this proposal will probably depend on the availability of convenient and unambiguous markers for the proposed species. Although, there is general agreement between the isoenzyme- and DNA-based approaches in regards to the relationships between the various biotypes, on occasion different molecular markers will result in different clade assignment.

Studies addressing correlations between virulence and genotype have not yielded consistent results. Some studies have found no differences in virulence between genotypes. Other studies have found higher levels of virulence associated with clade A, whereas other studies indicate clade B is more virulent.

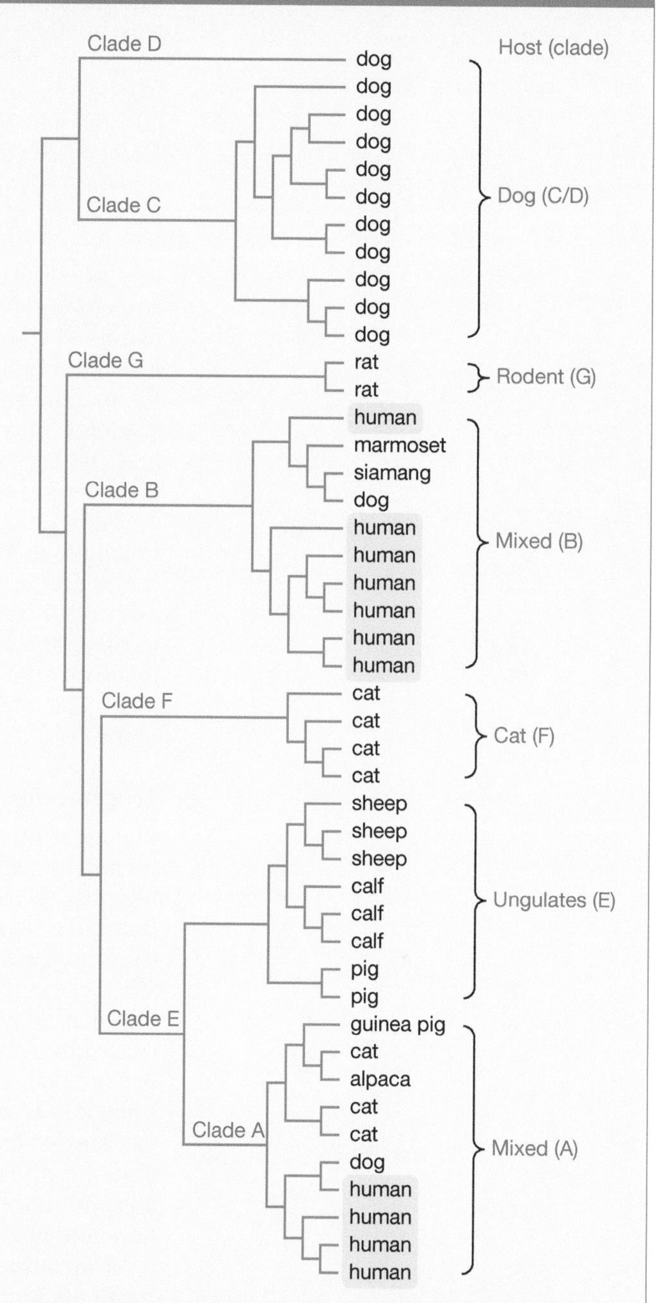

Figure 4A Phylogenetics of *Giardia* isolated from various mammals. Cladogram demonstrating relationships between *Giardia duodenalis* isolates from various mammalian hosts. In general, parasites isolated from a particular host species, or type of host, are related phylogenetically. These various parasite clades have previously been designated as assemblages A–G (Monis et al., 2009) as denoted by the letter in parentheses. Human isolates (shaded) exhibit in distinct biotypes within clades A and B.

1. Monis, P.T., Cacció, S.M. and Thompson, R.C.A. (2009) Variations in *Giardia*: towards a taxonomic revision of the genus. *Parasitol. Today* 25: 93–100.

Unique structural features of *Giardia*

The *Giardia* trophozoite exhibits a characteristic pear, or tear-drop, shape with bilateral symmetry when viewed from the top. It is typically 12–15 μm long, 5–10 μm wide, and 2–4 μm thick. Characteristic features of the stained trophozoite include: two nuclei, fibrils running the length of the parasite, and median bodies (Figure 4.1). The fibrils are called **axonemes** and are formed from the proximal regions of the flagella within the body of the trophozoite. *Giardia* trophozoites possess four pairs of flagella and are motile. Three pairs of flagella emerge from the dorsal surface (anterior, posterior-lateral, and caudal or extreme posterior end) and one pair emerges from the ventral surface. Trophozoites exhibit a distinctive erratic twisting motion, sometimes compared to that of a falling leaf. The **median bodies** are a pair of curved structures which lie posterior to the nuclei. At the ultrastructural level the median bodies contain an array of microtubules. The function of the median bodies is believed to be involved in the formation of the adhesive disk. This adhesive disk, not always visible by light microscopy, occupies the ventral side of the anterior end and mediates the attachment of trophozoites to epithelial cells of the small intestine.

The cysts of *Giardia* are oval shaped and typically measure 11–14 μm in length and 6–10 μm wide. *Giardia* cysts have a well-defined wall which is often set apart from the cytoplasm of the parasite. The axonemes, which extend across the length of the cyst, and the median bodies are also found in the cyst stages. These structures are quite distinctive and make *Giardia* relatively easy to identify and to distinguish from other protozoa found in feces. The mature cyst has four nuclei which are typically found at one end of the cyst.

Trophozoite physiology and metabolism

The ingested cyst passes through the stomach and excystation takes place in the duodenum. Excystation can be induced *in vitro* by a brief exposure of the cysts to acidic pH (~2), or other sources of hydrogen ions, and returning them to a neutral pH. Exposure to the acidic pH mimics the conditions of the stomach and probably functions as an environmental cue for the parasite. The return to neutral pH reflects the conditions of the small intestine. Flagellar activity begins within 5–10 minutes following this treatment and the trophozoite emerges through a break in the cyst wall. The breakdown of the cyst wall is believed to be mediated by parasite proteases (enzymes that break down other proteins). The quadrinucleated trophozoite will undergo cytokinesis (cell division without nuclear replication) within 30 minutes after emerging from the cyst resulting in binucleated trophozoites. While still within the small intestine the trophozoites will attached to the epithelium and begin the trophic stage of the life cycle.

After attaching to the intestinal epithelium the trophozoite absorbs nutrients from the intestinal lumen via pinocytosis from the dorsal surface. No specialized feeding organelles have been described in *Giardia*. As is the case with other luminal parasitic protozoa, *Giardia* exhibits an anaerobic metabolism and lacks traditional mitochondria. A remnant of the mitochondrion, called a **mitosome**, has been identified in *Giardia*. The surface of the trophozoite is covered with a protein from a family of structurally related proteins. Only a single gene from a repertoire of 150–200 variant surface protein (VSP) genes is expressed at any one time and switching of expression occurs approximately once every 10 generations. Since most of these VSPs are antigenically distinct, this switching of expression is presumed to be involved in avoidance of the immune system, thus allowing the parasite to maintain its presence within the host. The trophic stage is also characterized by asexual reproduction. Both nuclei divide at about the same time and cytokinesis restores the binucleated state. Each daughter cell receives one copy of each nucleus. Both nuclei appear equal in regard to gene expression and other properties.

Cyst formation

As an alternative to replication, the trophozoite can also encyst. During encystation the parasite rounds up, detaches from the intestinal epithelium, and secretes a cyst wall. Encystation can also be induced *in vitro*. Optimal cyst formation is obtained by depriving the trophozoites of bile at pH 7 followed by an exposure to high concentrations of bile at pH 7.8. The lack of bile at neutral pH mimics the conditions under the mucous blanket adjacent to the intestinal epithelial cells, whereas exposure to high concentrations of bile at slightly alkaline pH is analogous to the intestinal lumen. These studies highlight the extent to which *Giardia* has adapted to life within the gastrointestinal tract.

A major event in the encystation process is the formation of the cyst wall. The cyst wall is composed of four major cyst wall proteins which are transported to the cell surface by large secretory vesicles that appear during encystation. After secretion these proteins are probably cross-linked together via disulfide and other covalent bonds. Polysaccharides, and in particular polymers of acetylgalactosamine, are also incorporated into the cyst wall. After cyst wall formation the parasite undergoes one round of nuclear division without cytokinesis resulting in four nuclei. The flagella and adhesive disk are lost as the cyst matures but the axonemes and median bodies persist. The cysts are passed in the feces and can survive for months under appropriate temperature and moisture conditions. Mature cysts are infective to the next host that happens to ingest them, thus completing the life cycle.

The Adhesive Disk

A unique ultrastructural feature of *Giardia* is the **adhesive disk** (also called ventral disk, sucking disk, sucker, or striated disk). The adhesive disk is a concave structure which occupies approximately two-thirds of the anterior end of the ventral surface. On occasion the disk can be seen by light microscopy depending on the type of stain used. Scanning electron microscopy revels that the adhesive disk is formed from a concentric swirl originating in the center of the disk (Figure 4.2A). Part of the cell body of the trophozoite extends past the adhesive disk in a flattened protrusion called the ventrolateral flange. As the names imply, the adhesive disk plays a role in the attachment of the trophozoite to the intestinal epithelium and ultrastructural studies reveal close associations between the adhesive disk and the intestinal brush border (Figure 4.2B). In particular, the edges of the adhesive disk are often embedded within the microvilli of the intestinal epithelial cells.

Adhesive disks are composed of cytoskeletal elements and contractile proteins

The adhesive disk appears to be a relatively rigid structure and striations are evident by transmission electron microscopy (Figure 4.2B). These striations are formed from microtubules and a unique cytoskeletal element called a microribbon. Microribbons are long flattened structures and each microribbon is associated with a microtubule (Figure 4.2C). The combined microtubule–microribbon structures are arranged in concentric rows that form a flattened spiral with minimal overlap. The major components of microribbons are proteins called giardins. These giardins primarily play a structural role in the formation of the microribbons. The giardins show a limited homology to a protein called "striated fiber assemblin" from *Chlamydomonas* (a free-living, biflagellated unicellular algae). In *Chlamydomonas* this protein forms filamentous structures at the base of the flagella. The giardins have evolved to play a different functional role in *Giardia*, but are still associated with microtubule-based cytoskeletal elements.

The outer rim of the adhesive disk contains components of the actin–myosin cytoskeleton. Actin and myosin are proteins involved in the

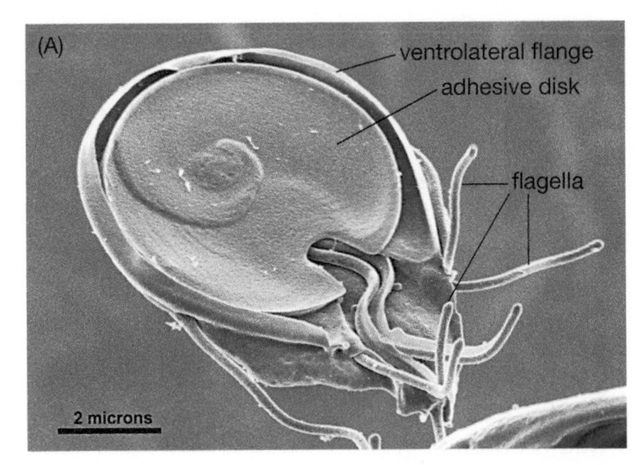

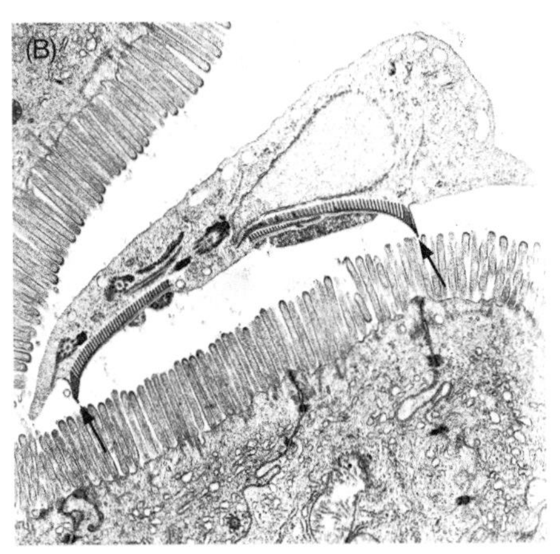

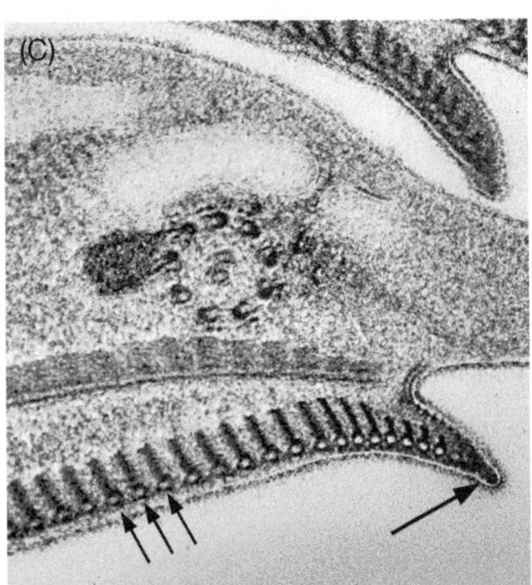

Figure 4.2 Ultrastructure of the adhesive disk. (A) Scanning electron micrograph of the ventral surface of a *Giardia* trophozoite showing the adhesive disk. (B) Transmission electron micrograph showing interaction of *Giardia* trophozoites with intestinal brush border. Arrows denote the outer edge of the adhesive disk and its interaction with the microvilli of the intestinal epithelial cell. Note also the striations in the adhesive disk. (C) Higher magnification showing detailed structure of microribbons and associated microtubules (small arrows). The outer edge of the adhesive disk contains actin and myosin (large arrow). (Micrographs A and B kindly provided by Stanley Erlandsen, University of Minnesota, Minneapolis, MN. Micrograph C kindly provided by J. Michael McCaffery, John Hopkins University, Baltimore, MD.)

generation of force and movement. This association of proteins involved in the generation of contractile force and other cytoskeletal elements in the adhesive disk suggests that attachment is mediated by mechanical forces generated by the parasite. The observation that imprints and circular dome-shaped lesions remain in the intestinal brush border (i.e., microvilli) following detachment of trophozoites is consistent with contractile forces playing a role in attachment (Figure 4.3). Inhibitors of microfilament function also inhibit attachment. In other words, the adhesive disk appears to be a unique contractile organelle which is capable of physically holding onto the intestinal epithelial cells. A mannose-binding lectin activity has also been identified on the surface of *Giardia* trophozoites. However, this lectin is found on the entire surface of the trophozoite and therefore may only participate in the initial interactions between the parasite and the intestinal epithelial cells.

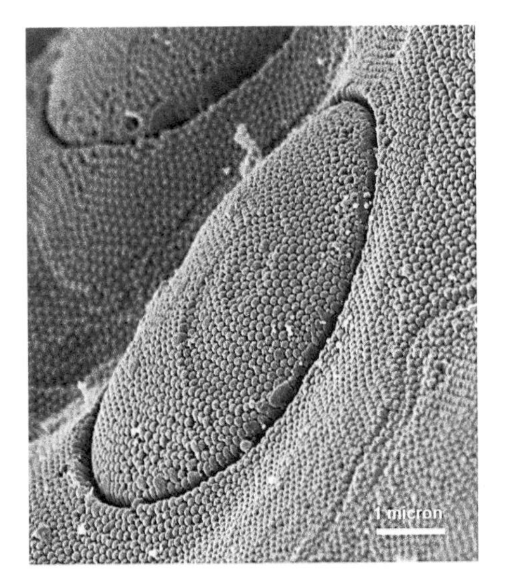

Figure 4.3 Impression in microvilli formed by adhesive disk. A scanning electron micrograph of the brush border following detachment of a *Giardia* trophozoite. (Micrograph provided by Stanley Erlandsen, University of Minnesota, Minneapolis, MN.)

Symptoms and Pathogenesis

The clinical features associated with *Giardia* infection range from total latency (i.e., asymptomatic), to acute self-resolving diarrhea, to chronic syndromes associated with nutritional disorders, weight loss, and failure to thrive. Children exhibit clinical symptoms more frequently than adults and subsequent infections tend to be less severe than initial infections. The incubation period is generally 1–2 weeks, but ranges of 1–75 days have been reported.

The first signs of acute giardiasis include nausea, loss of appetite, and an upper gastrointestinal uneasiness. These signs are often followed or accompanied by a sudden onset of explosive, watery, foul-smelling diarrhea. Stools associated with *Giardia* infection are generally described as loose, bulky, frothy and/or greasy with the absence of blood or mucus, which may help distinguish giardiasis from other acute diarrheas. Other gastrointestinal disturbances associated with giardiasis include: flatulence, bloating, anorexia, cramps, and foul sulfuric belching (Table 4.1). The acute stage usually resolves spontaneously in 3–4 days and is often not recognized as being giardiasis. Some of the individuals who resolve the acute symptoms do not clear the infection, but become asymptomatic cyst passers without clinical manifestations, whereas others may have a few sporadic recurrences of the acute symptoms. The acute infection can occasionally persist and lead to malabsorption, steatorrhea (excessive loss of fat in the feces), debility (loss of strength), and weight loss.

Acute infections can also develop into long-standing subacute or chronic infections, which in rare cases last for years. The typical chronic stage patient presents with recurrent brief episodes of loose foul stools which may be yellowish, frothy, and float, accompanied by intestinal gurgling, abdominal distention and flatulence. Between episodes the stools are usually mushy, but normal stools or constipation can also occur. Cramps are uncommon during chronic infections, but sulfuric belching is frequent. Anorexia, nausea, and epigastric uneasiness are additional frequent complaints during chronic infections. The development of chronic giardiasis has been associated with a lack of secretory IgA and an inability to clear the parasitemia. In the majority of chronic cases the parasites and symptoms spontaneously disappear.

Table 4.1 Common symptoms of acute giardiasis

Manifestation	%*
Diarrhea	89
Malaise	84
Flatulence	74
Greasy stools	72
Abdominal cramps	70
Bloating	69
Nausea	68
Anorexia	64
Weight loss	64

*Percentage of symptomatic patients exhibiting symptom

Factors contributing to *Giardia*-associated diarrhea

The specific mechanisms of *Giardia* pathogenesis leading to diarrhea and intestinal malabsorption are not completely understood and no specific virulence factors have been identified. In some patients the normal villus structure is affected. For example, villus blunting (atrophy), crypt cell hypertrophy, and an increase in crypt depth have been observed to varying degrees. Some studies suggest that increased rates of enterocyte apoptosis are observed in patients with giardiasis. The increase in crypt cell proliferation will lead to a repopulation of the intestinal epithelium by relatively immature enterocytes with reduced absorptive capacities, thus leading to osmotic diarrhea (Chapter 2). In addition, *Giardia* infection may cause a hypersecretion of chloride ions and water from the crypt cells. Therefore, a combination of malabsorption and secretion of electrolytes appears to be responsible for fluid accumulation in the intestinal lumen.

The attachment of trophozoites to the enterocytes could produce mechanical irritation or mucosal injury. Furthermore, the trophozoites can significantly cover localized areas of the intestinal epithelium (Figure 4.4) and thus contribute to inefficient absorption by serving as a physical barrier. However, extensive malabsorption cannot be explained simply by *Giardia* blocking uptake since the patches of trophozoites are quite small in comparison with the total surface area of the intestinal epithelium. Damage to the brush border and microvilli shortening is sometimes observed, and associated with this damage is a decrease in metabolite transport function and enzyme deficiencies—particularly disaccharidases (see lactose intolerance below). This reduced digestion and absorption of solutes, particularly sodium, glucose, and water, will lead to increased fluid retention in the intestinal lumen which in turn results in diarrhea.

The apoptosis and other physical damage to the enterocytes will also lead to loss of epithelial barrier function and this increases permeability between intestinal epithelial cells, thus possibly exposing host immune cells to antigens from the intestinal lumen. Indeed, an increased inflammatory cell infiltration in the lamina propria is occasionally observed in association with giardiasis. This inflammation is particularly common in patients with chronic giardiasis. Consequently, the activated lymphocytes and higher levels of cytokines associated with the inflammation may further increase the permeability between enterocytes resulting in further

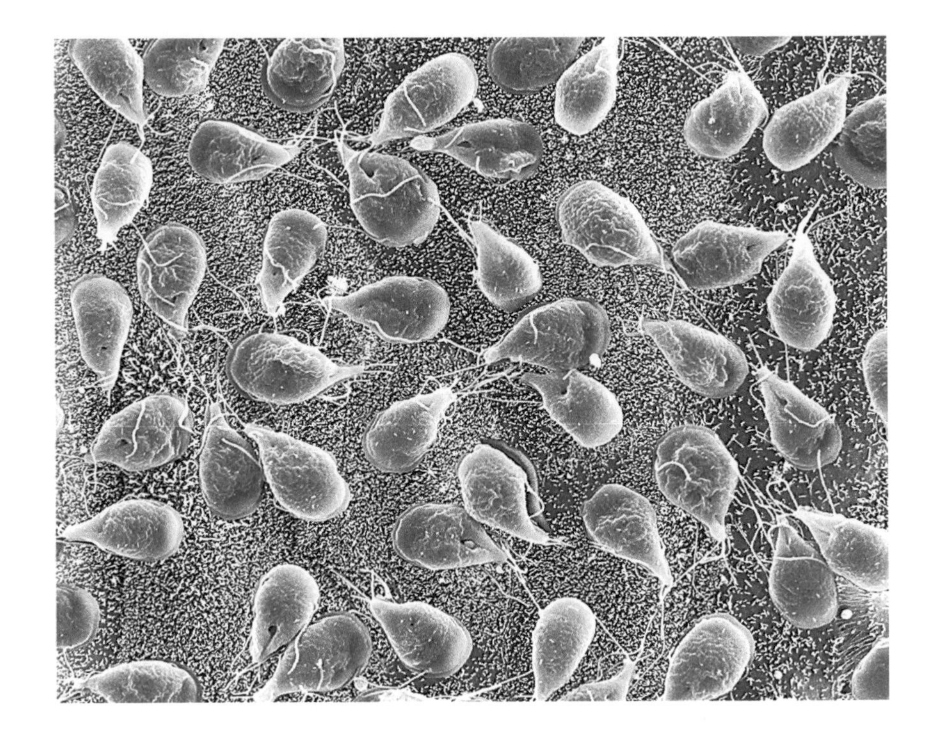

Figure 4.4 *Giardia* trophozoites attached to intestinal mucosa. A scanning electron micrograph showing a high density of trophozoites in a small area of the intestinal mucosa. Note also the loss of microvilli (i.e., brush border) on the right side. (Photograph provided by Marlene Benchimol, Universidade Santa Ursula, Rio de Janeiro, Brazil.)

fluid loss and exposure to luminal antigens. In addition, the inflammation can also cause or compound the damage to the intestinal brush border such as microvilli shortening and the accompanying loss in transporter and enzyme functions. Thus far, no single virulence factor or unifying mechanism explains the pathogenesis of giardiasis. But the combined effects of direct damage to the enterocytes caused by the parasite and the indirect effects due to apoptosis and inflammatory responses may lead to increased fluid in the intestinal lumen and the resulting diarrhea (Figure 4.5). Similar abnormalities in the intestinal epithelium are also observed during bacterial enteritis and inflammatory bowel disease.

Post-*Giardia* lactose intolerance

Some patients may present with a lactose intolerance during active *Giardia* infections which can persist after parasite clearance. This clinical manifestation is due to the parasite-induced lactase deficiency and is most common in ethnic groups with a predisposition for lactase deficiency. Lactase is an enzyme that breaks down lactose, a sugar found in milk, to

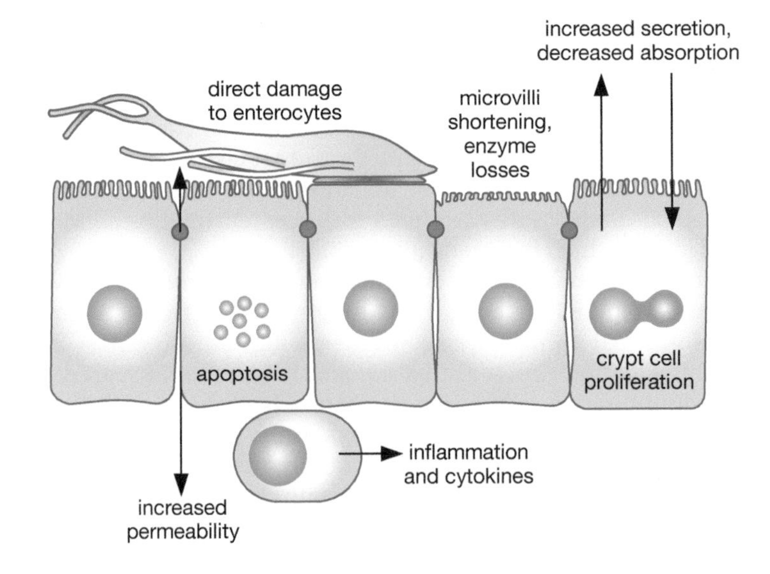

Figure 4.5 Potential mechanisms of pathogenesis. Schematic drawing highlighting some of the potential pathogenic mechanisms associated with giardiasis. Decreased absorption, increased secretion, and increased permeability can all lead to increased fluid in the intestinal lumen and result in diarrhea. This pathogenesis can result both from direct damage to the intestinal epithelium caused by the trophozoites and indirect effects, such as apoptosis and inflammation.

Table 4.2 Parasite detection	
Stool examination	3 nonconsecutive days
	Wet mount or stained
	IFA, copro-antigens
Duodenal aspirate or biopsy	Enterotest®
	Intubation or biopsy

monosaccharides which can be absorbed. Disaccharides, such as lactose, cannot be absorbed and will contribute to an osmotic type of diarrhea. Presumably this decrease in lactase, and other enzymes important for absorption, is due to epithelial cell damage caused directly or indirectly by the parasite. This lactose intolerance syndrome should be considered in persons who still present mushy stools and excessive gas following treatment, but have no detectable parasites.

Diagnosis

Diagnosis of giardiasis is confirmed by finding cysts or trophozoites in feces or in duodenojejunal aspirates or biopsies (Table 4.2). The unique features of the trophozoites and cysts make *Giardia* rather easy to identify and distinguish from other intestinal protozoa. However, detection of the parasites can be difficult since *Giardia* does not appear consistently in the stools of all patients. Some patients will express high levels of cysts in nearly all the stools, whereas others will only exhibit low parasite counts in some of the stools. A mixed pattern, characterized by periods of high cyst excretion alternating with periods of low excretion, has also been observed. In addition, parasites are easier to find during acute infections than chronic infections. Aspiration and biopsy may also fail to confirm the infection due to patchy loci of infection, and some practitioners question the usefulness of these invasive procedures.

Microscopic examination of stool specimens is the traditional and still widely used method for *Giardia* diagnosis. Three stools taken at intervals of at least 2 days should be examined. Watery or loose stools may contain motile trophozoites which are detectable by the immediate examination of wet smears. Otherwise the specimen should be preserved and stained because of trophozoite lability. The hardier cysts are relatively easy to recognize in either direct or stained smears. Numerous diagnostic kits and tests based on the detection of antigens in the stools (i.e., copro-antigens) are now available and are starting to replace microscopy as the standard method of diagnosis, especially in the United States. Tests using microtiter ELISA plates and immunochromatographic strips are available. Some of these tests simultaneously screen for *Cryptosporidium*. Immunofluorescence tests using antibodies against cyst wall proteins are also available.

Diagnosis can also be made by examining duodenal fluid for trophozoites. Duodenal fluid is obtained by either intubation or the Enterotest® (also called "string test"). The Enterotest® consists of a gelatin capsule containing a weighted nylon string of the appropriate length (Figure 4.6). The free end of the string is taped to the patient's face and the capsule is swallowed. After four hours to overnight the string is retrieved and the bile-stained mucus on the distal portion of the string is scraped off and examined for trophozoites by both wet mount and permanent staining. A small intestinal biopsy, preferably from multiple duodenal and jejunal sites, may also reveal trophozoites attached to the intestinal epithelium. A potential advantage of these more invasive procedures is that other enteric pathogens may be revealed.

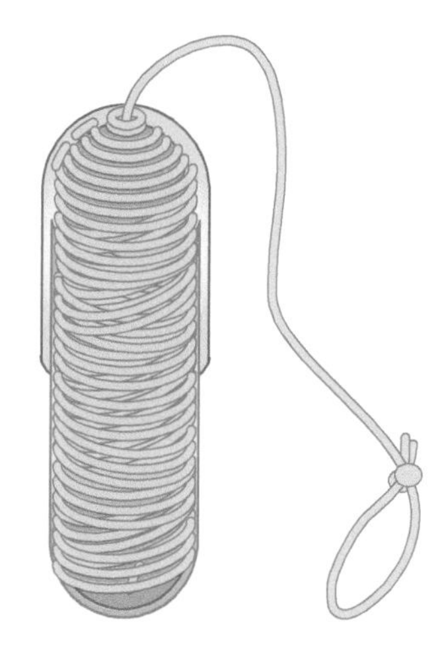

Figure 4.6 Enterotest®. Diagram showing the gelatin capsule with the weighted string that is used to diagnose *Giardia* and other infections of the small intestine. (Figure from Markell E.K., John D.T., and Krotoski W.A., 1999, *Markell and Voge's Medical Parasitology*, 8th Edn, Philadelphia. With kind permission from Elsevier.)

Treatment and Control

Generally infected individuals are treated because *Giardia* can persist and lead to severe malabsorption syndromes and weight loss. Several drugs are effective against *Giardia* (Table 4.3); treatment reduces morbidity and there are no sequelae. Metronidazole, although not licensed in the United States for giardiasis, effectively clears the parasite (cure rates approximately 85%) and is usually the drug of choice. Because of its short half-life the drug is usually given in multiple doses spread out over several days. Metronidazole is not recommended during pregnancy because it has teratogenic effects. Tinidazole is more effective than metronidazole when given as a single large dose and therefore some authorities consider this drug as the first line of therapy. Both metronidazole and tinidazole are nitroimidazoles that are activated via anaerobic metabolic pathways. Therefore, these drugs have a broad range of activity against anaerobic pathogens including *Entamoeba* (Chapter 3) and *Trichomonas* (Chapter 5). Side effects include metallic taste, nausea, headache, and vertigo and are usually dose dependent and self-limiting.

Nitazoxanide is another broad-spectrum drug effective against a wide range of intestinal bacteria, protozoa, and helminthes. It interferes with anaerobic metabolism by inhibiting oxidoreductase-dependent electron transfer. Quinacrine is an antimalarial that is also effective against *Giardia* by an unknown mechanism. Before the introduction of metronidazole, this drug was the usually the first choice for the treatment of giardiasis. It is quite effective, but exhibits more side effects than the other drugs. Furazolidone is a nitrofuran derivative that has the advantage of being available in a liquid formulation. Although less effective than the other drugs, it is more easily administered to infants and young children. Furazolidone is reduced to toxic nitro radicals by pathogen oxidases and, like other nitrofurans, acts by damaging important cellular components including DNA. Paramomycin is a broad-spectrum aminoglycoside antibiotic effective against many protozoa and helminthes and is proposed to inhibit protein synthesis. It is not absorbed and is recommended for the treatment of giardiasis during pregnancy.

Prevention and control

The widespread distribution of *Giardia* and the infectivity of the cysts make it unlikely that human infection will be completely eliminated. Control measures to prevent or reduce *Giardia* infection will depend on the specific circumstances of transmission, but in general involve measures which prevent the ingestion of substances contaminated with fecal material (Chapter 2). Health promotion and education aimed at improving personal hygiene, and emphasizing hand washing, sanitation, and food handling,

Table 4.3 Drugs for treatment of giardiasis

Drug	Trade name	Comments
Metronidazole	Flagyl®	Mainstay therapy for > 40 years
Tinidazole	Fasigyn®	Single dose treatment effective
Nitazoxanide	Alinia®	Most recently approved drug
Quinacrine	Atabrine®	Significant side effects
Furazolidone	Furoxone®	Available in liquid formulation
Paramomycin	Humatin®	Recommended during pregnancy

are effective control activities for the reduction of person-to-person transmission. Special attention to personal hygiene in high-risk situations such as day-care centers and other institutions is needed. Treatment of asymptomatic household members prevents reinfection in nonendemic areas. However, the value of treating asymptomatic carriers in hyperendemic communities is questionable since reinfection rates are high.

The socioeconomic situation in many developing countries makes it difficult to prevent infection. Public health measures to protect water supplies from contamination are required to prevent epidemics and to reduce endemicity. Conventional water treatment employing coagulation, sedimentation, and filtration does remove *Giardia* cysts. Tourists should not drink tap water without additional treatment in places where water purity is questionable. Boiling or iodine treatment kills *Giardia* cysts, but standard chlorination does not. There are no safe or effective chemoprophylactic drugs for giardiasis.

Summary and Key Concepts

- *Giardia duodenalis* (synonyms *lamblia* and *intestinalis*) is a common flagellated protozoan that resides in the small intestine and attaches to the epithelial layer via a specialized structure known as the adhesive disk.
- The infection is acquired via the ingestion of cysts in a typical fecal–oral transmission cycle and humans are the predominant source of infection.
- Clinical manifestations associated with infection can range from asymptomatic, to an acute self-resolving diarrhea, to chronic syndromes.
- Acute symptoms are typically a sudden onset of explosive, watery, foul-smelling diarrhea along with other gastrointestinal disturbances.
- The chronic disease is characterized by anorexia, weight loss, and a failure to thrive.
- Diagnosis primarily depends on detection of cysts or trophozoites in the feces which can be difficult because of inconsistent excretion in the feces.
- Giardiasis is effectively treated with metronidazole and other drugs.
- Control and prevention of giardiasis generally involve activities that are effective against other fecal–oral transmitted diseases and in particular improving personal hygiene.

Further Reading

Adam, R.D. (2001) The biology of *Giardia* spp. *Microbiol. Rev.* 55: 706–732.

Buret, A.G. (2008) Pathophysiology of enteric infections with *Giardia duodenalis*. *Parasite* 15: 261–265.

Cacció, S.M. and Ryan, U. (2008) Molecular epidemiology of giardiasis. *Mol. Biochem. Parasitol.* 160: 75–80.

Elmendorf, H.G., Dawson, S.C. and McCaffery, J.M. (2003) The cytoskeleton of *Giardia lamblia*. *Int. J. Parasitol.* 33: 3–28.

Escobedo, A.A. and Cimerman, S. (2007) Giardiasis: a pharmacotherapy review. *Expert Opin. Pharmacother.* 8: 1885–1902.

Huang, D.B. and White, A.C. (2006) An updated review on *Cryptosporidium* and *Giardia*. *Gastroenterol. Clin. North Am.* 35: 291–314.

Lauwaet, T., Davids, B.J., Reiner, D.S. and Gillin, F.D. (2007) Encystation of *Giardia lamblia*: a model for other parasites. *Curr. Opin. Microbiol.* 10: 554–559.

Morrison, H.G., McArthur, A.G., Gillin, F.D. et al. (2007) Genomic minimalism in the early diverging intestinal parasite *Giardia lamblia*. *Science* 317: 1921–1926.

Thompson, R.C.A. and Monis, P.T. (2004) Variations in *Giardia*: implications for taxonomy and epidemiology. *Adv. Parasitol.* 58: 69–137.

Traub, R.J., Minis, P., Robertson, I., Irwin, P., Mencke, N. and Thompson, R.C.A. (2004) Epidemiological and molecular evidence support the zoonotic transmission of *Giardia* among humans and dogs living in the same community. *Parasitology* 128: 53–62.

Trichomonas vaginalis and Other Trichomonads

Disease(s)	Trichomoniasis
Etiological agent(s)	*Trichomonas vaginalis*
Major organ(s) affected	Urogenital tract, especially vagina, urethra, prostate and epididymis
Transmission mode or vector	Direct contact during sexual intercourse
Geographical distribution	Worldwide and most prevalent nonviral STD
Morbidity and mortality	Generally asymptomatic or mild vaginitis or urethritis. Never fatal
Diagnosis	Culture of parasite from vaginal, urethral, prostatic secretions
Treatment	Metronidazole
Control and prevention	Safe sex practices

The trichomonads are a group of flagellated protozoa. Most of the members of this group are commensals or parasites and only a few free-living species have been identified. Generally the trichomonads are nonpathogenic commensals and only a few species are of veterinary or medical importance. Four species of trichomonads infect humans (Table 5.1). Among these only *Trichomonas vaginalis* is clearly pathogenic and it is usually of low virulence. The others exhibit a questionable pathogenicity or are of low virulence or low prevalence. The most notable among the veterinary trichomonads is probably *Tritrichomonas foetus* which is sexually transmitted among bovines and can result in abortion. The estimated annual loss for the cattle industry is estimated at $650 million.

The various trichomonads of humans inhabit different anatomical locations. *T. vaginalis* is a common sexually transmitted disease found in the urogenital tract. *T. tenax* is a commensal of the human oral cavity, found particularly in patients with poor oral hygiene and advanced periodontal disease. *T. tenax* has also occasionally been reported in the lungs. Such cases have been primarily in patients with underlying cancers or other lung diseases or following surgery. *Pentatrichomonas hominis*, formerly *Trichomonas hominis*, is a nonpathogenic commensal of the large intestine. *Dientamoeba fragilis* was originally believed to be an intestinal ameba based on its morphology and lack of flagella. Now it is known to be related to the trichomonads.

Table 5.1 Trichomonads of humans

Species	Location
Trichomonas vaginalis	Urogenital tract
Trichomonas tenax	Oral cavity
Pentatrichomonas hominis	Intestines
Dientamoeba fragilis	Intestines

Morphological and Metabolic Features of Trichomonads

The trichomonads only exhibit trophozoite stages and reproduce asexually by binary fission. A distinctive morphological feature of the group is an axostyle which runs the length of the organism and appears to protrude from the posterior end (Figure 5.1). The **axostyle** is a cytoskeletal element

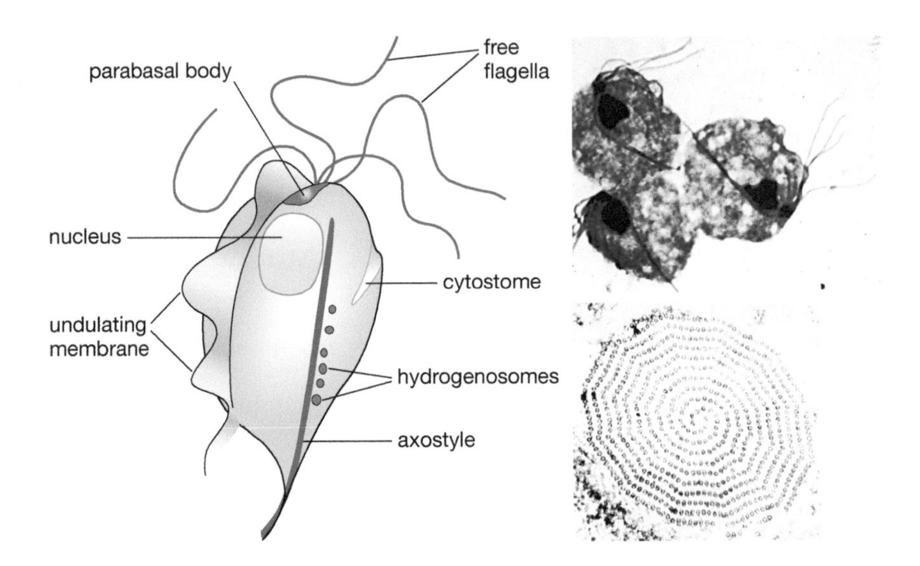

Figure 5.1 Morphological features of trichomonads. (Left) Schematic representation showing the major structural features of trichomonads. (Upper right) Giemsa-stained trophozoites of *T. vaginalis* from *in vitro* culture. (Micrograph from Thomas Orihel of Tulane University, School of Public Health.) (Lower right) Electron micrograph of axostyle cross section showing concentric layers of microtubules.

composed of concentric rows of microtubules and is believed to function in the attachment of the parasite to epithelial cells. A single nucleus is found at the anterior end of the parasite and 4–6 flagella emerge from the anterior end. One of the flagella is attached to the body of the organism and forms a posteriorly directed undulating membrane, whereas the remaining flagella are free. The number of free flagella is used as a criterion in designating genera. For example, three free flagella are classified as *Tritrichomonas*, four free flagella as *Trichomonas*, and five free flagella as *Pentatrichomonas*. Occasionally the cytostomal groove is also observed.

Another structural feature of trichomonads visible with the light microscope in stained specimens is the **parabasal body**. Most authors feel that the parabasal body is a type of Golgi apparatus which is linked to the basal bodies via fibers. This parabasal body and associated cytoskeletal elements define a group of protozoa called the parabasalids which includes the trichomonads and a group formerly called the hypermastigids. The hypermastigids are found in the guts of termites and other insects and are characterized by numerous flagella.

The trichomonads, like many other intestinal protozoa, exhibit an anaerobic metabolism and lack mitochondria. Part of energy metabolism of trichomonads involves a unique organelle called the **hydrogenosome**. The hydrogenosome has a double membrane and, like mitosomes in other anaerobic symbionts, is likely a highly derived mitochondrion. However, the hydrogenosome lacks DNA, cytochromes, and many typical mitochondrial functions such as enzymes of the tricarboxylic acid cycle and oxidative phosphorylation. The primary function of the hydrogenosome is the metabolism of pyruvate, produced during glycolysis within the cytosol, to acetate and carbon dioxide with the concomitant production of ATP (Figure 5.2). The electrons released from the oxidation of pyruvate are transferred to hydrogen ions to produce molecular hydrogen, hence the name hydrogenosome. This is in contrast to aerobic metabolism carried out by the mitochondria in which the released electrons are transferred to oxygen to form water. In addition, nitroimidazoles, such as metronidazole, are reduced to cytotoxic free radicals by the ferredoxin and the pyruvate:ferredoxin oxidoreductase (PFO) of the hydrogenosome indicating that this unique organelle is a drug target.

Trichomonas vaginalis

Trichomonas vaginalis was first described by Alfred Donné in 1836 from purulent vaginal discharges and by the early part of the 20th century it was recognized as an etiological agent of **vaginitis**. Trichomoniasis is a

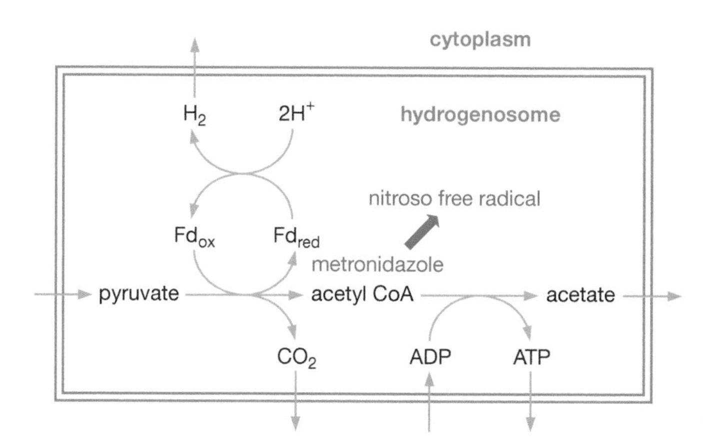

Figure 5.2 Abbreviated metabolism of the trichomonad hydrogenosome. Pyruvate produced by glycolysis in the cytoplasm is metabolized to acetate and carbon dioxide within the hydrogenosome. Pyruvate decarboxylation is catalyzed by a pyruvate:ferredoxin oxidoreductase (PFO) and the electrons released from pyruvate are transferred to ferredoxin (Fd), a low molecular weight electron carrier protein. The ferredoxin is reoxidized with protons (H^+) as terminal electrons acceptors through the action of a hydrogenase (H2ase) and thereby producing molecular hydrogen (H_2). ATP synthesis is coupled to the conversion of acetyl coenzyme A to acetate. Metronidazole is also metabolically activated to toxic free radical compounds by the ferredoxin and the pyruvate:ferredoxin oxidoreductase within the hydrogenosome.

common sexually transmitted disease with a worldwide distribution and an estimated 167 million people becoming infected per year worldwide and an estimated 5 million new infections per year in the United States. Prevalence tends to be higher in resource-limited settings. Trichomoniasis is believed to be the most common nonviral sexually transmitted disease and the prevalence is similar in magnitude to chlamydia and substantially greater than that of gonorrhea. Despite the frequency of trichomoniasis, in the past it has been considered more of a nuisance parasite rather than a major pathogen. However, it is now recognized as a factor in promoting HIV transmission (Box 5.1), possibly causing low-weight and premature births, and predisposing women to substantial discomfort and stress.

T. vaginalis, despite its name, infects both men and women. In females the organism primarily inhabits the vagina and in males it is usually found in the urethra, prostate, or epididymis. The trophozoites live closely associated with, or attached to, the epithelium of the urogenital tract, where they replicate by binary fission. The life cycle consists only of a trophozoite stage which is transmitted by direct contact during sexual intercourse. Nonvenereal transmission is rare, but possible, since the trophozoites can survive 1–2 days in urine and 2–3 hours on a wet sponge. In addition, neonates have been infected during the birth process. The risk factors for

Box 5.1 Trichomoniasis and HIV

The pathology caused by *Trichomonas* may enhance the efficiency of HIV transmission [1, 2]. *T. vaginalis* infection typically elicits a local cellular immune response with inflammation of the vaginal epithelium and cervix in women and the urethra of men. This inflammatory response includes the infiltration of potential HIV target cells such as $CD4^+$ bearing lymphocytes and macrophages. In addition, *T. vaginalis* can cause punctate hemorrhages on the vaginal walls and cervix. This leukocyte infiltration and the genital lesions may increase the number of target cells for the virus and allow direct viral access to the bloodstream through open lesions. In addition, the hemorrhages and inflammation can increase the level of virus in body fluids and the numbers of HIV-infected lymphocytes and macrophages present in the genital area in persons already infected with HIV. This increase of free virus and virus-infected leukocytes can increase the probability of HIV exposure and transmission to an uninfected partner and some studies suggest that concurrent *Trichomonas* infections can increase the risk of HIV transmission by 1.5–3 fold. Furthermore, increased cervical shedding of HIV has been shown to be associated with cervical inflammation, and substantially increased viral loads in semen have been documented in men with trichomoniasis. Moreover, since many patients with *Trichomonas* infection are asymptomatic, or only mildly symptomatic, they are likely to remain sexually active in spite of infection. In consideration of the high prevalence of trichomoniasis throughout the world even a modest increase in the risk of HIV transmission can translate into a large number of cases.

1. Sorvillo, F., Smith, L., Kerndt, P. and Ash, L. (2001) *Trichomonas vaginalis*, HIV, and African-Americans. *Emerg. Infect. Dis.* 7: 927–932.

2. McClelland, R.S., Sangare, L., Hassan, W.M., et al. (2007) Infection with *Trichomonas vaginalis* increases the risk of HIV-1 acquisition. *J. Infect. Dis.* 197: 698–702.

acquiring *T. vaginalis* infections are similar to those of other sexually transmitted infections. For example, a higher prevalence of infection is found in patients visiting sexual health clinics and among sex workers and drug users. However, in contrast to other sexually transmitted infections, the prevalence of *T. vaginalis* infection tends to increase with age among sexually active males and females.

Symptoms and pathogenesis

T. vaginalis causes different clinical manifestations in males and females (Table 5.2) and the severity of symptoms may be related to the degree of parasitemia. Women are more likely to exhibit symptoms which tend to persist longer. The incubation period typically ranges from 4 to 28 days. In females the infection can present as a mild inflammation of the vagina (i.e., vaginitis), an acute or chronic vulvovaginitis, or urethritis. The onset or exacerbation of symptoms commonly occurs during or immediately after menstruation. The most common complaint associated with *T. vaginalis* infection is a persistent mild vaginitis associated with a vaginal discharge that is often accompanied by burning or itching (pruritus). The discharge can range from thin and scanty to copious and foul-smelling and it is most often gray, but can be yellow or green and is occasionally frothy or blood tinged. As the infection becomes more chronic the discharge diminishes. Many women also experience painful or difficult coitus (dyspareunia). Urethral involvement occurs in some cases and is characterized by painful urination (dysuria) and frequent urination.

The vaginal epithelium is the primary site of infection. Thus the vaginal walls are usually erythematous (i.e., red) and may show petechial (a small non-raised spot) hemorrhages. Punctate hemorrhages of the cervix, called strawberry cervix, are observed in approximately 2% of the cases. This strawberry cervix is a distinctive pathological observation associated with trichomoniasis not seen with other sexually transmitted diseases. Males are likely to be asymptomatic (50–90%) and the infection tends to be self-limiting. The urethra and prostate are the most common sites of infection. Common symptoms include: urethral discharge (ranging from scant to purulent), dysuria (painful urination), and urethral pruritus (itching). Some men experience a burning immediately after coitus.

Little is known about the pathophysiology associated with *T. vaginalis* infection, but it is presumably due to interactions between the parasite and host epithelial cells. *In vitro* studies indicate that *T. vaginalis* can destroy cells in a contact-dependent manner. Therefore, adhesion of the trophozoites to the epithelium is believed to be a major factor in the pathogenesis. The trophozoite takes on a more ameboid appearance when it binds to a host cell (Figure 5.3). Several potential adhesion proteins have been identified on the surface of the trophozoites. In addition, secreted proteases which could play a role in pathogenesis have also been identified. Both cellular and humoral immune responses are evident in response to *T. vaginalis* infection, but are probably not protective.

Table 5.2 Clinical manifestations

Females	Males
Asymptomatic (15–20%*)	Asymptomatic (50–90%*)
Vaginal discharge (50–75%*)	Urethral discharge (50–60%**)
Dyspareunia (50%*)	Dysuria (12–25%**)
Pruritus (25–50%*)	Urethral pruritus (25%**)

*% of infected; **% of symptomatic

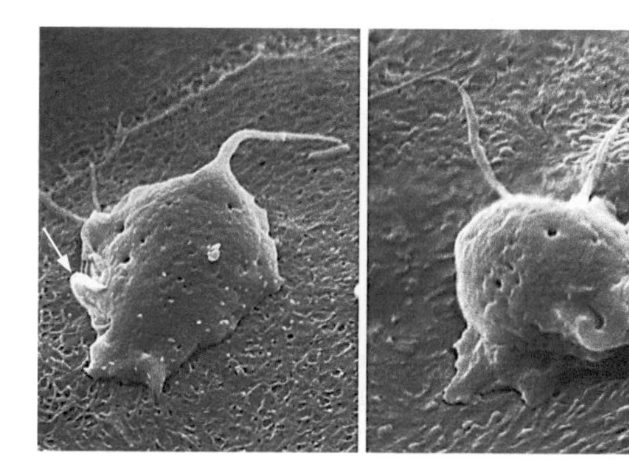

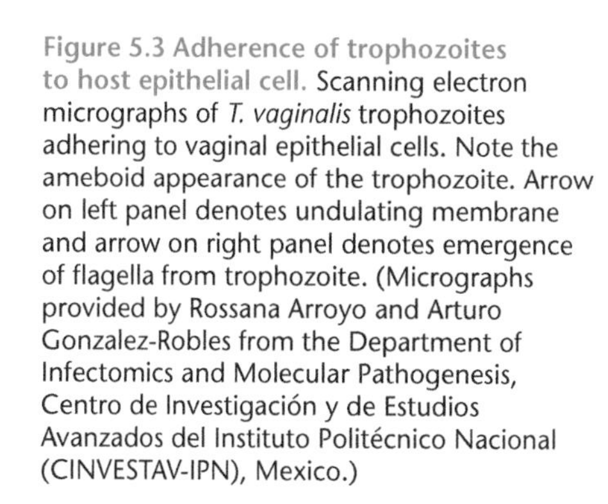

Figure 5.3 Adherence of trophozoites to host epithelial cell. Scanning electron micrographs of *T. vaginalis* trophozoites adhering to vaginal epithelial cells. Note the ameboid appearance of the trophozoite. Arrow on left panel denotes undulating membrane and arrow on right panel denotes emergence of flagella from trophozoite. (Micrographs provided by Rossana Arroyo and Arturo Gonzalez-Robles from the Department of Infectomics and Molecular Pathogenesis, Centro de Investigación y de Estudios Avanzados del Instituto Politécnico Nacional (CINVESTAV-IPN), Mexico.)

Diagnosis

In general, the clinical manifestations are not reliable as the sole means of diagnosis since the clinical presentation is similar to other sexually transmitted diseases. In addition, many patients have mild or no symptoms. Diagnosis is confirmed by the demonstration of trophozoites in vaginal, urethral, or prostatic secretions, or urine sediment (following prostate massage). Traditionally, microscopic examination of wet mounts of fresh vaginal discharge, preferably collected with a speculum on a cotton-tipped applicator, is the most practical method of diagnosis. Specimens should be diluted in saline and examined immediately. *T. vaginalis* is recognized by its characteristic morphological features and its rapid jerky motility. Specimens can also be fixed and stained with Giemsa or fluorescent dyes. However, the organism may be difficult to recognize on stained slides.

The reliability of direct observation ranges from 40 to 80%. Therefore, *in vitro* culture is considered the gold standard for diagnosis despite some limitations, such as a need for laboratory facilities and that organisms require 2–7 days of growth before they are detected. The accessibility issue is partly resolved by the InPouch™ TV culture system (Biomed Diagnostics). This is a commercially available self-contained system for the detection of *T. vaginalis* in clinical specimens. Approved immunochromatographic antigen detection systems (i.e., dipsticks) are available and can provide a diagnosis in less than 1 hour. Reliable PCR primers for DNA-based tests have been described but are not yet approved for diagnosis.

Treatment

Generally trichomoniasis is readily treatable as long as sexual partners are treated at the same time to prevent reinfection. Metronidazole (Flagyl®) and other nitroimidazoles, such as tinidazole (Tindamax®), are highly effective against trichomoniasis. Nitroimidazoles are activated by the hydrogenosome to nitroso radical ion intermediates which are cytotoxic. Metronidazole has been the predominant drug used for the treatment of trichomoniasis for several decades as either a single 2 g dose (85–92% cure rate) or 250 mg three times daily for 7–10 days (>95% cure rate). However, several clinical trials have demonstrated that a single 2 g dose of tinidazole is equally or more efficacious and it has a half-life of approximately double that of metronidazole. Nonetheless, cost issues often favor the use of metronidazole. Generally both metronidazole and tinidazole are well tolerated and adverse effects are usually limited to nausea, vomiting, constipation, cramping, and metallic taste. In addition, some patients are allergic to metronidazole. Generally the side effects of tinidazole are less than metronidazole. But there is no evidence indicating that tinidazole is safe for patients with a metronidazole allergy.

It is estimated that 2.5–5% of all cases of trichomoniasis exhibit some degree of resistance to metronidazole. Resistant parasites express lower levels of the hydrogenosomal enzymes (Figure 5.2). Because the metronidazole competes for the protons in the ferredoxin redox cycles, this results in less metronidazole being activated to cytotoxic radicals. Metronidazole resistance can usually be overcome with higher doses and some evidence suggests that tinidazole can be used to treat *Trichomonas* infections that are resistant to metronidazole.

Control

The epidemiology of trichomoniasis exhibits features similar to other sexually transmitted diseases and incidence generally correlates with the number of sexual partners. In addition, co-infection with other sexually transmitted infections, particularly gonorrhea, is common. It is estimated that up to 25% of sexually active women will become infected at some point during their lives and the disease will be transmitted to 30–70% of their male partners. Measures used in the control of other sexually transmitted infections, such as limiting number of sexual partners and use of condoms, are also effective in preventing trichomoniasis. Primary prevention should also include health education with the aim of modifying behavior.

Two possible complications in controlling trichomoniasis are the potentially long durations of untreated infections and the high reinfection rates. Longitudinal cohort studies indicate that *T. vaginalis* can persist as an asymptomatic infection in untreated women for at least 3 months and probably longer. Similarly, several studies have shown a high rate of reinfection with *T. vaginalis* for up to a year following treatment, particularly among high risk groups. These two phenomena can result in a large reservoir of infected individuals and a high prevalence within the community or within high risk groups. Control in such situations (e.g., sex workers) may require improved access to health care and screening and treatment of asymptomatic individuals.

Dientamoeba fragilis

Dientamoeba fragilis was first described in 1918 by Dobell. It was originally classified as an ameba based on its morphology and motility. However, later it was recognized to exhibit morphological features more similar to the turkey parasite *Histomonas meleagridis* than to other ameba. *Histomonas* trophozoites can exhibit both ameba-like forms and flagellated forms with a single flagellum. Subsequent ultrastructural studies of *Dientamoeba* revealed parabasal filaments and an extranuclear spindle apparatus as found in the parabasalids including the trichomonads. In addition, *Dientamoeba* has hydrogenosomes similar to the trichomonads. Early immunological studies also revealed that *Trichomonas*, *Histomonas*, and *Dientamoeba* all share some structurally related antigens with *Dientamoeba* and *Histomonas* being more closely related to each other than either to *Trichomonas*. More recent molecular sequence analyses have confirmed this close phylogenetic relationship between *Dientamoeba* and *Histomonas* and a more distal relationship to *Trichomonas*. Phylogenetic analyses also suggest that *Dientamoeba* and *Histomonas* share a common ancestor with a more complex cytoskeletal structure and that both species have arisen through a secondary loss of these structures. In other words, *Dientamoeba* is a trichomonad that has permanently lost its flagella and other cytoskeletal elements characteristic of the trichomonads.

Transmission and morphology

As with other trichomonads, *Dientamoeba* only exhibits a trophozoite stage. This raises some questions about the mode of transmission in that a cyst stage is usually involved in fecal–oral transmission. In addition, the trophozoites of *Dientamoeba* can only survive outside of the body for a short time. *H. meleagridis* also lacks a cyst stage and has been demonstrated to be transmitted via the eggs of a nematode. Because of the close phylogenetic relationship between *Histomonas* and *Dientamoeba*, it has been proposed that *Dientamoeba* is also transmitted via helminth eggs. Several studies have examined the potential role of the pinworm *Enterobius vermicularis* as the carrier for *Dientamoeba*. However, no clear correlations between *Dientamoeba* and pinworms have been demonstrated. However, numerous studies have shown that a large proportion of patients are infected with other intestinal protozoa supporting a direct transmission by the fecal–oral route. Other intestinal protozoa, most notably *Pentatrichomonas hominis*, also appear to be transmitted by a fecal–oral route without cyst stages. Thus, these two trichomonads may share a similar means of transmission which has yet to be elucidated.

The trophozoite of *Dientamoeba* is pleomorphic and typically ranges in size from 5 to 15 μm. This typical ameboid appearance and overlapping size range with other intestinal amebas (Chapter 2) can complicate the diagnosis. A notable feature of *Dientamoeba* is that the trophozoites are often binucleated (Figure 5.4). The proportion of uninucleated and binucleated trophozoites can vary considerably from patient to patient and even from the same patient on different days. The nucleoli (i.e., karyosome) often appear as 4–8 granules and there is no peripheral chromatin. Thus the key features used to correctly identify *D. fragilis* are the high frequency of binucleated trophozoites, the lack of cyst stages, and the fragmented karyosome.

Clinical features

Historically *Dientamoeba* has been considered as a nonpathogenic commensal. However, clinical symptoms often correlate with the presence of large numbers of trophozoites and treatment of the infection resolves these symptoms. The prevalence of symptoms is estimated at 15–30% of infected individuals. Clinical symptoms associated with *Dientamoeba* include intermittent diarrhea, abdominal pain, flatulence, nausea, and fatigue. Little is known about the pathogenesis and *Dientamoeba* probably acts as a low-grade irritant of intestinal mucosal surfaces that may lead to some inflammation. Diagnosis is usually through the examination of permanently stained fecal smears. No commercially available antigen detection tests are available, but reliable PCR tests have been described. Iodoquinol is generally the drug of choice for the treatment of *Dientamoeba*. Tetracycline, paromomycin, and metronidazole are also effective. Although the exact mode of transmission is not known, prevention is presumably the same as other infections transmitted by the fecal–oral route (Chapter 2).

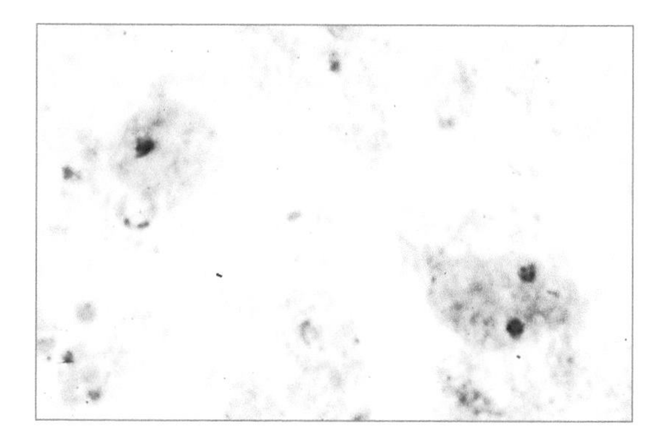

Figure 5.4 *Dientamoeba fragilis.* Iron-hematoxylin fecal smear showing a uninucleated (left) and binucleated (right) trophozoite. Note the ameba-like morphology and fragmented appearance of the nuclei. (Micrograph from Thomas Orihel of Tulane University, School of Public Health.)

Summary and Key Concepts

- *Trichomonas vaginalis* is a protozoan of the urogenital tract transmitted during sexual intercourse and is probably the most common nonviral sexually transmitted disease.
- The life cycle, as with other trichomonads, consists only of a trophozoite stage.
- The most common manifestation associated with *T. vaginalis* infection is a persistent mild vaginitis associated with a vaginal discharge that is often accompanied by burning or itching.
- Infections in men involve the urethra and prostate, and tend to exhibit milder symptoms and a shorter persistence of the infection.
- The basis of pathogenesis is unknown but likely involves the adherence of trophozoites to epithelial cells.
- Diagnosis is based on the microscopic detection or the *in vitro* cultivation of trophozoites from clinical specimens (e.g., vaginal swabs).
- Treatment with metronidazole or tinidazole is highly effective in curing trichomoniasis if both sexual partners are treated.
- *Dientamoeba fragilis* is found in the intestinal tract and causes some mild gastrointestinal symptoms.

Further Reading

Benchimol, M. (2009) Hydrogenosomes under microscopy. *Tissue Cell* 41: 151–168.

Huppert, J.S. (2009) Trichomoniasis in teens: an update. *Curr. Opin. Obstet. Gynecol.* 21: 371–378.

Johnson, E.H., Windsor, J.J. and Clark, C.G. (2004) Emerging from obscurity: biological, clinical, and diagnostic aspects of *Dientamoeba fragilis*. *Clin. Microbiol. Rev.* 17: 553–570.

Johnston, V.J. and Mabey, D.C. (2008) Global epidemiology and control of *Trichomonas vaginalis*. *Curr. Opin. Infect. Dis.* 21: 56–64.

Lehker, M.W. and Alderete, J.F. (2000) Biology of trichomonosis. *Curr. Opin. Infect. Dis.* 13: 37–45.

Schwebke, J.R. and Burgess, D. (2004) Trichomoniasis. *Clin. Microbiol. Rev.* 17: 794–803.

Stark, D.J., Beebe, N., Marriott, D., Ellis, J.T. and Harkness, J. (2006) Dientamoebiasis: clinical importance and recent advances. *Tr. Parasitol.* 22: 92–96.

Wendel, K.A. and Workowski, K.A. (2007) Trichomoniasis: challenges to appropriate management. *Clin. Infect. Dis.* 44: S123–S129.

Balantidium coli and *Blastocystis hominis*

Disease(s)	Balantidiosis, balantidiasis, balantidial dysentery
Etiological agent(s)	*Balantidium coli*
Major organ(s) affected	Colon
Transmission mode or vector	Fecal–oral and possible zoonotic cycle involving pigs
Geographical distribution	Worldwide, but more prevalent in the tropics. Prevalence rates rarely exceed 1%
Morbidity and mortality	Generally mild gastrointestinal symptoms with occasional dysentery and rarely spread to other organs. Rarely fatal
Diagnosis	Detection of the parasite in the feces
Treatment	Tetracycline
Control and prevention	Avoid fecal contamination of food or water

The majority of intestinal protozoa are classified as flagellates, amebas, or coccidia. Two exceptions to this are *Balantidium coli* and *Blastocystis hominis. Balantidium* is a **ciliate** which exhibits a typical fecal–oral life cycle and can cause a dysenteric syndrome. *Blastocystis* is a **stramenopile**. Stramenopiles are a diverse group of protozoa with many members found in planktonic environments. Thus far, *Blastocystis* is the only known parasitic stramenopile. Presumably *Blastocystis* is transmitted by a fecal–oral route, but many features about *Blastocystis*, such as mode of transmission and pathogenic potential, have not been conclusively demonstrated. The ciliates and stramenopiles are distally related to each other. The ciliates, along with the dinoflagellates and apicomplexans form a group called the alveolates. The alveolates and stramenopiles are both related to a group of algae known as Chromista (Chapter 1).

General Ciliate Biology

Ciliates are a large and diverse group of protozoa. Most ciliates are free-living and are found in a variety of habitats. Well-known ciliates include *Paramecium* species and *Ichthyophthirius multifiliis. Paramecium* species are found in ponds throughout the world and are often used in high school biology classes resulting in *Paramecium* probably being the quintessential ciliate. *I. multifiliis* is an ectoparasite of fish that causes white spot disease (also called "ick") and is well known to aquarium enthusiasts. As the name implies, virtually all ciliates possess cilia at some point during their life cycles. The cilia are generally arranged in longitudinal rows, called kineties, and typically cover the surface of the organism (Figure 6.1). The cilia generally function in motility, but can also play a role in the acquisition of food. Most ciliates are phagotrophic and ingest nutrients through a specialized organelle called the cytostome. Contractile vacuoles, which primarily function in osmotic regulation, are also usually evident.

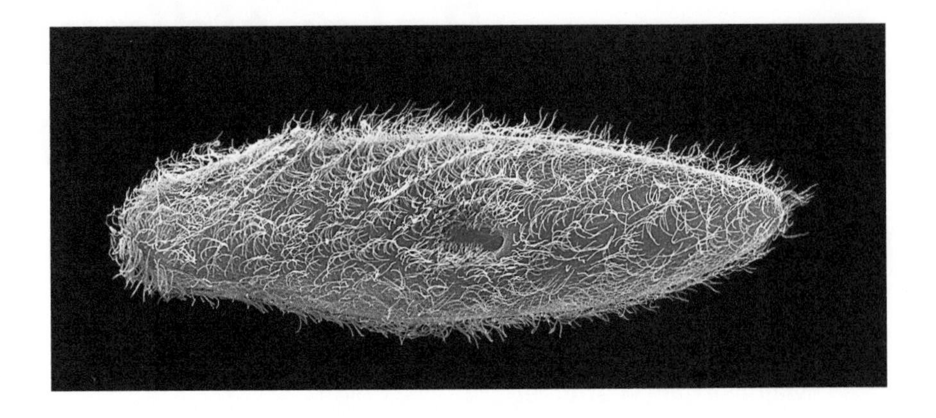

Figure 6.1 Scanning electron micrograph of *Paramecium caudatum*. (Provided by Wilhelm Foissner, Organismische Biologie, Universität Salzburg, Austria.)

The defining attribute of the ciliates is nuclear dimorphism characterized by two distinct nuclei. The large, often kidney-shaped, **macronucleus** is involved in the "housekeeping" or metabolic functions of the cell, whereas the smaller spherical **micronucleus** contains the complete genome and is primarily involved in the genetics and sexual recombination of the organism. The macronucleus typically contains thousands of copies of transcriptionally active "minichromosomes" representing 10000–20000 different DNA molecules. These minichromosomes are formed by a fragmentation of the chromosomal DNA into small pieces containing one or a few genes. DNA sequences corresponding to genes not expressed at high levels during the normal asexual cycle are lost and the remaining DNA fragments are amplified. Telomeres (chromosome ends) are also added to the ends of the minichromosomes. In fact, telomeres were first described in ciliates and the large number of telomeres resulted in ciliates being an early model system for the study of telomeres and telomerase (the enzyme that synthesizes telomeres).

Ciliates undergo both an asexual reproduction (i.e., binary fission) and a sexual reproduction involving **conjugation** (Figure 6.2). During conjugation, two ciliates of opposite mating types pair and exchange genetic material. Conjugal contact triggers meiosis in the micronuclei resulting in haploid micronuclei. In addition, the macronucleus breaks down and disappears. The exact fate of the micronuclei and the details of the sexual cycle vary according to species. In many species three of the micronuclei disintegrate and the remaining micronucleus divides again. Each of the conjugating organisms donates a micronucleus to its mate via a cytoplasmic bridge that connects them. The donated micronucleus is often referred to as either the gametic or male micronucleus, whereas the micronucleus remaining with the trophozoite is called either the stationary or female micronucleus. The gametic micronucleus fuses with the stationary micronucleus forming

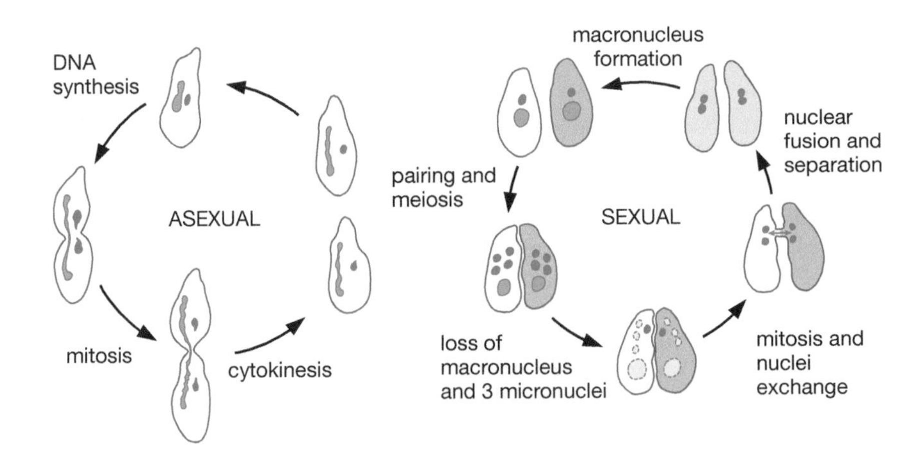

Figure 6.2 Schematic representation of asexual and sexual reproduction in ciliates. (Left) Trophozoites replicate by binary fission in which both the macro- and micronucleus undergo mitosis and each of the daughter cells receives copies of both. (Right) Trophozoites pair and exchange nuclei during sexual reproduction.

the diploid zygotic micronucleus. The conjugating pair separates and the zygotic nuclei undergo another round of mitosis resulting in two micronuclei. One of these micronuclei develops into the macronucleus, thus completing the cycle.

Balantidium coli

Balantidium coli is found worldwide, but like many other fecal–orally transmitted diseases, infections are more prevalent in the tropics. However, prevalence rates rarely exceed 1%. *B. coli* also infects a wide variety of mammals and is especially common in monkeys and pigs. Prevalence in pigs ranges from 20 to 100% and human balantidiosis usually exhibits an increased prevalence in communities that live in close association with pigs. For example, in Papua New Guinea, where pigs are the principal domestic animals, the prevalence among swine herders and slaughterhouse workers has been reported to exceed 25%. Contamination of the water supply with swine feces is believed to be a major source of infection. Human-to-human transmission has also been documented and this mode of transmission is likely to occur in similar situations as other fecal–oral infections (Chapter 2).

Life cycle and morphology

Balantidium exhibits a typical fecal–oral life cycle consisting of trophozoite and cyst stages. The trophozoite is oval shaped and has an average size of 70 × 45 μm, but can range upwards to 150–200 μm. The cyst has a distinctive cyst wall and is more spherical than the trophozoite with an average diameter of 55 μm. *Balantidium* exhibits the typical ciliate subcellular features such as nuclear dimorphism, rows of cilia, cytostome, and contractile vacuole (Figure 6.3). In stained specimens the most obvious internal structure is the large macronucleus. The micronucleus is usually adjacent to the macronucleus and may not always be apparent in stained preparations because of its close association with the macronucleus. The rows of cilia give the parasite surface a fuzzy appearance and are less pronounced in the cyst stage. Contractile vacuoles and the cytostome are often visible. The replicating trophozoites divide transversely forming asymmetric daughter cells in that one daughter retains the oral apparatus and cytostome and the other must regenerate the oral apparatus. Conjugation has been reported for *Balantidium* but the details of the nuclear events are not known.

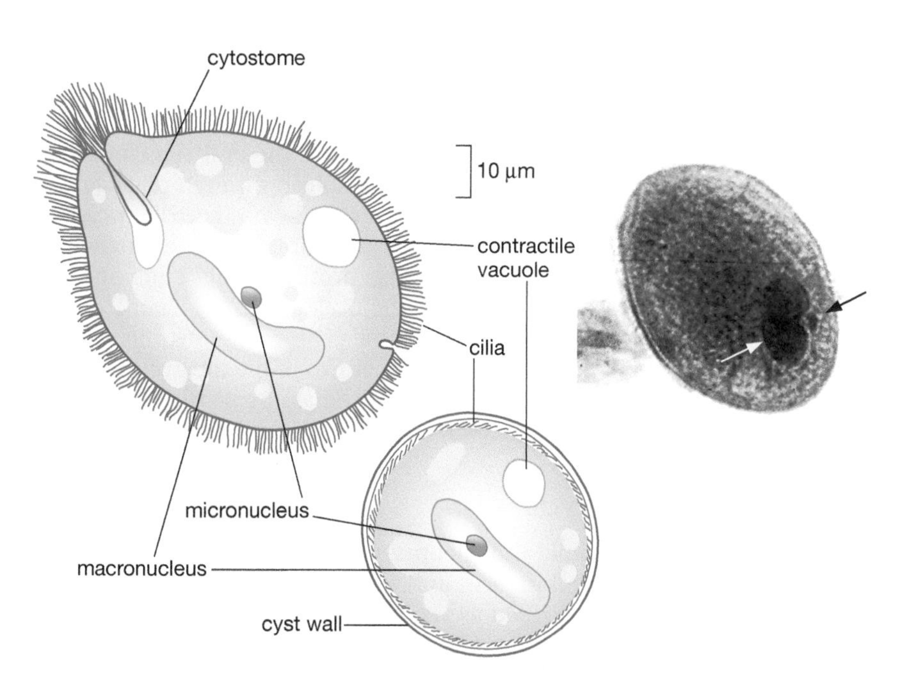

Figure 6.3 Morphological features of *Balantidium*. (Left) Schematic representation of trophozoite and cyst showing key morphological features. (Right) Micrographs of eosin-stained fecal sample showing micronucleus (black arrow) and macronucleus (white arrow).

Balantidium produces a range of symptoms

Some individuals infected with *B. coli* are completely asymptomatic or only exhibit mild symptoms. Symptoms typically include diarrhea or dysentery, tenesmus, nausea, vomiting, anorexia, and headache. Mild or chronic infections are characterized by intermittent diarrhea and constipation, weight loss, and abdominal pain. The diarrhea can persist for weeks to months before developing into dysentery. Superficial inflammation of the colonic mucosa may occur which can result in diarrhea and colicky pain. On rare occasions the trophozoites will invade the intestinal epithelium and produce ulceration of the colon. Clinically this results in an acute diarrhea with mucus and blood (i.e., dysentery). This balantidial dysentery is similar to the dysentery produced by *Entamoeba histolytica* (see Chapter 3) and a few deaths associated with balantidiosis have been recorded. Rare extra-intestinal infections involving lungs, vagina, ureter, and urinary bladder, and intestinal perforations leading to peritonitis, have been reported. Severe disease and death have usually been associated with other potentially compromising situations such as cancer, alcoholism, and poor health, suggesting *Balantidium* may be somewhat opportunistic.

Laboratory diagnosis is made by identifying the organism in feces. Because of their large size and spiraling motility the trophozoites are readily detectable in wet mount preparations from freshly collected stools. Permanent stains used for other intestinal protozoa can also be used. The large size and unique morphological features of *Balantidium* preclude its confusion with any other protozoa found in human feces. However, heavily stained cysts can be confused with helminth eggs. The treatment of choice is tetracycline given at 500 mg doses four times per day for 10–14 days. Iodoquinol is the recommended alternate drug. Metronidazole has not produced consistent results, but has been found effective in some studies. Preventive measures are similar to other diseases transmitted by the fecal–oral route (Chapter 2). The noted exception is that special care should be taken by persons working with swine and the contamination of drinking water and food with pig wastes needs to be avoided.

Blastocystis hominis

Blastocystis hominis is a common organism found in human stools. Prevalence in developed countries is typically 1–1.5% and in developing countries ranges from 30 to 50%. In addition, *Blastocystis* species have been isolated from a wide range of hosts including nonhuman primates, pigs, rodents, birds, reptiles, amphibians, insects, and annelids. Since its initial description approximately 100 years ago, *Blastocystis* has been variously classified as an ameba, a yeast, a sporozoan, and the cyst stage of a flagellate. Its phylogenetic position was finally resolved through phylogenetic analysis and molecular sequence data, which indicate that *Blastocystis* is most closely related to the stramenopiles, a complex and heterogeneous assemblage of unicellular and multicellular protists. Other stramenopiles include diatoms, brown algae, and water molds. Many of the characteristics of *Blastocystis* are unknown or controversial. The mode of transmission, mechanism of cell replication, and other features of the life cycle have not been conclusively demonstrated. Similarly, the status of *Blastocystis* as a pathogen, commensal, or opportunistic organism is unknown.

Morphology and life cycle

Blastocystis is polymorphic and pleomorphic in that a variety of morphological forms are found in feces and *in vitro* culture. Distinct forms of *Blastocystis* have been described as vacuolar, granular, ameboid, and cyst forms. The most widely recognized form is vacuolar form which is generally spherical and ranging in size from 10 to 15 μm in diameter with a large central vacuole (Figure 6.4). This large vacuole pushes the nuclei and other organelles to the periphery of the cell. The vacuole is sometimes filled with a granular material (i.e., granular form). Small resistant cyst-like forms have been identified from *in vitro* cultures and occasionally observed in feces. These presumed cysts are approximately 5 μm and surrounded by a multilayered wall. Furthermore, the cysts do not lyse when placed in water suggesting that they are resistant to environmental conditions.

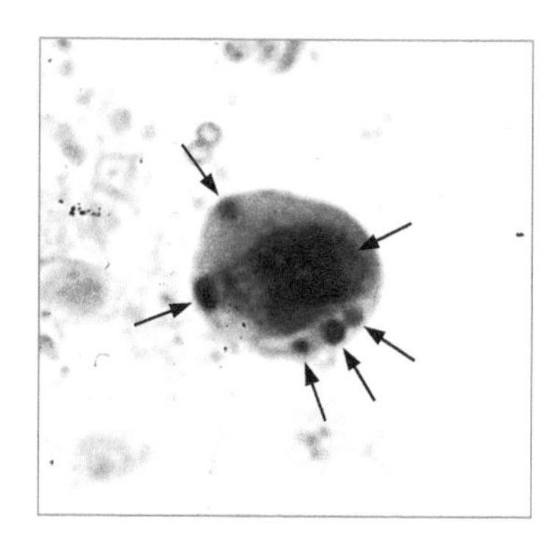

Figure 6.4 Vacuolated form of *Blastocystis hominis.* Arrows denote nuclei found at the periphery of the organism surrounding the large central vacuole.

There is considerable disagreement concerning the life cycle and mechanisms of cell division in *Blastocystis*. Presumably *Blastocystis* is transmitted via a fecal–oral route and a few studies have implicated contaminated water as a source of infection. *Blastocystis* is also found in other hosts ranging from worms to primates. This complicates the designation of species and historically human isolates have been designated as *B. hominis*, and isolates from other hosts as *Blastocystis* species corresponding to the particular host. Furthermore, *Blastocystis* exhibits a wide range of molecular diversity and the genetic distance between *Blastocystis* isolates is greater than the genetic distance between *Entamoeba histolytica* and *E. dispar* (Chapter 3). At least seven distinct clades among mammalian and avian isolates have been defined. Phylogenetic analysis also reveals that there are no exclusively human clades and human isolates are found in all of the clades. These studies raise the possibility that there are several distinct species of *Blastocystis*, none of which are highly host specific, and transmission cycles can involve human-to-human, animal-to-animal, animal-to-human, and human-to-animal routes.

Clinical aspects

As there is uncertainty about the taxonomic status and life cycle of *Blastocystis*, it is currently unclear whether *Blastocystis* represents a commensal, opportunistic organism, or a pathogen. There have been several reports suggesting *Blastocystis* causes disease, as well as many reports suggesting the opposite. Symptoms generally associated with *Blastocystis* infections are generally nonspecific and include diarrhea, cramps, nausea, vomiting, and abdominal pain. Asymptomatic infections are also quite common and large numbers of *Blastocystis* have been observed in feces of persons exhibiting no symptoms. In addition, some studies have shown that treatment alleviates the symptoms and clears the organisms suggesting that *Blastocystis* causes disease. However, the drugs used against *Blastocystis* (e.g., metronidazole) also work against many other intestinal protozoa and bacteria. The inability to rule out other organisms as the source of symptoms and the observation that many infected persons exhibit no symptoms makes it difficult to draw any definitive conclusions about the pathogenicity of *Blastocystis*. Furthermore, it could be that *Blastocystis* is opportunistic and exhibits virulence under specific host conditions such as concomitant infections, poor nutrition, or immunosuppression. In addition, the wide range of genetic diversity might explain the controversy concerning the pathogenicity of *Blastocystis* in that some genotypes may be more virulent than others. However, thus far the studies addressing this issue are inconclusive. Resolution of the confusion about the taxonomy, transmission, and virulence of *Blastocystis* will require additional studies.

Summary and Key Concepts

- *Balantidium coli* is the only ciliate that infects humans.
- Transmission of *B. coli* is via a fecal–oral route and may involve a zoonotic cycle with pigs.
- Clinical symptoms associated with balantidiosis can range from an asymptomatic carrier state to a dysenteric syndrome.

- *Blastocystis hominis* is a stramenopile presumably transmitted by the fecal–oral route.
- The exact taxonomic status, mode of transmission, and pathogenic potential have not been conclusively demonstrated for *B. hominis*.

Further Reading

Garcia, L.S. (2008) *Balantidium coli*. In *Emerging Protozoan Pathogens* (ed. N.A. Khan). Taylor and Francis Group, New York, pp. 353–366.

Schuster, F.L. and Ramirez-Avila, L. (2008) Current world status of *Balantidium coli*. *Clin. Microbiol. Rev.* 21: 626–638.

Stenzel, D.J. and Boreham, P.F.L. (1996) *Blastocystis hominis* revisited. *Clin. Microbiol. Rev.* 9: 563–584.

Tan, K.S.W. (2004) *Blastocystis* in humans and animals: new insights using modern methodologies. *Vet. Parasitol.* 126: 121–144.

Yoshikawa, H. Morimoto, K., Wu, Z., Singh, M. and Hashimoto, T. (2004) Problems in speciation in the genus *Blastocystis*. *Tr. Parasitol.* 20: 251–255.

Kinetoplastids

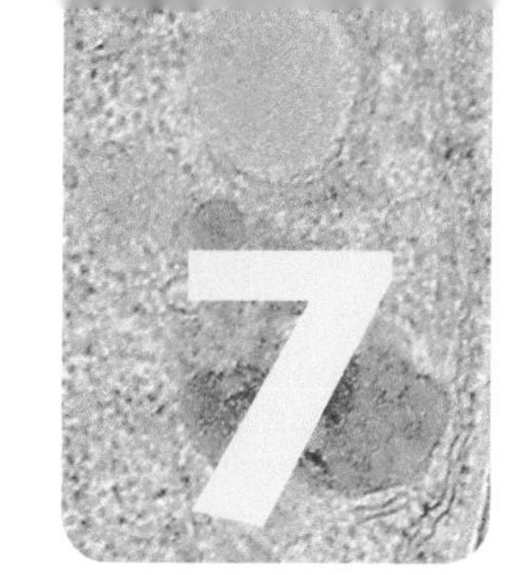

The kinetoplastids are a widespread protozoan group. Members of this group parasitize virtually all animal groups—ranging from fish to humans—as well as plants and insects. There are also free-living kinetoplastids which feed on bacteria in aquatic, marine, and terrestrial environments. The kinetoplastids are a monophyletic group and are distally related to the jakobids. Jakobids are small free-living bacteria-eating flagellates and the ancestors of the kinetoplastids were probably similar to the jakobids in terms of both their morphology and ecology. Parasitism may have evolved at least four different times within the natural history of the kinetoplastids.

Three distinct kinetoplastids cause human disease: African trypanosomes (Chapter 8), *Trypanosoma cruzi* (Chapter 9), and *Leishmania* species (Chapter 10). All three parasitize the blood or tissues of the human host and are transmitted by insect vectors (Table 7.1). They all exhibit complex and distinct life cycles in regard to both the human host and the arthropod vector. The three protozoa also exhibit distinct disease manifestations.

A major distinguishing feature of the kinetoplastids is a subcellular structure known as the **kinetoplast**. The kinetoplast is a structure within the cell that is visible by light microscopy and readily detected after staining with Giemsa, but is distinct from the nucleus (Figure 7.1). The kinetoplast is found near the basal body which is located at the base of the flagellum. Because of this location near the flagellum, it was previously believed that the kinetoplast was somehow associated with cell movement—hence the name. However, the staining of the kinetoplast is due to one of the dyes within the Giemsa stain mix binding to the mitochondrial DNA and is not directly involved in motility. In fact, the existence of extranuclear (i.e., organellar) DNA was first demonstrated in the kinetoplastids.

Kinetoplastids have a single mitochondrion that generally runs the length of the organism and a mass of DNA is localized to a discrete domain within the mitochondrion. The kinetoplast is physically associated with a particular region of the inner surface of the mitochondrial membrane and the basal body is associated with the same region on the outer surface. Thus the kinetoplast and basal body are indirectly connected and are always located close to each other within the cell body. The physical size of the kinetoplast varies according to species and this is reflective of the amount of mitochondrial DNA in any particular species.

Basic Morphology

Generally the kinetoplastids exhibit a long slender morphology with a free flagellum located at the anterior end. Anterior and posterior refer to the direction of movement, and thus the flagellum pulls the organism. A microtubule-based cytoskeleton lying beneath the plasma membrane (aka, pellicle) and running the length of the organism is a major determinant of the

Table 7.1 Kinetoplastids causing human disease

Organism	Disease	Vector
African trypanosomes	Human African trypanosomiasis	Tsetse fly
Trypanosoma cruzi	Chagas' disease	Triatomine bugs
Leishmania species	Leishmaniasis	Sand fly

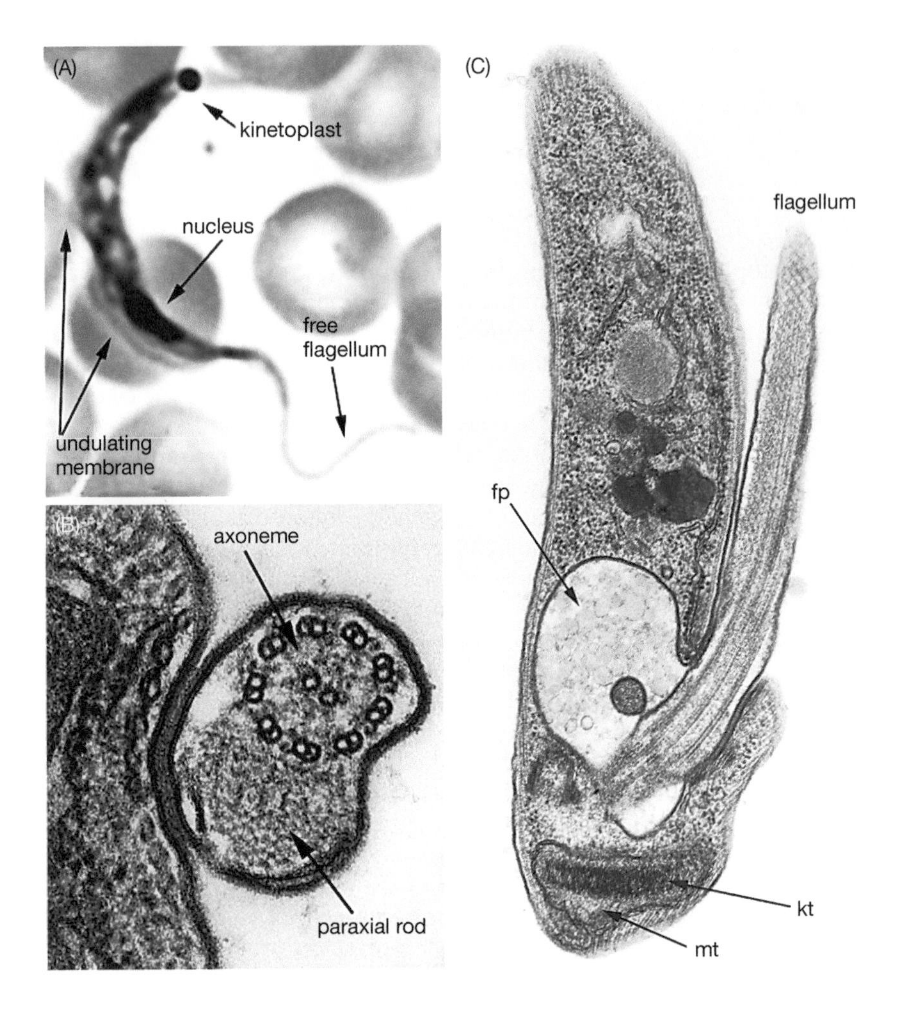

Figure 7.1 Basic morphological features of kinetoplastids. (A) Light micrograph of trypomastigote form showing the basic morphological features of kinetoplastids. (B) Transmission electron micrograph showing cross section of the undulating membrane enclosing the axoneme and paraxial rod. (C) Transmission electron micrograph showing the kinetoplast (kt) within the mitochondrion (mt) at the base of the flagellum emerging from the flagellar pocket (fp). (Electron micrographs provided by Paul Webster of the House Ear Institute, Los Angeles.)

shape of the organism. This subpellicular cytoskeleton is a common feature of many other protozoa as well as plant cells. The membrane surrounding the flagellum of the kinetoplastids is sometimes attached to the body of the organism forming an **undulating membrane** (Figure 7.1). This undulating membrane functions like a fin and increases the motility of the organism. Also enclosed within the undulating membrane and lying between the axoneme and the body of the organism is a structure called the paraxial rod.

Kinetoplastids can exhibit several different morphological forms. These various morphological forms are associated with different life cycle stages in the various species (to be discussed in subsequent chapters). The four major morphological forms found in kinetoplastids of medical importance are **trypomastigotes**, **epimastigotes**, **promastigotes**, and **amastigotes** (Figure 7.2). Other morphological forms have been described in other species. The different morphological forms are distinguished by the position of the kinetoplastid in relation to the nucleus and the presence or absence of an undulating membrane. In trypomastigotes the kinetoplast is located on the posterior end of the parasite. The flagellum emerges from the posterior end but folds back along the parasite's body, thus forming an undulating membrane that spans the entire length of the parasite. The kinetoplast is more centrally located in epimastigotes and is usually just anterior to nucleus. The flagellum emerges from the middle of the parasite and forms a shorter undulating membrane than observed in trypomastigote forms. The kinetoplast in promastigotes is usually located more toward the anterior end and the flagellum with no undulating membrane emerges at the anterior end. In all three forms the free flagellum is at the front of the parasite (i.e., anterior end) in respect to the direction of movement. The amastigote is smaller and

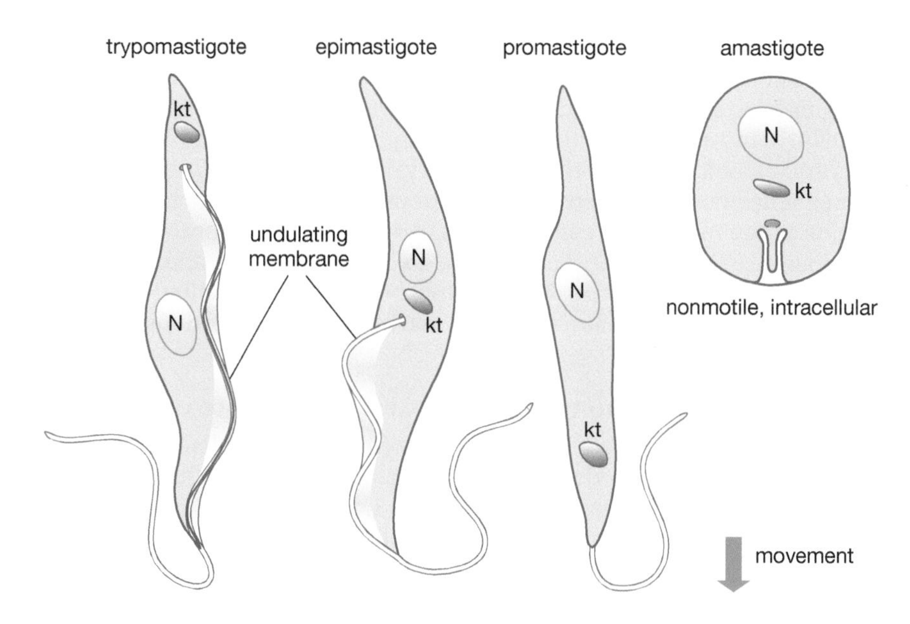

Figure 7.2 Morphological forms of kinetoplastids. Schematic representation of the morphological forms observed in medically important kinetoplastids. The position of the kinetoplast (kt) in relation to the nucleus (N) and the presence and length of the undulating membrane is used to distinguish the extracellular motile forms. Amastigotes are nonmotile intracellular stages.

more spherical in shape and has no free flagellum. However, a basal body and the base of the flagellum can be seen with the electron microscope. The kinetoplast is usually detectable via light microscopy as a darkly staining body near the nucleus. This form is a nonmotile intracellular stage.

Unique Features of Kinetoplastids

The kinetoplastids exhibit several unique cellular and molecular features. For this reason the kinetoplastids, and especially the African trypanosomes, are used as model systems for the study of basic biological phenomena. For example, glycosylphosphatidylinositol (GPI) membrane anchors were first described in African trypanosomes and have been extensively characterized from these organisms. Most notable is the GPI anchor of the variant surface glycoprotein discussed in detail in Chapter 8. GPI is a distinctive glycolipid attached to the C-terminus of membrane proteins in which the fatty acid chains of the lipid portion insert into the lipid bilayer of the membrane and anchor the protein onto the outer surface of the plasma membrane. Initially it was believed that GPI anchors were unique to the kinetoplastids. However, it was soon realized that such membrane anchors are quite ubiquitous among eukaryotes. Other unique features or phenomena first described in kinetoplastids include: mitochondrial DNA architecture, mitochondrial RNA editing, arrangement of genes in large polycistronic clusters, *trans*-splicing of mRNA transcripts, novel modifications of nucleotides, compartmentalization of glycolysis, and novel thiol-based redox mechanisms.

Kinetoplast DNA

The ability to detect kinetoplast DNA in stained samples is due to its relative abundance compared with mitochondrial DNA in other organisms. Kinetoplast DNA (ktDNA) is composed of two classes of circular DNA called **mini-circles** and **maxi-circles**. These two types of ktDNA form a concatenated mass of DNA in which the individual circular DNA molecules are interconnected like chain links into a complex network in which each circular DNA molecule is connected to several other circular DNA molecules. Thus this large network of interconnected DNA molecules is readily detected by light microscopy when stained with DNA binding dyes.

Maxi-circles range in size from 20 to 38 kb and encode several genes that are typically found on mitochondrial genomes from other organisms such as rRNA genes and genes encoding proteins associated with metabolic processes that take place in mitochondria. Approximately 20–50 copies are found per mitochondrion. Generally 5000–10 000 copies of the mini-circles, ranging in size from 0.5 to 1.5 kb, are found within the concatenated mass of ktDNA. In most kinetoplastid species the mini-circles are quite heterogeneous and exhibit a high level of sequence variability with only small regions of sequence homology between the various mini-circles. The homologous region of the mini-circle can be used as a sensitive probe for the detection of kinetoplastids by DNA methods due to its high copy number. Similarly, the high level of mini-circle DNA heterogeneity is often exploited in genetic fingerprinting methods as a genotypic marker to distinguish different parasite isolates and strains.

Kinetoplastid mitochondrial gene transcripts require editing

Several of the genes located on the maxi-circles are incomplete in that transcripts do not exactly code for the proteins. In other words, the transcripts need to be edited before being translated. Editing of these "cryptogenes" (i.e., hidden genes) is carried out by a complex of proteins referred to as the **editosome** and a special type of RNA found on the mini-circles called **guide RNA** (gRNA). During editing, uridines (nucleotides found only in RNA) are added, or sometimes removed, from the primary transcript. The positions of the uridine additions or deletions are specified by gRNAs. The gRNAs are small RNA transcripts that are complementary to the cryptogenes near the sites of editing.

The first step of RNA editing is the formation of a duplex between the primary mRNA transcript and its cognate gRNA in the complementary regions (Figure 7.3). An endonuclease will then cleave the pre-mRNA molecule in the region adjacent to complementary region between the pre-mRNA and the gRNA. In the case of insertion editing, a terminal uridine transferase adds uridines to the pre-RNA, whereas in deletion editing a uridine-specific exonuclease removes the extra uridines in the pre-mRNA. The number of uridines added or removed is determined by the sequence of the gRNA. In both cases the two fragments of the pre-mRNA are then rejoined by the action of DNA ligase to form the mature mRNA.

Kinetoplastids exhibit atypical transcription and mRNA processing

The 5′ end of all nuclear-encoded mRNA molecules from kinetoplastids consists of a highly conserved spliced leader sequence. The DNA sequence for this spliced leader is not contiguous with the protein coding portion of the gene, but is added post-transcriptionally. In other words, the mature mRNA of kinetoplastids is a chimera of the spliced leader transcript and

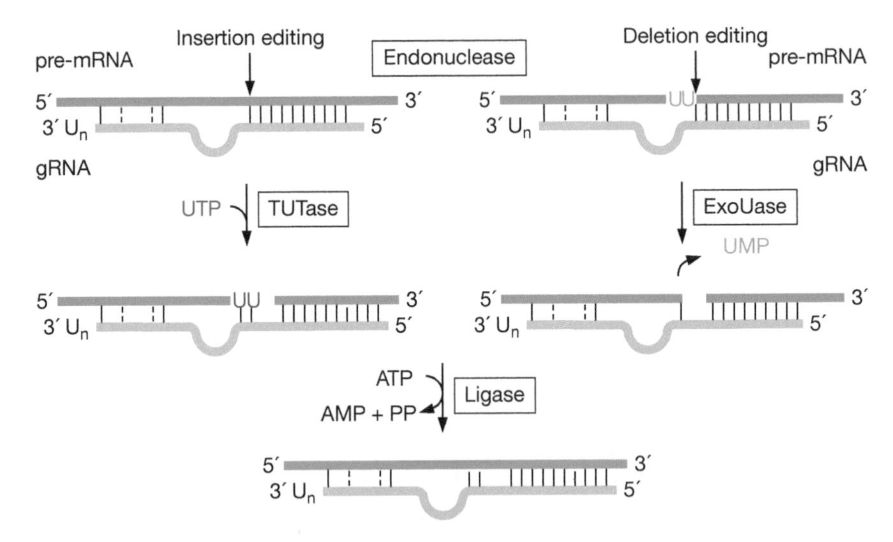

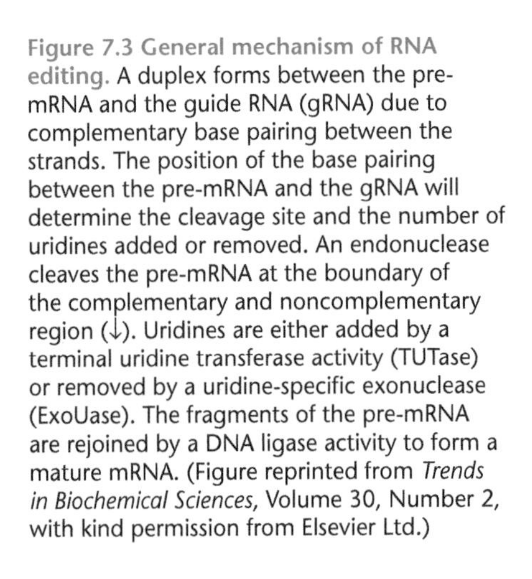

Figure 7.3 General mechanism of RNA editing. A duplex forms between the pre-mRNA and the guide RNA (gRNA) due to complementary base pairing between the strands. The position of the base pairing between the pre-mRNA and the gRNA will determine the cleavage site and the number of uridines added or removed. An endonuclease cleaves the pre-mRNA at the boundary of the complementary and noncomplementary region (↓). Uridines are either added by a terminal uridine transferase activity (TUTase) or removed by a uridine-specific exonuclease (ExoUase). The fragments of the pre-mRNA are rejoined by a DNA ligase activity to form a mature mRNA. (Figure reprinted from *Trends in Biochemical Sciences*, Volume 30, Number 2, with kind permission from Elsevier Ltd.)

the protein encoding transcript produced by *trans*-splicing. Furthermore, the protein encoding genes are found in dense arrays and are transcribed as extremely long pre-mRNA molecules containing numerous genes. These long polycistronic RNA molecules are then processed into mRNA molecules which encode for the individual proteins. This is in contrast to the transcription normally seen in most eukaryotes which involves the transcription of just a single gene into an mRNA molecule.

Processing of this polycistronic RNA into individual mRNA molecules involves the *trans*-splicing of the leader sequence to the 5′ ends of the individual genes within the polycistronic RNA. The *trans*-splicing is carried out by similar machinery as found in other eukaryotes for the removal of introns by *cis*-splicing. The net result of this process is a cleavage of the polycistronic RNA at specific sites in the space between the genes and the addition of the splice leader sequence to the 5′ end of one gene and the addition of a poly-A tail to the 3′ end of the preceding gene (Figure 7.4). This process is repeated for each of the genes in the polycistronic RNA resulting in mature mRNA molecules corresponding to the individual genes. The phenomenon of polycistronic transcription and *trans*-splicing was initially discovered in the kinetoplastids and subsequently shown to occur in widely divergent phyla, including euglenids, cnidarians, rotifers, nematodes, flatworms, and chordates.

Glycolytic enzymes are concentrated in the glycosome

Enzymes associated with glycolysis are usually found in the cytoplasm of eukaryotic cells. In kinetoplastids many of the glycolytic enzymes are located in a specialized peroxisome-like compartment called the **glycosome**. Like peroxisomes, the glycosome is bound by a single membrane and has a protein-dense matrix. Enzymes typically associated with peroxisomes are also found in the glycosomes and the targeting sequences for glycosomes are the same as peroxisomal targeting sequences. Thus the glycosomes are divergent peroxisomes with a distinct function. In particular, the seven enzymes involved in the conversion of glucose to 3-phosphoglycerate are present within the glycosome (Figure 7.5), whereas the remaining enzymes of the glycolytic pathway—responsible for converting 3-phosphoglycerate to pyruvate—are found within the cytosol.

The compartmentalization of the glycolytic enzymes may account in part for the high rates of glycolysis observed during some life cycle stages of the kinetoplastids (e.g., bloodstream form of African trypanosomes). The high concentration of enzymes and metabolites within this specialized organelle likely promote higher rates of glycolysis. In addition, the glycosomal membrane exhibits a low permeability leading to a segregation of metabolites between the glycosome and cytosol. This segregation of metabolites may prevent metabolic interference between competing pathways.

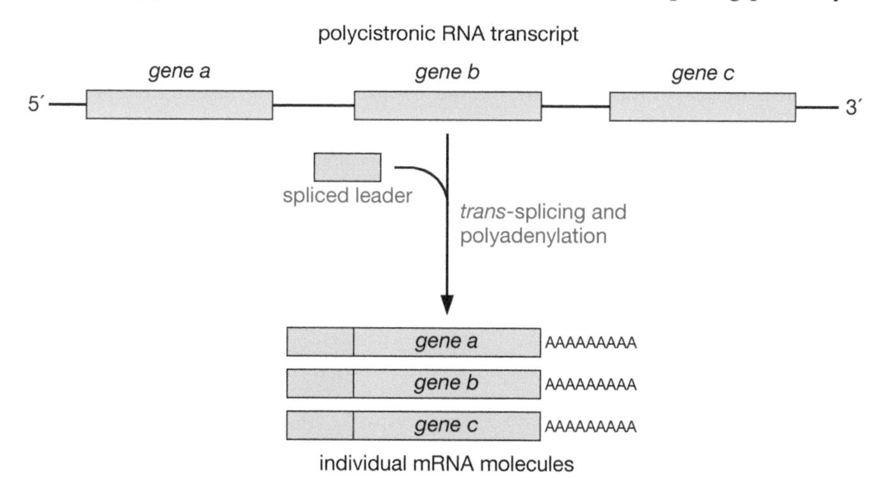

Figure 7.4 Kinetoplastid RNA processing. DNA is transcribed into long polycistronic RNA molecules containing numerous genes and then subsequently processed into the individual mature mRNA molecules by *trans*-splicing and polyadenylation.

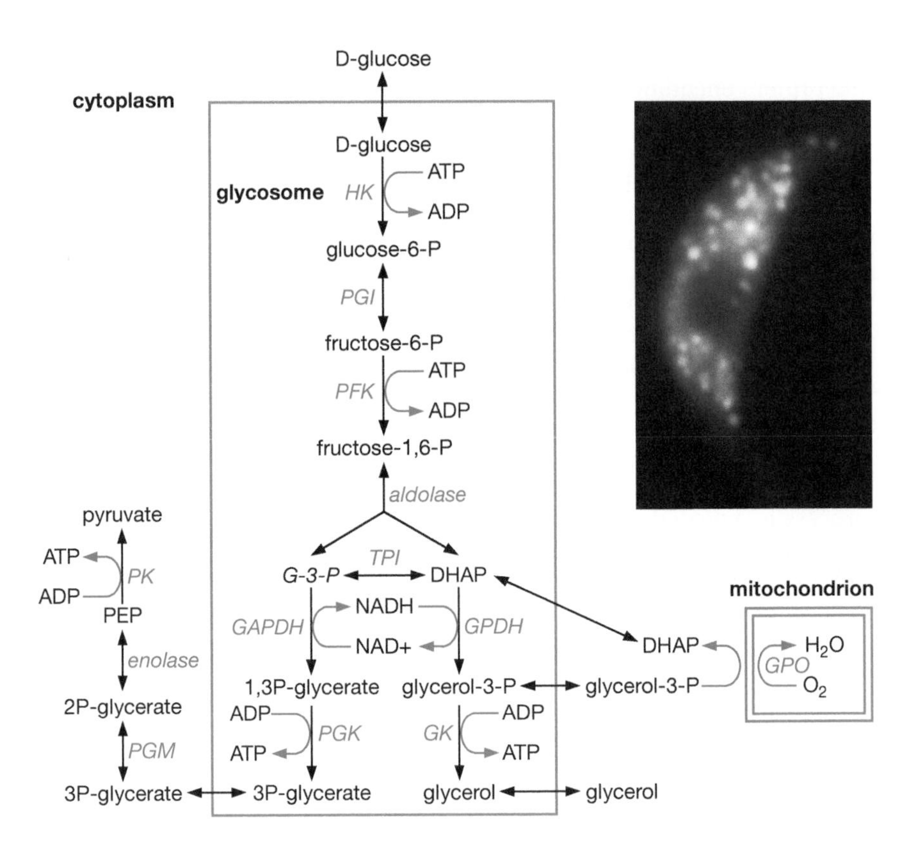

Figure 7.5 Glycolysis and the glycosome. The metabolism of glucose within the kinetoplastids involves the glycosome, the cytosol, and the mitochondrion. Shown is the compartmentalization of the enzymes involved in the metabolism of glucose between these compartments. Abbreviations: DHAP (dihydroxyacetone phosphate), G-3-P (glyceraldehyde-3-phosphate), GAPDH (glyceraldehyde-3-phosphate dehydrogenase), GK (glycerol kinase), GPDH (glycerol-3-phosphate dehydrogenase), GPO (glycerol phosphate oxidase), HK (hexose kinase), PFK (phosphofructose kinase), PGI (glucose phosphate isomerase), PGK (phosphoglycerate kinase), PGM (phosphoglycerate mutase), PK (pyruvate kinase), and TPI (triose phosphate isomerase). The inset in the upper right is a fluorescent image of a procyclic trypomastigote form of a trypanosome expressing a glycosomal protein tagged with a fluorescent marker. (Micrograph provided by Tracy Saveria and Marilyn Parsons of the Seattle Biomedical Research Institute, Seattle, WA.)

The exact composition of the glycosomes appears to differ in the various kinetoplastid species as well as between life cycle stages. Thus the glycosome promotes more efficient glycolysis, or other metabolic processes, and allows for metabolic specialization according to the exact needs of the parasite in its various life cycle stages.

Redox metabolism includes a unique thiol-based compound

Reactive oxygen intermediates, such as superoxide and peroxide, are produced as by-products of aerobic metabolism. These reactive oxygen intermediates are toxic in that they are highly reactive and damage the DNA, protein, and lipids within the cell and can lead to the death of the organism. Parasitic protozoa not only have to deal with these reactive by-products resulting from their own metabolism, but also need to deal with reactive oxygen intermediates produced as a result of the host metabolism as well as the oxidative respiratory burst associated with the host immune system. For example, macrophages and other immune effector cells produce reactive oxygen intermediates as a defensive strategy to kill pathogens.

Thiol-based redox reactions based on glutathione play key roles in the detoxification of reactive oxygen intermediates in virtually all organisms. However, the kinetoplastids utilize a unique compound consisting of two glutathione moieties joined together via a spermidine (i.e., bis-glutathionylspermidine) which is called trypanothione. The trypanothione system replaces the nearly ubiquitous glutathione system for providing reducing equivalents. For example, reduced trypanothione is used as a cofactor in various processes such as detoxification of peroxides and heavy metals, as well as the other metabolic processes such as the reduction of nucleotides for DNA synthesis. Reduced trypanothione is then regenerated by an NADPH-dependent trypanothione reductase analogous to the generation of reduced glutathione by glutathione reductase (Figure 7.6). Trypanothione reductase is an essential enzyme and several compounds that specifically

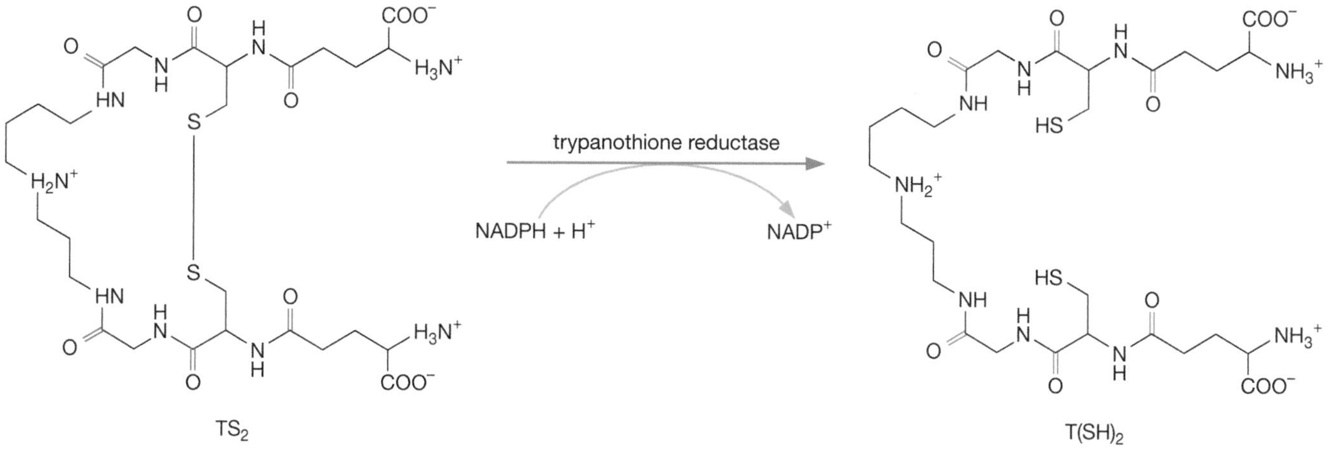

Figure 7.6 Trypanothione. Trypanothione consists of two glutathione molecules joined together by a spermidine molecule. Oxidized trypanothione (TS_2) is reduced by trypanothione reductase. The reduced trypanothione ($T(SH)_2$) is then used as a cofactor in many cellular reactions to provide reducing equivalents.

inhibit trypanothione reductase, but not human glutathione reductase, have been identified. It is hoped that the uniqueness of the trypanothione system may provide a sensitive drug target.

Summary and Key Concepts

- The kinetoplastids are a widespread group of flagellated protozoa which exhibit many unique biological features.
- The defining characteristic is a Giemsa-staining structure called the kinetoplast which is the result of a concatenated mass of mitochondrial DNA.
- Human diseases caused by kinetoplastids are African sleeping sickness, Chagas' disease, and leishmaniasis.
- Distinct morphological forms of the kinetoplastids are defined by the position of the kinetoplast in relation to the nucleus and the presence or absence of a flagellum with or without an undulating membrane.

Further Reading

Krauth-Siegel, R.L., Meiering, S.K. and Schmidt, H. (2003) The parasite-specific trypanothione metabolism of trypanosoma and leishmania. *Biol. Chem.* 384: 539–549.

Liang, X., Haritan, A., Uliel, S. and Michaeli, S. (2003) *Trans* and *cis* splicing in trypanosomatids: Mechanism, factors, and regulation. *Eukaryot. Cell* 2: 830–840.

Ogbadoyi, O.E., Robinson, D.R. and Gull, K. (2003) A high-order trans-membrane structural linkage is responsible for mitochondrial genome positioning and segregation by flagellar basal bodies in trypanosomes. *Mol. Biol. Cell* 21: 1769–1779.

Parsons, M. (2004) Glycosomes: parasites and the divergence of peroxisomal purpose. *Mol. Microbiol.* 53(3): 717–724.

Simpson, A.G.B., Stevens, J.R. and Lukeš, J. (2006) The evolution and diversity of kinetoplastid flagellates. *Tr. Parasitol.* 22: 168–174.

Stuart, K.D., Brun, R., Croft, S., Fairlamb, A., Gürtler, R.E., McKerrow, J., Reed, S. and Tarleton, R. (2008) Kinetoplastids: related protozoan pathogens, different diseases. *J. Clin. Invest.* 118: 1301–1310.

Stuart, K.D., Schnaufer, A., Ernst, N.L. and Panigrahi, A.K. (2005) Complex management: RNA editing in trypanosomes. *Tr. Biochem. Sci.* 30: 97–105.

African Trypanosomiasis

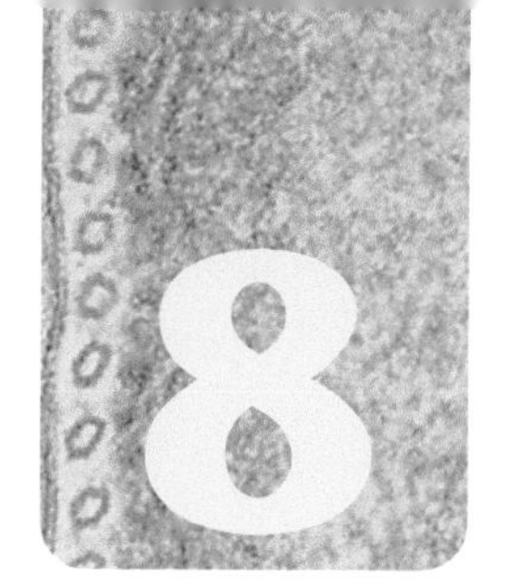

Disease(s)	African trypanosomiasis, African sleeping sickness
Etiological agent(s)	*Trypanosoma gambiense, Trypanosoma rhodesiense*
Major organ(s) affected	Blood, lymphatics, CNS
Transmission mode or vector	Tsetse fly
Geographical distribution	Foci throughout central Africa
Morbidity and mortality	An acute phase due to a blood infection advancing to neurological symptoms associated with invasion of the CNS. Fatal if not treated
Diagnosis	Detection of parasite in blood, lymphatics, or CNS
Treatment	Pentamidine or suramin during the acute stage and melarsoprol or eflornithine after involvement of the CNS
Control and prevention	Protective clothing and insect repellents to prevent infection and traps to reduce number of tsetse

Human African trypanosomiasis, also known as African sleeping sickness in its late stage, is found in equatorial Africa and exhibits a patchy distribution depending upon specific topographical features and the presence of the vector. African trypanosomes, the causative agent of the disease, are transmitted by several *Glossina* species, commonly known as **tsetse** flies (Figure 8.1). The control of African trypanosomiasis is complicated by poverty, political instability, and civil wars which are often ongoing in areas endemic for the disease. These factors contribute to disruption of the disease surveillance and control activities. In addition, climatic changes affecting vegetation and reservoir populations may be playing a role in focal resurgences of the disease. Human African trypanosomiasis is endemic in much of sub-Saharan Africa with an estimated 60 million people in 36 nations at risk of infection. The incidence of the disease had been increasing from the mid-1960s to the end of the 20th century with 300 000–500 000 cases in 1998. The true number of cases is difficult to determine since the disease is not

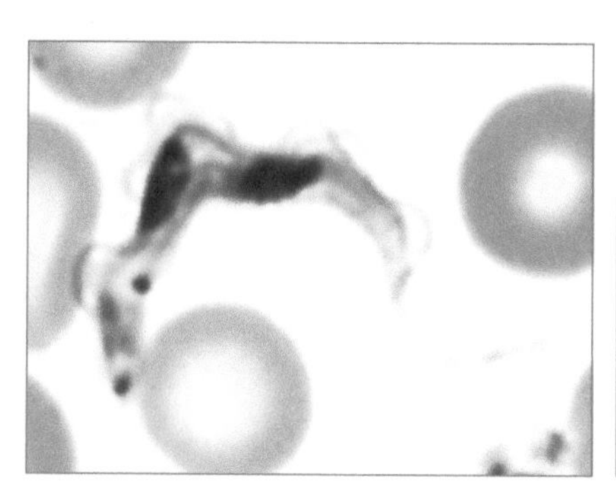

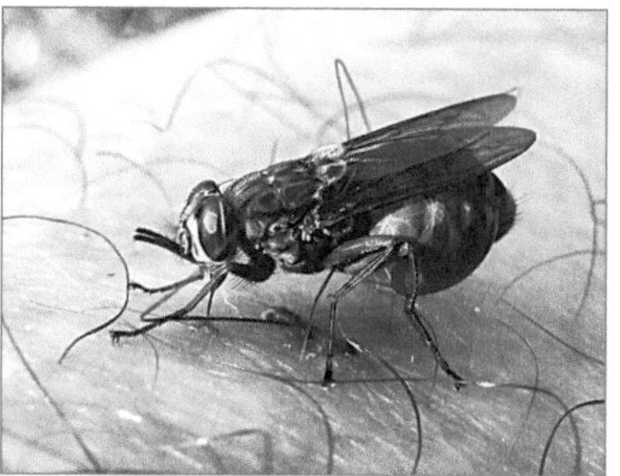

Figure 8.1 Agent and vector of human African trypanosomiasis. (Left) Micrograph of a Giemsa-stained blood smear showing a trypanosome in the process of replicating. Note the two nuclei, two kinetoplasts, and two flagella with undulating membranes. (Right) The blood-feeding tsetse fly transmits African trypanosomes. Shown is *Glossina pallidipes* in the act of feeding. (Photograph from David Bygott, Tucson, AZ.)

always easy to accurately diagnose and afflicts primarily the rural poor and individuals beyond the reach of national healthcare programs. Increased awareness of the disease and surveillance and control activities in the early part of the 21st century, as well as the availability of a new drug, have led to a precipitous drop in the number of cases and currently 50 000–70 000 people are infected per year.

Causative Agent

African trypanosomes are kinetoplastids and share many features with other *Trypanosoma* and *Leishmania* species (Chapter 7). Two distinct species of African trypanosomes infect humans: *Trypanosoma gambiense* and *T. rhodesiense* (Table 8.1). These two species are morphologically identical to each other and to *T. brucei*, which infects several animal species, but not humans. These three species are sometimes considered as subspecies of *T. brucei*. Various molecular and biochemical markers indicate the *T. gambiense* is relatively homogeneous across wide geographical ranges and can be readily distinguished from *T. brucei* and *T. rhodesiense*. On the other hand, both *T. brucei* and *T. rhodesiense* exhibit a high level of heterogeneity and in some cases more homology is observed between *T. rhodesiense* and *T. brucei* than between isolates of the same species obtained from different geographical regions. Furthermore, genetic crosses between *T. brucei* and *T. rhodesiense* have been successfully carried out. Thus, one might consider *T. rhodesiense* as a host-range variant of *T. brucei*. The inability of *T. brucei* to infect humans is due to a **trypanosome lytic factor** (TLF) found in human sera and the presence of a serum resistance associated (SRA) gene in *T. rhodesiense* (Box 8.1).

T. brucei is a major pathogen of wild and domestic animals and has far reaching effects on raising livestock. It and two other morphologically distinct trypanosome species, *T. vivax* and *T. congolense*, that are transmitted by the tsetse cause a disease in animals known as nagana (Table 8.1). Nagana is characterized by fever, weakness, lethargy, weight loss, and anemia and is usually fatal unless treated. In fact, although trypanosomiasis can be a devastating human disease, the greatest impact on human health is probably at the agricultural level. Large areas of Africa are unsuitable for raising livestock and other domestic animals due to the presence of the tsetse and the transmission of trypanosomes. This inability to raise cattle and other livestock contributes to protein-deficient diets among the indigenous population.

As the names imply, *T. gambiense* and *T. rhodesiense* are distinguished by their geographical distributions (Rhodesia is the former name for Zimbabwe). *T. rhodesiense* is found in east and southern Africa and *T. gambiense* is found in west and central Africa. Uganda is the only country in which both forms of human African trypanosomiasis exist, and even there the two forms are separated geographically. However, human African trypanosomiasis is not widespread throughout the endemic regions, but

Table 8.1 Tsetse-transmitted trypanosomes in Africa

Species	Disease
Trypanosoma gambiense	West and central African sleeping sickness
Trypanosoma rhodesiense	East African trypanosomiasis
Trypanosoma brucei	Nagana (wild animals and livestock)
Trypanosoma vivax	Nagana (wild animals and livestock)
Trypanosoma congolense	Nagana (wild animals and livestock)

Box 8.1 Trypanosome lytic factor

Humans and some primates are naturally resistant to the cattle parasite *T. brucei* because of a serum factor that lyses the parasite [1, 2]. This trypanolytic activity has been demonstrated to be associated with a sub-fraction of high density lipoprotein (HDL) particles and is called trypanosome lytic factor (TLF). TLF consists of lipids and several proteins collectively known as apolipoproteins. HDL particles containing the ubiquitous apolipoprotein A1 (ApoA1) and the primate-specific apolipoprotein L1 (ApoL1) and haptoglobin-related protein (HRP) provide innate protection against *T. brucei* infection. African trypanosomes take up these particles via receptor-mediated endocytosis and this uptake is important in regard to the toxicity. TLF is activated in the acidic lysosome and leads to the disruption of the lysosomal membrane.

Initially it was believed that HRP was responsible for trypanolytic activity and the mechanism involved oxidative damage to the lysosomal membrane. However, no direct evidence for a role of peroxidative mechanisms could be shown for TLF-mediated lysis. ApoL1 contains a pore-forming domain and within the acidic lysosome ApoL1 forms pores on the lysosomal membrane leading to the uptake of chloride ions and fatal osmotic fluxes. Furthermore, studies using transgenic mice indicate that ApoL1 is necessary and sufficient for killing trypanosomes and HRP does not cause trypanosome lysis [3]. Thus, it appears that ApoL1 may be the primary toxic factor. However, maximal TLF activity is found with the combination of HRP, ApoA1, and ApoL1. HRP facilitates the uptake of HDL particles and therefore may indirectly influence the lytic activity [4].

The human parasites—*T. gambiense* and *T. rhodesiense*—are resistant to the trypanosome lytic factor found in human serum. In *T. rhodesiense* a serum resistance associated (SRA) gene has been identified. SRA is a truncated variant surface glycoprotein (VSG) that binds to ApoL1. This interaction is believed to neutralize the toxic effect of ApoL1 and/or to prevent the trafficking of ApoL1 to the lysosome [1, 2]. However, the SRA gene is only found in *T. rhodesiense*, not *T. gambiense*, suggesting that there may be multiple mechanisms by which the parasite avoids the lytic activity of human sera.

1. Pays, E., Vanhollebeke, B., Vanhamme, L., Paturiaux-Hanocq, F., Nolan D.P. and Perez-Morga, D. (2006) The trypanolytic factor of human serum. *Nat. Rev. Microbiol.* 4: 477–486.

2. Shiflett, A.M., Faulkner, S.D., Cotlin, L.F., Widener, J., Stephens, N. and Hajduk, S.L. (2007) African trypanosomes: intracellular trafficking of host defense molecules. *J. Eukaryot. Microbiol.* 54: 18–21.

3. Molina-Portela, M.P., Samanovic, M. and Raper, J. (2008) Distinct roles of apolipoprotein components within the trypanosome lytic factor complex revealed in a novel transgenic mouse model. *J. Exp. Med.* 205: 1721–1728.

4. Vanhollebeke, B., Nielsen, M.J., Watanabe, Y., Truc, P., Vanhamme, L., Nakajima, K., Moestrup, S.K. and Pays, E. (2007) Distinct roles of haptoglobin-related protein and apolipoprotein L-I in trypanolysis by human serum. *Proc. Natl Acad. Sci.* 104: 4118–4123.

tends to be found in specific foci. The restricted distributions of the African trypanosomes are determined in part by their specific vectors. In addition to being transmitted by different vectors, *T. gambiense* and *T. rhodesiense* are distinguished by animal reservoirs, epidemiology, and disease virulence (Table 8.2).

Table 8.2 Distinguishing features of human trypanosomes

	T. rhodesiense	*T. gambiense*
Tsetse vector	*G. morsitans* group	*G. palpalis* group
Ecology	Dry bush or woodland	Rainforest, riverine, lakes
Geographical range	East and central Africa	West and southern Africa
Transmission cycle	Ungulate–fly–human	Human–fly–human
Nonhuman reservoir	Wild and domestic animals	Rare, if ever
Epidemiology	Sporadic, safaris	Endemic, some epidemics
Disease progression	Rapid progression to death	Slow (~1 yr) acute → chronic
Parasitemia	High	Low
Asymptomatic carriers	Rare	Common

The vectors of *T. rhodesiense* are *G. morsitans* and related species. They are found in woodland or dry bush environments. The natural vertebrate hosts of *T. rhodesiense* are antelope and other wild ungulates (hoofed animals). *T. rhodesiense* also infects domestic animals which serve as reservoirs in some situations. Disease transmission depends upon coming in contact with infected vectors and can be associated with activities such as safaris which bring humans into contact with infected vectors. In addition, molecular epidemiological studies indicate that cattle are an important reservoir and some localized epidemics have been caused by the movement of cattle herds. Thus, *T. rhodesiense* is primarily a zoonosis and little human–fly–human transmission occurs. The disease typically progresses rapidly and is almost always fatal if not treated.

T. gambiense is transmitted by members of the *G. palpalis* group. These vectors are found along rivers and lakes which are often in close proximity to human habitation. Consequently, the disease tends to be endemic with occasional epidemics. *T. gambiense* is also found in wild and domestic animals. In principle such animals may serve as a source for sporadic epidemics, although this has never been definitively demonstrated. In contrast to *T. rhodesiense, T. gambiese* appears to be transmitted primarily via a human–fly–human transmission cycle. The disease is less virulent and is characterized by a slow progression from an acute disease to a chronic disease. The higher levels of contact between *T. gambiense* and humans may account for the lower virulence of *T. gambiense* as compared with *T. rhodesiense.*

Life Cycle

African trypanosomes alternate between a mammalian host and a tsetse fly vector (Figure 8.2). Tsetse flies are pool feeders which lacerate the skin with their mouth parts and then ingest the blood and lymph that accumulates in the lesion. Saliva containing **metacyclic trypomastigotes** is injected into the bite site. This gives the trypanosome access to the blood and lymph of the mammalian host. The parasite retains the trypomastigote morphology and replicates as an extracellular parasite within the bloodstream and tissue

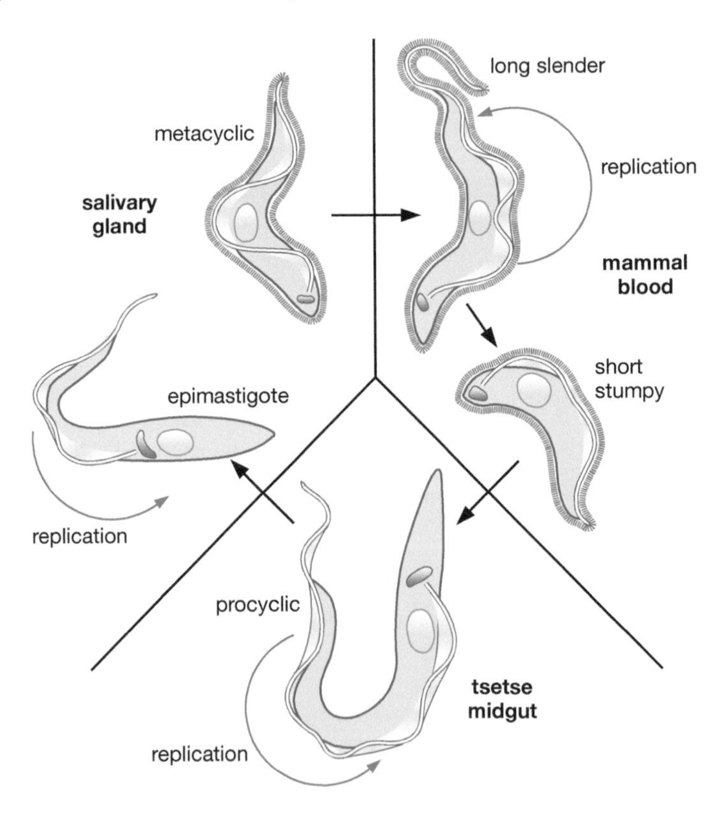

Figure 8.2 Life cycle of African trypanosomes. The infective stage for the human host is the metacyclic trypomastigote which is introduced with the saliva of the tsetse during feeding. The parasite maintains its trypomastigote morphology and replicates within the blood and tissues of the host. Metacyclic and bloodstream forms are covered with a surface coat. When taken up by a tsetse the parasite differentiates to a procyclic trypomastigote and replicates in the gut of the vector before migrating to the salivary glands and converting to an epimastigote form. The epimastigotes convert to metacyclic trypomastigotes to complete the life cycle.

fluids. Replication proceeds by binary fission and the trypomastigotes can exhibit a doubling time of 6 hours. This rapid rate of replication allows the parasite to establish itself within the mammalian host.

Mammal–vector transition requires complex metabolic changes

The replicating forms exhibit a long slender morphology. As the infection proceeds some of these long slender trypomastigotes cease to divide and exhibit a short stumpy morphology. The ratio of the long slender forms to the short stumpy forms varies with each wave of parasitemia and proportionately more short stumpy forms are seen later in the infection. These short stumpy forms also undergo metabolic changes that are consistent with a pre-adaptation for the tsetse fly. For example, glucose is abundant within the bloodstream of the mammalian host and the parasite metabolizes glucose solely via glycolysis; it does not exhibit a tricarboxylic acid cycle or oxidative phosphorylation. Consistent with this lack of normal aerobic metabolism, the mitochondria of the long slender forms are acristate, whereas those stages found in the vector contain cristae. The cristae are folds of the inner mitochondrial membrane that contain the proteins involved in electron transport (i.e., cytochromes) and ATP synthesis. Acristate mitochondria are usually associated with anaerobic metabolism and the lack of a tricarboxylic cycle and oxidative phosphorylation. During the conversion to the short stumpy forms the parasite begins to exhibit a metabolism that is intermediate between the bloodstream form and the insect form.

Bloodstream forms are taken up by the tsetse and differentiate into **procyclic trypomastigotes** within the gut of the tsetse. This differentiation is characterized by changes in the expression of surface proteins as well as changes in metabolism. The environment within the vector is quite different than that of the mammalian bloodstream and therefore the parasite must quickly change its metabolism. In particular, the parasite is moving from the glucose-rich mammalian bloodstream to a glucose-poor environment. Thus the parasite needs to be more efficient in the use of glucose for the generation of ATP and switches to an aerobic metabolism. In addition, the parasite switches from using glucose as its principal source of energy to using the amino acid proline. The tsetse uses proline as its energy source for flight and proline is abundant in the tsetse hemolymph. Coincident with this switch to aerobic metabolism, the mitochondria develop tubular cristae and mitochondrial enzymes are substantially more active in the procyclic forms than the bloodstream forms. In addition, the volume of the mitochondrion increases substantially, and changes in the morphology and enzyme composition of the glycosome (Chapter 7) also occur during the differentiation to procyclic trypomastigotes.

Differentiation to procyclic trypomastigotes is also accompanied by a loss of the surface coat on the membrane and replacement of the variant surface glycoprotein (VSG) with another membrane surface protein called procyclin. The thick surface coat of VSG protects the parasite from complement-mediated lysis in the bloodstream of the mammalian host. This surface coat is no longer needed during the insect stage and the VSG is removed from the surface of the parasite and replaced with procyclin. The procyclin is resistant to digestion with proteases and probably protects the parasite while it is living within the gut of the fly.

Completion of the cycle requires migration to the salivary glands

The focus in medical parasitology courses tends to be on the complex interactions between the parasite and the human host which result in pathology. However, parasites also interact with, and undergo complex developmental processes in, the vector. One problem for the trypanosome is that it must move from the gut to the salivary glands of the tsetse. In addition, blood meals are surrounded by a chitinous peritrophic membrane that the parasite must traverse. The exact mechanism by which the parasite migrates from the tsetse gut to the salivary glands is not known. Because of these various potential barriers the majority of infections in the tsetse flies fail. Thus the flies are somewhat refractory to trypanosome infection and the infection rates in nature are quite low.

Approximately 4 days after ingestion by the tsetse, procyclics are observed in the space between the peritrophic membrane and the gut epithelium. It appears that the trypomastigotes directly penetrate the peritrophic membrane rather than exit at the open end of the hindgut. During the following week the ectoperitrophic space becomes filled with dividing trypanosomes. Some procyclics appear capable of invading midgut epithelial cells and traversing the midgut epithelium. However, the hemolymph of the tsetse contains trypanocidal factors. Thus penetration of the midgut wall and passage through the hemocoel does not appear to be the normal route to the salivary glands. Rather the procyclics appear to backtrack through the esophagus and mouthparts of the tsetse and then migrate up the salivary duct to the salivary glands.

After reaching the salivary glands the procyclic trypomastigotes transform into **epimastigotes** and attach to epithelial cells via their flagella. The epimastigotes undergo further replication within the salivary gland. In addition, a sexual recombination may occur during the epimastigotes stage. This sexual recombination is nonobligatory and is relatively rare. The epimastigotes are noninfective for the mammalian host and must mature into metacyclic trypomastigotes before they are capable of infecting the mammalian host. During this maturation, the mitochondria lose their cristae and the glycosomes revert back to their blood stage morphology. In addition, the VSG surface coat also reappears. All of these changes are in preparation for life within the mammalian bloodstream. The mature metacyclic trypomastigote detaches from the salivary gland epithelium and is free within the lumen of the salivary gland waiting to be transferred to a vertebrate host when the tsetse feeds again, thus completing the life cycle. The complete maturation of the trypanosome in the tsetse vector takes 20–30 days and the fly remains infective for the rest of its life, which is generally 1–3 months.

Antigenic Variation

African trypanosomes establish a long-term chronic infection in the mammalian host. This is common among the vector-transmitted protozoa since it increases the prospects of the infection being transmitted to another vector. Establishing a long-term chronic infection involves special adaptations to avoid the immune system. For example, the surface of the blood stage trypomastigote of the African trypanosome is composed of a 12–15 nm thick surface coat which protects the trypanosome against complement-mediated lysis. The host does eventually make antibodies against the proteins of the surface leading to an antibody-mediated lysis of the trypanosome. However, the trypanosome continually switches these proteins to antigenically distinct forms to stay one step ahead of the host immune system. This results in cyclical peaks of parasitemia with each wave of parasitemia characterized by an antigenically distinct population of trypanosomes (Figure 8.3).

Variant genes encode variant forms of the surface coat

The dense surface coat of the blood stage trypomastigotes is formed primarily from a single protein called the **variant surface glycoprotein** (VSG). African trypanosomes contain over a thousand copies of VSG genes and VSG pseudogenes. Comparing the sequences of the various VSG genes reveals an N-terminal variable domain and a C-terminal conserved domain (Figure 8.4). The C-terminus has a glycosylphosphatidylinositol (GPI) anchor which

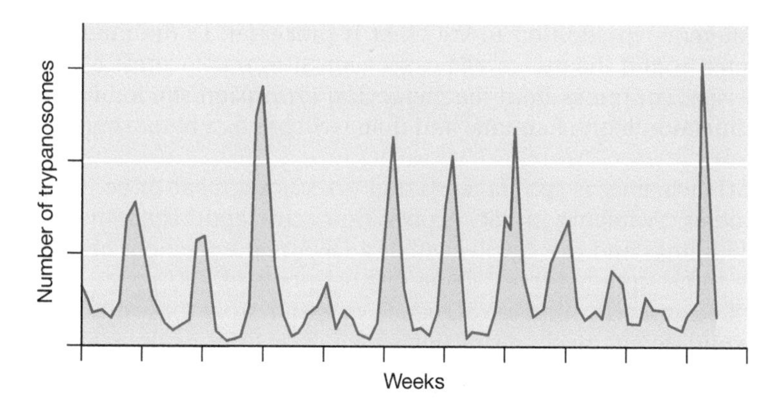

Figure 8.3 Fluctuating parasitemia associated with African trypanosomiasis. Infection with African trypanosomes is characterized by a rapid rise in parasitemia followed by parasite clearance. The trypanosomes found in each of the peaks of parasitemia are antigenically distinct. Disease symptoms are usually most pronounced during the peaks of parasitemia.

is imbedded into the lipid bilayer of the plasma membrane. Approximately 10^7 copies of VSG are expressed and they form a densely packed monolayer over the surface of the trypanosome. This dense packing prevents access to the conserved C-terminal region by serum antibodies. Thus only the variable N-terminal region is exposed to the host immune system.

An individual trypanosome only expresses a single VSG gene at any one time. The host mounts an immune response against this particular VSG and the resulting antibodies will lead to the clearance of parasites. However, some of the trypanosomes will switch to the expression of a different VSG gene leading to the replacement of the surface coat with a protein not recognized by the antibodies currently present in the serum. These trypanosomes with the new surface coat will then rapidly increase in number since they are not recognized by the antibodies. The immune system will then learn to recognize this variant of VSG and make antibodies that again lead to parasite clearance. However, some of the trypanosomes will again change their VSG coat and cause a new wave of parasitemia. Therefore, the recurring waves of parasitemia are the result of repeated cycles of the host killing the parasites only to have them reemerge due to this antigenic variation resulting from the switch in VSG gene expression.

VSG switching involves several molecular mechanisms

VSG expression is carried out at specific promoter sites located at the telomeres (i.e., ends) of chromosomes. There are two types of expression sites corresponding to the two life cycle stages in which VSG is expressed on the parasite surface: metacyclic expression sites and bloodstream form expression sites. There are 15–20 of each of these two types of expression sites and only one of these approximately 40 sites is active at any one time. The bloodstream form expression sites produce a polycistronic RNA containing

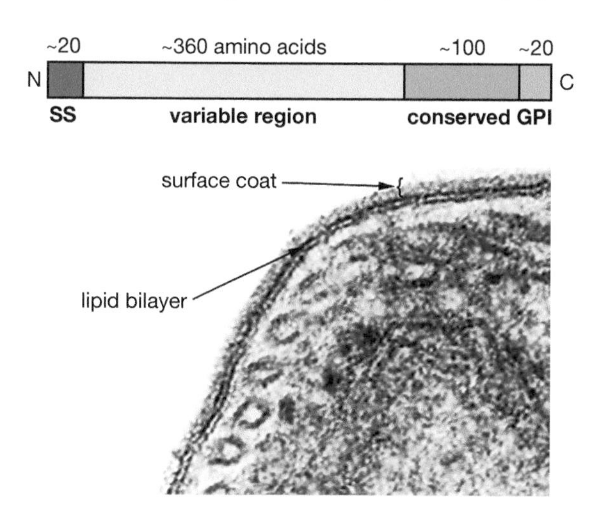

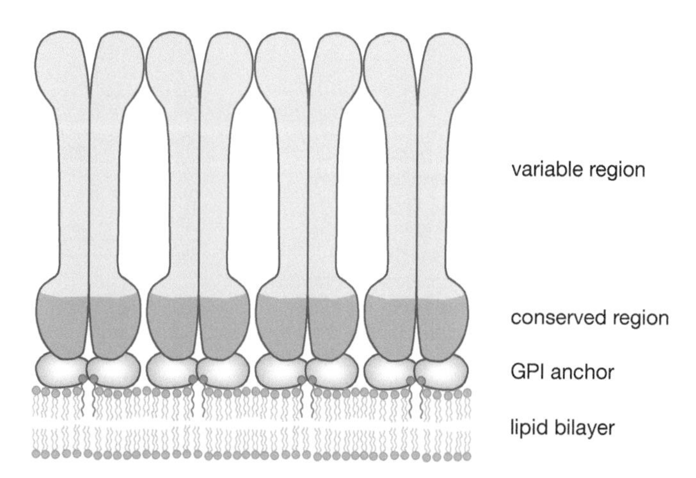

Figure 8.4 Structure of VSG and the surface coat. (Upper left) Schematic representation of the VSG gene showing an N-terminal signal sequence (SS) to target the protein to the membrane, followed by a variable region, a conserved region, and a glycosylphosphatidylinositol (GPI) signal sequence. (Lower left) Electron micrograph showing the cross section of a trypomastigote with the surface coat and lipid bilayer denoted. (Right) Schematic representation of the arrangement of VSG on the surface of the trypanosome with the variable region exposed on the outermost surface and the protein attached to the lipid bilayer via the GPI anchor. (Electron micrograph provided by Paul Webster, House Ear Institute, Los Angeles, CA.)

several genes in addition to VSG that is processed to the mature mRNA (Chapter 7) and the metacyclic expression sites only express a VSG gene. Expression continues from the metacyclic expression site for a short time after infection of the mammal and then switches to a bloodstream expression site.

VSG switching is spontaneous and does not appear to be induced by antibodies. Switching in VSG expression occurs approximately every 100 cell doublings and can involve either a DNA rearrangement or regulation at the transcriptional level. Three distinct pools of VSG genes have been identified and activation of VSG gene expression from these different pools involves different mechanisms (Figure 8.5). The three distinct pools of VSG genes are: (i) those already located at expression sites near telomeres; (ii) those located on minichromosomes; and (iii) those found as long tandem arrays in subtelomeric locations on the chromosomes.

One mechanism of switching VSG gene expression involves turning off the currently active expression site and initiating expression from another expression site. For example, the switch from a metacyclic expression site to a bloodstream form expression site involves this *in situ* switch often called allelic exclusion. In other words, the various forms of the VSG genes are alleles and only one allele is expressed at a time. This allelic exclusion appears to be regulated by an epigenetic mechanism and there is evidence suggesting that the active expression site is located in a distinct region of the nucleus. This specific nuclear domain can only accommodate a small segment of DNA from a single chromosome. Therefore, only a single VSG allele is expressed per trypanosome at any given time. The *in situ* switch in expression is associated with a nuclear reorganization in which a previously silent VSG-containing expression site replaces the currently expressed VSG-containing expression site in this specific nuclear domain. A different VSG allele will then be transcribed and translated and the resulting protein will replace the VSG currently on the surface of the trypanosome.

In addition to VSG genes located in these expression sites, there are VSG genes located elsewhere in the genome. For example, approximately 200 VSG genes are found at the telomeres of the minichromosomes. The minichromosomes are 50–100 kilobases in size and are unique to African

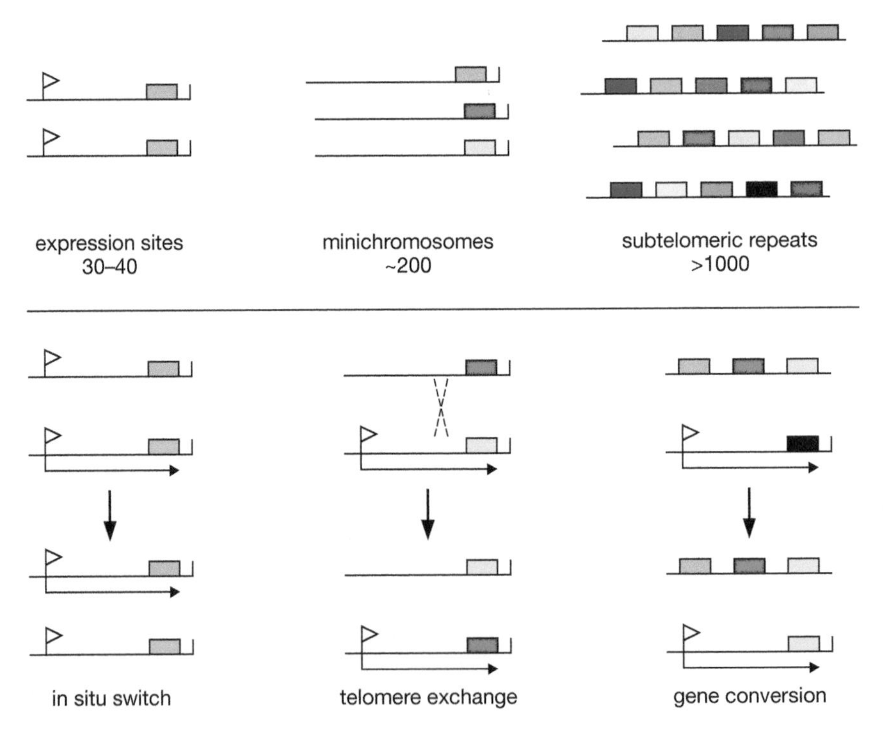

Figure 8.5 VSG genes and mechanisms of switching. Three distinct pools of VSG genes are found in the genome of African trypanosomes (upper panel). Expression of VSG genes from these different pools is activated by three different mechanisms (lower panel). The boxes represent the various VSG genes and the flags represent promoters at expression sites. Transcription is denoted by the arrows running from the promoter through the gene. See text for details. (Modified from Taylor and Rudenko. 2006. Switching trypanosome coats: what's in the wardrobe? *Trends in Genetics* 22: 614-620,with permission from Elsevier Ltd.)

trypanosomes. There are approximately 100 minichromosomes per trypanosome and these minichromosomes only code for VSG. However, no promoter is associated with the VSG genes on these minichromosomes and thus the genes cannot be transcribed. Expression of the VSG genes located on minichromosomes involves a recombination event between the telomere with the active expression site and a telomere of a minichromosome. This recombination is similar to the crossing-over process that occurs between chromosomes and involves breaking the DNA at specific sites and then rejoining the DNA fragments so that the gene on the minichromosome is now associated with an expression site and *vice versa*. Following telomere exchange the previously silent VSG gene is now located in the active expression site and the previously expressed VSG gene is now located on a minichromosome.

The majority of VSG genes are found in long tandem arrays at subtelomeric locations on the chromosomes. However, most of these genes are pseudogenes (see below). Expression of these genes involves yet a third mechanism of VSG switching called duplicative transposition or gene conversion. During duplicative transposition a copy of the target VSG gene is made and then this copied gene replaces the gene in the active expression site via DNA recombination. Thus the gene being copied remains intact in its original location on the chromosome and can be viewed as an archival copy of the gene. During the acute stage of the infection telomere exchange appears to be the more predominant mechanism for the switch in gene expression, whereas gene conversion is probably more important during the chronic stage of the infection.

The processes involving DNA rearrangements and recombination (i.e., telomere exchange and gene conversion) only appear to be functioning in the blood stage trypomastigote and not in the metacyclic trypomastigote. Only the *in situ* activation of a new expression site appears to be functioning during the formation of the metacyclic trypomastigote. Thus each metacyclic trypanosome only has the ability to express one of the 15–20 different VSG genes found in the metacyclic expression sites and not the entire VSG repertoire.

VSG pseudogenes may be involved in generation of new variants

Sequencing of the genome of African trypanosomes revealed that relatively few (approximately 7%) of the VSG genes are intact functional genes, whereas the majority of the VSG genes have frame shift errors or in-frame stop codons. These pseudogenes can undergo an intragenic recombination with other VSG genes and form chimeric genes which now have an intact coding sequence. The formation of these chimeric genes also leads to the formation of new VSG alleles. This gives the African trypanosome an enormous potential to generate diversity in that the repertoire of VSG genes is continuously changing during the course of the infection. Not only does the trypanosome have a large repertoire of VSG genes to choose from, but new variants are continuously being created and recently expressed VSG variants are potentially being lost due to gene conversion. Thus, the African trypanosomes confront their hosts with a high degree of heterogeneity.

Disease Course and Symptoms

Infection with African trypanosomes can result in disease manifestations ranging from asymptomatic or mild to a severe fulminating disease. *T. gambiense* tends to cause a slow progressing disease which may take months or years to develop into a chronic disease involving the central nervous system (CNS). *T. rhodesiense* is more likely to cause a rapidly progressing and fulminating disease that invades the CNS in weeks to months. The disease is often broken down into two stages. Stage 1, also called the hemolymphatic stage or early stage, is characterized by parasites spreading via the bloodstream

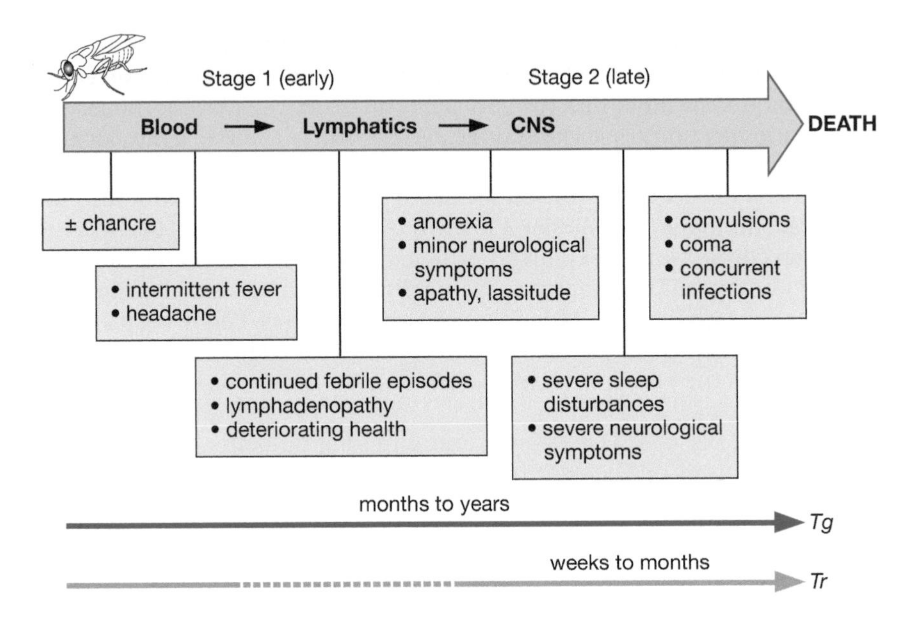

Figure 8.6 Progression of African trypanosomiasis. The symptoms associated with the disease progressively worsen as the parasites move from the blood to the lymphatics to the CNS. The early part of the infection, or stage 1, involves the blood and lymphatics. The late stage, or stage 2, is associated with the invasion of the CNS.

and lymphatics to systemic organs including liver, spleen, heart, endocrine system, and eyes. Parasites crossing the blood–brain barrier and entering the CNS mark the beginning of stage 2, also called the meningoencephalitic stage or late stage. During the course of the infection the symptoms associated with African trypanosomiasis progressively worsen (Figure 8.6). It is generally believed that untreated infections will progress to death. Reports of spontaneous parasite clearance are extremely rare.

Acute African trypanosomiasis is characterized by febrile attacks

The infection is initiated when metacyclic trypomastigotes are introduced from the saliva of the tsetse into the bite wound. Generally there is an asymptomatic incubation period of 1–2 weeks in which the trypomastigotes are replicating within the tissue near the site of the bite. A local inflammatory nodule known as a "trypanosomal chancre" is sometimes observed at this site. In Europeans and others from nonendemic areas the frequency of the chancre is 25–40%, whereas it is seldom observed in Africans. Chancres are usually tender and painful and ulceration may occur. Trypanosomes can be recovered from fluid aspirated from the ulcer.

The trypanosomes invade the capillaries and enter the circulatory system during the incubation period and continue to replicate within the blood of the human host. The establishment of this acute blood stage infection is characterized by irregular episodes of fever and headache. The time between the febrile episodes can be a few days or up to 1 month. In the case of *T. gambiense* the number of parasites in the blood tends to be very low and often the infected person exhibits no symptoms, whereas most persons infected with *T. rhodesiense* will exhibit higher parasitemias and a more pronounced fever which is sometimes associated with rigor. In addition, the febrile episodes are more frequent in *T. rhodesiense* infections than *T. gambiense* infections.

The trypanosomes circulate throughout the body and live as extracellular parasites in the blood and tissue fluids. Invasion of the lymphatics is associated with enlarged lymph nodes, particularly on the back of the neck, and continued febrile attacks. The febrile attacks will usually last a few days to a week followed by an asymptomatic period of a few weeks. An erythematous rash, sometimes accompanied by itching and swelling, may also appear during the attacks. The rash is more prominent and frequent in Caucasian patients. Symptoms associated with specific organs may also be evident during this period. The febrile attacks and other overt symptoms are generally associated with higher levels of parasitemia. These fluctuations in parasitemia and clinical symptoms are due to the antigenic variation exhibited by the African trypanosomes (see above). Involvement of the lymphatics occurs less frequently and is less pronounced in *T. rhodesiense* infections than *T. gambiense* infections.

Neurological manifestations of African trypanosomiasis

Disease progression is associated with a worsening of the overall condition of the patient. Patients begin to exhibit signs of anorexia, generalized weakness, and wasting as the disease ensues. Neurological symptoms also become apparent as the infection becomes more chronic and the trypanosomes invade the CNS. Generally it is 6–12 months (or even years) after the infection before the neurological symptoms start to become apparent in the case of *T. gambiense*. Neurological manifestations can occur within weeks or months after *T. rhodesiense* infection.

Trypanosomes crossing the blood–brain barrier result in a generalized meningoencephalitis characterized by progressively worsening symptoms. The onset of CNS involvement is insidious and can exhibit a wide range of clinical manifestations that include psychiatric, motor, and sensory abnormalities and sleep disturbances. Indications of nervous impairment include: lassitude, apathy, fatigue, confusion, somnolence, and motor changes, such as tics, slurred speech, incoordination, and other behavioral changes (Table 8.3). Increasing apathy and lassitude tends to mark the beginning of the neurological involvement and the symptoms progressively worsen. A notable feature of chronic African trypanosomiasis is somnolence, hence the common name African sleeping sickness. These changes in sleep patterns are often characterized by extreme fatigue during the day and agitation at night. In the later stages of the disease the patients become more difficult to rouse. In contrast to the rest of the body, the face becomes edematous and the facial features become blank and expressionless (Figure 8.7).

There may be profound changes in the personality and extreme mental deterioration in the final stages of the disease. If untreated, the CNS stage of the disease will almost always progress to include convulsions or coma followed by death in both *T. gambiense* and *T. rhodesiense* infections. The gradual deterioration of *T. gambiense*-infected persons and the associated anorexia will lead to a weakened state and immunosuppression. Quite often the terminal stages of the disease are associated with concurrent infections and in many cases death is due to malaria, dysentery, or pneumonia, aided by starvation.

Table 8.3 Behavioral changes associated with African trypanosomiasis
Lack of interest and disinclination to work
Morose and melancholic attitude alternating with exaltation
Lethargy and lassitude
Slow and shuffling gait
Low and tremulous speech
Tremors and chorea (rapid jerky movements)
Altered reflexes

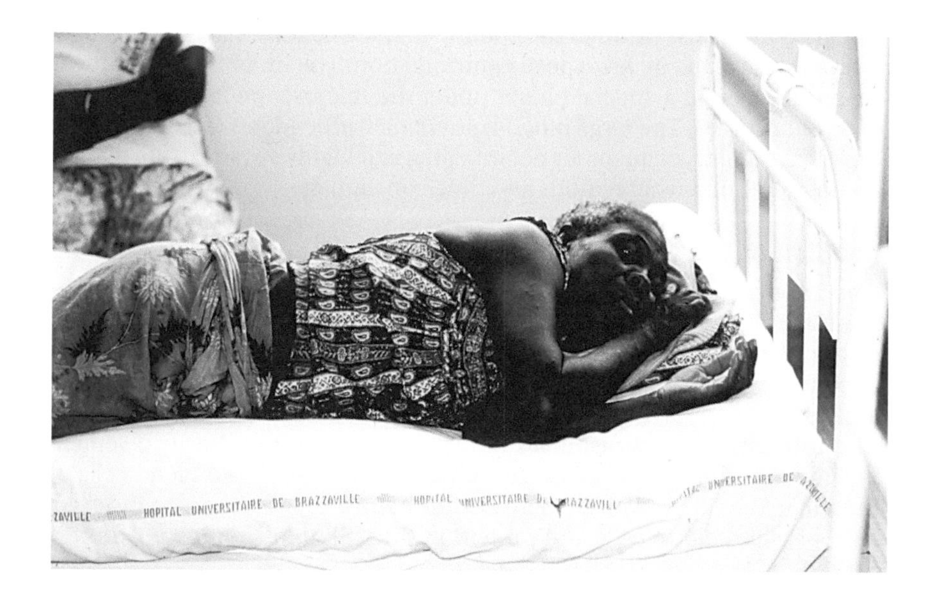

Figure 8.7 Patient in later stages of African sleeping sickness. Note the edematous face and blank expression. (Photograph from Antonio D'Alessandro, Tulane University, Department of Tropical Medicine.)

Diagnosis

Diagnosis relies upon a combination of the clinical symptoms, geographical presence of the parasite, and diagnostic tests. Appropriate clinical symptoms and living in or visiting an endemic area is suggestive of African trypanosomiasis. However, the nonspecific nature of the clinical symptoms and the presence of other infectious agents that exhibit similar symptoms require confirmation of the parasite. In cases where the parasite cannot be confirmed, serological tests can be used to complement clinical diagnosis. For example, the card agglutination trypanosomiasis test (CATT) is a frequently used method for the detection of antibodies against *T. gambiense.* It is rapid and easy to carry out. However, serological tests cannot distinguish between current and past infections and the CATT is not acceptable for the diagnosis of *T. rhodesiense.*

Confirmed diagnosis depends upon the detection of trypanosomes in the blood, lymph node aspirations, or spinal fluid. Typically few trypanosomes are detected in the blood or other bodily fluids during *T. gambiense* infections. *T. rhodesiense* generally exhibits higher parasitemias making detection and diagnosis easier. The trypanosomes are more likely to be detected during symptomatic periods (e.g., febrile episodes). In the absence of detectable parasites, travel or residence in an endemic area combined with the symptoms can be used as a presumptive diagnosis.

Enrichment of trypanosomes can help confirm diagnosis

Because of the low parasitemias exhibited during *T. gambiense* infections it is often not possible to detect parasites by standard thin and thick blood smears. Techniques to increase the sensitivity of detection are often needed (Table 8.4). For example, looking at fresh whole blood mounts may increase the sensitivity due to the distinctive movement of the trypomastigote. Another method to increase sensitivity is to centrifuge the blood in a microhematocrit tube. The parasites are enriched in the "buffy coat" which is the band of white cells directly above the packed erythrocytes. The tube is then broken at the buffy coat and Giemsa-stained blood smears are prepared from the cells found in the buffy coat. Inoculation of mice or rats with patient blood and looking for the development of parasitemia is also possible. This generally works better for *T. rhodesiense* infections than *T. gambiense* infections.

The miniature anion-exchange chromatography technique (mAECT) is another method used in the diagnosis of African sleeping sickness. Blood is passed through an anion exchange column. Blood cells are more negatively charged than the trypanosomes and they are retained on the column. The trypanosomes pass through the column and are collected at the bottom of a sealed glass tube by low-speed centrifugation. The tip of the glass tube is then examined in a special holder under the microscope for the presence of trypanosomes. The large blood volume (300 μl) enables the detection of fewer than 100 trypanosomes per ml. Although highly sensitive, the manipulations are somewhat tedious and time consuming.

Table 8.4 Methods for detection of African trypanosomes
Giemsa-stained thin and thick blood smears
Fresh whole blood (characteristic movement of trypomastigotes)
Examine buffy coat (microhematocrit)
Inoculate mice or rats
Mini-anion exchange chromatography technique (mAECT)

Invasion of the CNS is an important component of diagnosis

One important issue in diagnosis of African trypanosomiasis is to distinguish the encephalitic stage of the disease from the blood and lymphatic stages of the disease. This is important since the treatment is different depending on whether there is CNS involvement (see below). Patients who are suspected of having CNS involvement should undergo a lumbar puncture to examine cerebrospinal fluid (CSF). Criteria for CNS involvement include detection of parasites in the CSF or elevated white blood cells or protein in the CSF. However, there is some controversy in regard to the exact level of white blood cells and protein in the CSF that should constitute a classification of CNS involvement in the absence of detectable parasites. There is an urgent need for a reliable test that can be performed in the field for diagnosis of late-stage African trypanosomiasis.

Treatment

Drug treatment of African trypanosomiasis is usually effective when it is started relatively early in the infection during the blood and lymphatic stages. However, the prognosis is not so good once the trypanosomes have invaded the CNS. Another problem associated with the treatment of African trypanosomiasis is that the currently available drugs are somewhat toxic and have other drawbacks (Table 8.5). In addition, since treatment failures do occur, all patients need to be followed up at 6-month intervals for 2 years after successful treatment. And even patients successfully cleared of the parasites may still experience long-term neurological sequelae such as weakness, muscular incoordination, cognitive impairment, epilepsy, or psychiatric disorders.

The recommended drugs for treatment if there is no evidence for CNS involvement are pentamidine isethionate (Pentacarinat®, Pentam 300® or NebuPent®) or suramin (Germanin®). Pentamidine is less toxic than suramin and is the preferred drug for treatment of *T. gambiense*. However, pentamidine is not effective against *T. rhodesiense* and suramin must be used. Pentamidine and suramin do not cross the blood–brain barrier and therefore melarsoprol (Mel-B®) is recommended for the late stage of the infection. Melarsoprol, an arsenic-based drug, has been used for more than 50 years despite its relatively high toxicity. Although usually effective, melarsoprol is followed by a severe post-treatment encephalopathy in about 10% of the patients and about half of those die. Furthermore, the treatment course is long (minimum 10 days) and requires hospitalization, thus adding significantly to the costs. There is clearly an urgent need for the development of better antitrypanosomal drugs. However, since the victims of African sleeping sickness are primarily the rural poor, there is little financial incentive for research and development.

Table 8.5 Drugs for African trypanosomiasis

Drug	Use	Drawbacks
Pentamidine	Effective against early-stage *T. gambiense* disease	Adverse side effects
		Non-oral route
Suramin	Effective against early-stage *T. gambiense* and *T. rhodesiense* disease	Adverse side effects
		Non-oral route
Melarsoprol	First-line drug for late-stage *T. gambiense* and *T. rhodesiense* disease involving CNS	Adverse side effects, especially encephalopathy
		Fatal in 1–5% of cases
		Parasite resistance
		Non-oral route
Eflornithine ± nifurtimox	Effective against late-stage *T. gambiense* disease involving CNS (soon may be the first-line therapy)	High cost
		Not effective against *T. rhodesiense*
		Non-oral route

Eflornithine (Ornidyl®), an ornithine decarboxylase inhibitor, is an effective antitrypanosomal drug that has been recently developed. It has been nicknamed the "resurrection drug" because of its spectacular effect on comatose patients in late-stage African sleeping sickness. It is highly effective against *T. gambiense* with a cure rate of 99% and exhibits relatively low toxicity. However, it not as effective against *T. rhodesiense* and higher doses resulting in significant side effects are required. Eflornithine is expensive and the standard treatment is 14 consecutive daily injections. Recent results suggest that a combination of eflornithine and nifurtimox, a drug used in the treatment of Chagas' disease (Chapter 9), is a promising treatment for late-stage *T. gambiense* infections and may soon be the recommended first line of therapy.

Prevention and Control

Wearing protective clothing or using insect repellents are the recommended prophylactic measures against African trypanosomiasis. Other measures to lessen contact with the day-biting tsetse, such as avoidance of streams and water holes during the warm dry season, can also be taken. Prophylactic drugs are contraindicated since they may mask latent CNS infections and promote drug resistance. In addition, the available drugs are quite toxic and require hospitalization for administration.

Control activities are primarily focused on *T. gambiense* and involve reducing the number of infected humans as well as reducing the vector. Systematic surveillance and treatment of infected persons is an effective control measure because humans are the primary reservoir and the disease in the early stages is often asymptomatic or mild and progresses slowly. Control of the riverine tsetse flies includes destruction of their habitats and breeding places by clearing stream banks of trees and shrubs. Widespread application of insecticides can also be used to reduce the number of tsetse.

Traps and targets may be cost-effective ways to control tsetse

The use of traps and targets is another potential means to reduce the number of tsetse in a localized area. Traps and targets function by attracting the tsetse to a contraption that collects or kills them and thereby reduces the tsetse population. A common type of trap consists of a pyramid of mosquito netting with alternating blue and black sheets forming quadrants (Figure 8.8). Also within the pyramid is a plastic bag filled with kerosene. The flies are attracted to the trap and fly into the kerosene bag and are killed. Traps can also be used for entomological surveillance, as well as for control. Targets are screens that are impregnated with insecticides in order to kill any flies that land on them. The strategy is to place the traps or targets around a village or other areas such as along riverbanks where bathing and washing activities take place or along paths. This leads to a reduction in the number of tsetse in the vicinity. Traps and targets are also advantageous in that there is less environmental impact than widespread insecticide spraying or clearing brush from large areas. Although this strategy is relatively low-tech, it does require community awareness and participation since the traps and targets need to be maintained and replenished with attractants and insecticides.

Summary and Key Concepts

- African trypanosomes are transmitted by tsetse flies.
- Two species of African trypanosomes infect humans: *Trypanosoma gambiense* and *Trypanosoma rhodesiense.*
- *T. gambiense* is transmitted in riverine environments and tends to be an endemic disease that slowly progresses to a chronic disease.

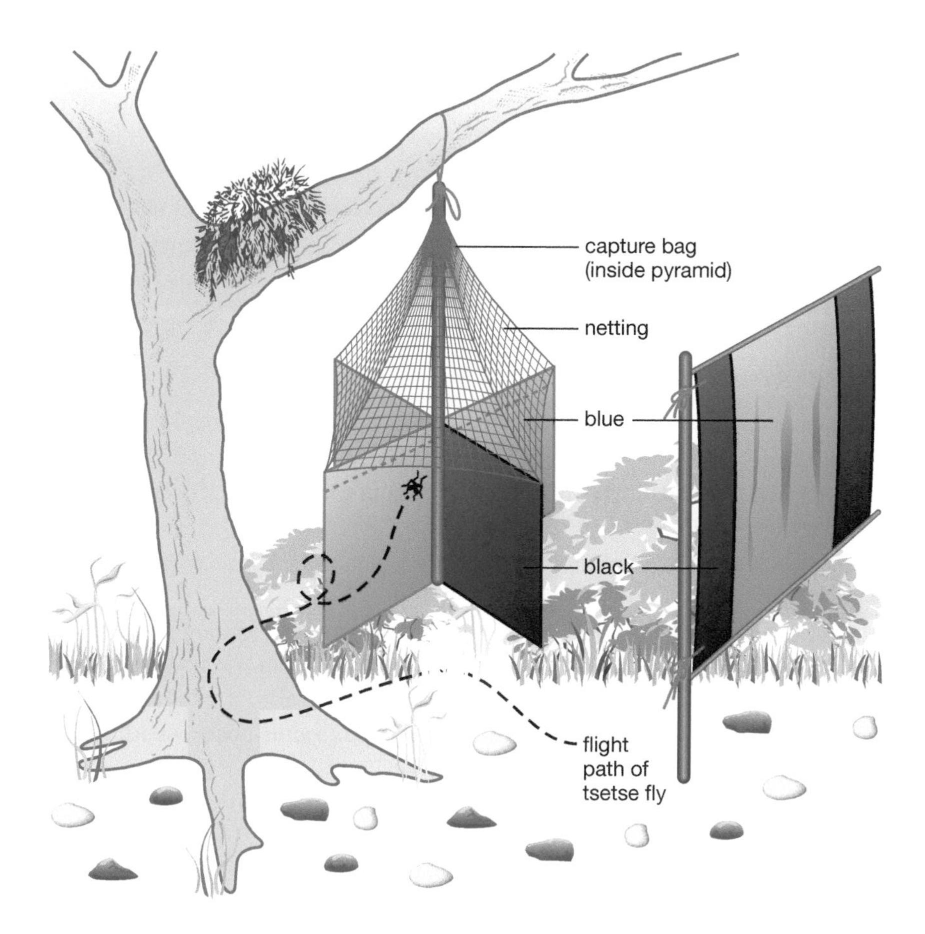

Figure 8.8 Examples of traps and targets used to control tsetse flies. The traps (left) and targets (right) are made of blue and black cloth. These colors attract the tsetse and the flies are captured and killed by the traps and targets. (Adapted from figures by the World Health Organization.)

- *T. rhodesiense* is transmitted in the savannas and human infection is more sporadic and causes a more rapidly progressing disease than *T. gambiense.*
- African trypanosomes live and replicate within the bloodstream and body fluids of the human host.
- Antigenic variation in which the trypanosome changes the expression of the protein making up its surface coat allows the establishment of long-term chronic infections.
- The early phases of the disease are characterized by intermittent febrile episodes corresponding to the waves of parasitemia associated with the antigenic variation.
- Invasion of the central nervous system by the trypanosomes results in various neurological manifestations including disturbances in sleep patterns.
- Geographical presence of the parasite, symptoms, and immunological tests can be used in making a presumptive diagnosis.
- Confirmed diagnosis depends on detection of trypanosomes in blood, lymph, or cerebrospinal fluid.
- A limited number of drugs with major drawbacks are available for the treatment of African trypanosomiasis.
- Preventive measures include protective clothing and insect repellants and other means to avoid tsetse bites.
- Effective control activities include active case detection and treatment and/or means to reduce tsetse–human contact.

Further Reading

Aksoy, S., Gibson, W.C. and Lehane, M.J. (2003) Interactions between tsetse and trypanosomes with implications for the control of trypanosomiasis. *Adv. Parasitol.* 53: 1–83.

Barrett, M.P. (2006) The rise and fall of sleeping sickness. *Lancet* 367: 1377–1378.

Checchi, F., Filipe, J.A.N., Barrett, M.P. and Chandramohan, D. (2008) The natural progression of Gambiense sleeping sickness: What is the evidence? *PLoS Negl. Trop. Dis.* 2(12): e303.

Gibson, W. (2002) Will the real *Trypanosoma brucei rhodesiense* please step forward? *Tr. Parasitol.* 18: 486–490.

Hide, G. and Tait, A. (2009) Molecular epidemiology of African sleeping sickness. *Parasitology* 136. Published online by Cambridge University Press, 26 June 2009.

Jannin, J. and Cattand, P. (2004) Treatment and control of human African trypanosomiasis. *Curr. Opin. Infect. Dis.* 17: 565–570.

Kennedy, P.G.E. (2008) The continuing problem of human African trypanosomiasis (sleeping sickness). *Ann. Neurol.* 64: 116–127.

Tait, A., MacLeod, A., Tweedie, A., Masiga, D. and Turner, C.M.R. (2007) Genetic exchange in *Trypanosoma brucei*: evidence for mating prior to metacyclic stage development. *Mol. Biochem. Parasitol.* 151: 133–136.

Taylor, J.E. and Rudenko, G. (2006) Switching trypanosome coats: what's in the wardrobe? *Tr. Genetics* 22: 614–620.

Vickerman, K. (1985) Developmental cycles and biology of pathogenic trypanosomes. *Br. Med. Bull.* 41: 105–114.

Trypanosoma cruzi and Chagas' Disease

Disease(s)	Chagas' disease, American trypanosomiasis
Etiological agent(s)	*Trypanosoma cruzi*
Major organ(s) affected	Blood, heart, esophagus, colon
Transmission mode or vector	Triatomine bugs
Geographical distribution	Foci throughout South and Central America
Morbidity and mortality	A generally asymptomatic acute stage is followed by decades of latency before chronic symptoms involving the heart or gastrointestinal system emerge. A large portion of infected individuals will die of congestive heart failure
Diagnosis	Detection of the parasite in the blood during the acute stage and various serological tests during the chronic stage
Treatment	Nifurtimox and benznidazole are generally effective during the acute phase of the disease, but questionable during the chronic stage. Supportive care during the chronic stage may also be necessary
Control and prevention	Elimination of triatomines through insecticides and improvement of human dwellings combined with education

Trypanosoma cruzi infects a wide range of vertebrates. It is transmitted by blood-sucking triatomine bugs and causes a human disease commonly referred to as **Chagas' disease**. Carlos Chagas first discovered the trypanosome while studying the vectors and named it after his mentor Oswaldo Cruz. Chagas determined the life cycle and found that the parasite could infect a wide range of mammals, including humans, and subsequently described the salient features of the disease. Chagas' disease exhibits a patchy distribution throughout the Americas ranging from southern United States to southern Argentina. It is predominantly found in poor rural areas of Central and South America and is the leading cause of heart disease in Latin America.

Life Cycle

T. cruzi is transmitted to humans via hematophagous (i.e., blood feeding) insects called **triatomine bugs** (Figure 9.1). Numerous species of triatomines are capable of transmitting *T. cruzi*. However, the most important vectors for human transmission are *Triatoma infestans* and *Rhodnius prolixus*. Some common names of triatomine bugs include assassin bugs, kissing bugs (in reference for its tendency to bite around the face), and conenose bugs (in reference to their pointed heads). In addition, they are called reduviid bugs in reference to being in the family Reduviidae or triatomine bugs in reference to the subfamily Triatominae. The life cycle consists of five nymphal

Figure 9.1 Adult triatomine bug. *Rhodnius prolixus* taking a blood meal. Note the feces in the lower left corner. (Photograph from WHO/TDR/STAMMERS TDR Image Library and the Liverpool School of Tropical Medicine.)

instars followed by sexually mature and winged adults. Triatomines feed on blood throughout their life and all stages can become infected with *T. cruzi*.

Triatomines are relatively large insects (5–30 mm) that generally feed at night while the person is sleeping. As it is taking a blood meal, the triatomine bug defecates and the feces contain **metacyclic trypomastigotes** which are infective to humans or other mammalian hosts (Figure 9.2). Infection is generally associated with rubbing the infected fecal material into the bite wound or eyes while the person sleeps. The motile trypomastigotes gain access to the blood system via the bite wound. Trypomastigotes are also capable of penetrating mucous membranes, especially the thin membrane which covers the eyelid and eye (i.e., conjunctiva). Thus the eyes are another common entry point for the metacyclic trypomastigotes. The transmission of *T. cruzi* via feces is referred to as "hindgut station" or "stercorarian" and is considerably less efficient than the salivarian transmission exhibited by African trypanosomes and several other vector-transmitted protozoa. This is because the infectious stages are not directly injected into the skin, but need to find their own entry points into the host. Animals can become infected by ingesting infected triatomine bugs, which may represent the normal route of infection for many animals. In this case the trypomastigotes penetrate mucosal membranes around the mouth or gastric epithelium.

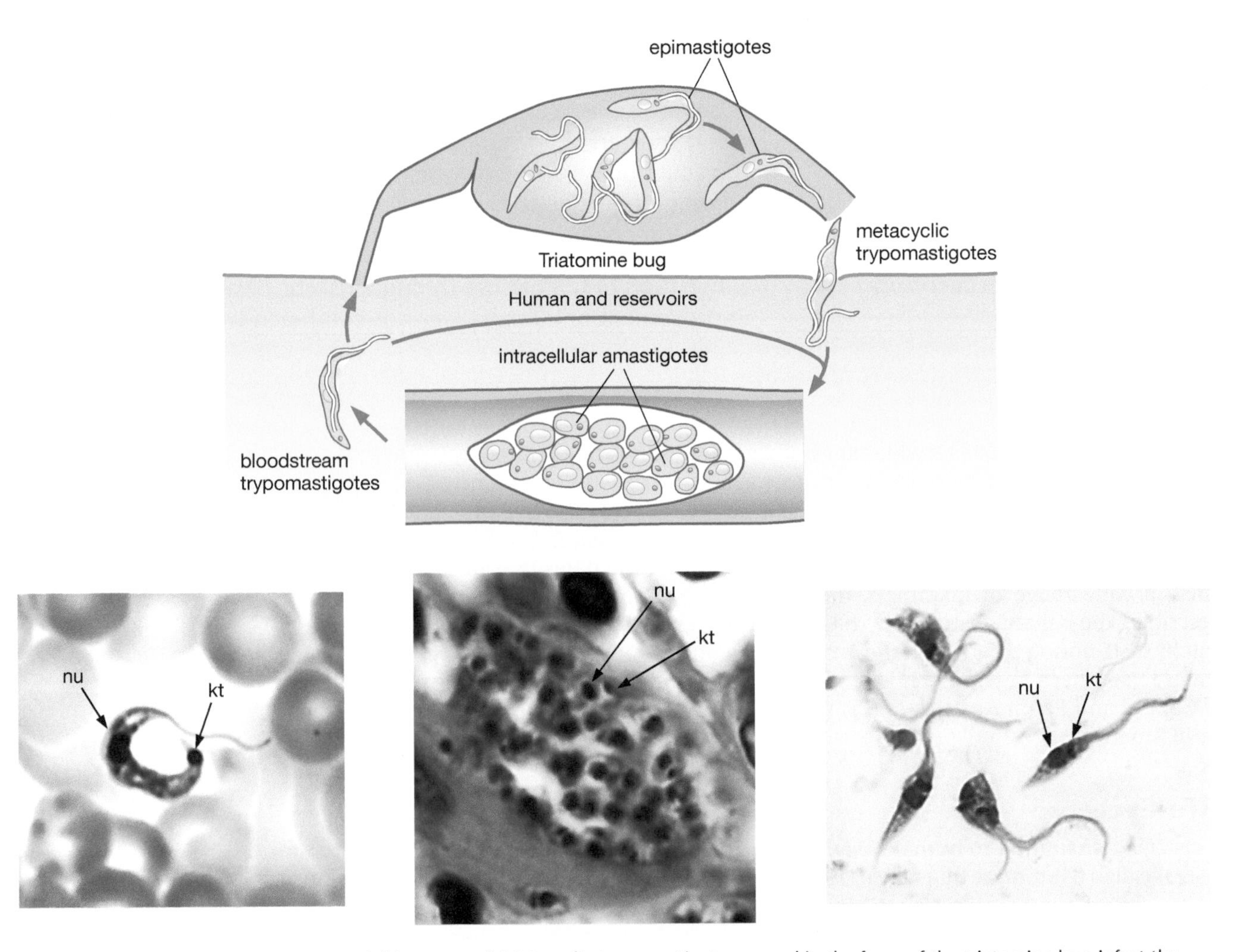

Figure 9.2 Life cycle of *Trypanosoma cruzi*. (Upper panel) Metacyclic trypomastigotes passed in the feces of the triatomine bug infect the vertebrate host and invade a wide range of cell types. Within the host cells the parasites convert to amastigote forms and replicate. The amastigotes convert back into trypomastigotes and are released from the host cell. The trypomastigotes can reinvade other host cells to reinitiate the replication cycle or be taken up by the triatomine bug and convert into replicating epimastigotes within the midgut of the vector. Maturation to the infective metacyclic trypomastigote occurs in the hindgut. (Lower panels) Trypomastigotes found circulating in blood (left). Amastigotes in muscle cell (middle). Epimastigotes from *in vitro* culture (right). Nuclei (nu) and kinetoplasts (kt) are denoted with arrows.

T. cruzi replicates as an intracellular parasite

After gaining entry into the vertebrate host, the trypomastigote invades a host cell and becomes intracellular. *T. cruzi* is capable of invading and replicating within many different cell types which plays a key role in disease progression. The invasion of host cells is distinct from phagocytic pathways exploited by some intracellular pathogens such as *Leishmania*. However, host cell invasion does involve the recruitment of host cell lysosomes to a site on the host cell plasma membrane adjacent to the trypomastigote after initial contact between the parasite and host cell (Figure 9.3). These lysosomes fuse with the plasma membrane and the trypomastigote enters into this forming a parasitophorous vacuole. Lysosomes continue to fuse with the nascent parasitophorous vacuole as parasite entry proceeds. This entry does not involve an actin-mediated phagocytosis process. Instead, the recruitment of lysosomes appears to be part of a process by which cells repair lesions on the plasma membrane. Consistent with this membrane repair process is an elevation of intracellular calcium levels associated with the invasion process. Thus, the parasite is exploiting the plasma membrane wound repair mechanism to gain entry into the host cell.

After completion of parasite entry the trypomastigote lies within a parasitophorous vacuole. Entry independent of lysosomes has also been described in which the parasite is engulfed by the plasma membrane of the host cell without the coincident fusion of the lysosomes. However, lysosomes still fuse with this parasitophorous vacuolar membrane after the entry is completed. Thus, regardless of the entry method the parasitophorous vacuole is analogous to the phagolysosome. Approximately 24 hours after cell invasion the parasite escapes from the parasitophorous vacuole. This escape is mediated by a pore-forming protein secreted by the parasite into the parasitophorous vacuole. The acidic environment of the lysosome activates this pore-forming protein which then disrupts the parasitophorous vacuolar membrane. A neuraminidase activity on the surface of the parasite may also facilitate the escape from the parasitophorous vacuole to the cytoplasm of the host cell.

Once within the cytoplasm of the host cell the trypomastigote converts into the **amastigote** form. The amastigotes replicate by binary fission within the cytoplasm of the host cell and the progeny fill up the host cell.

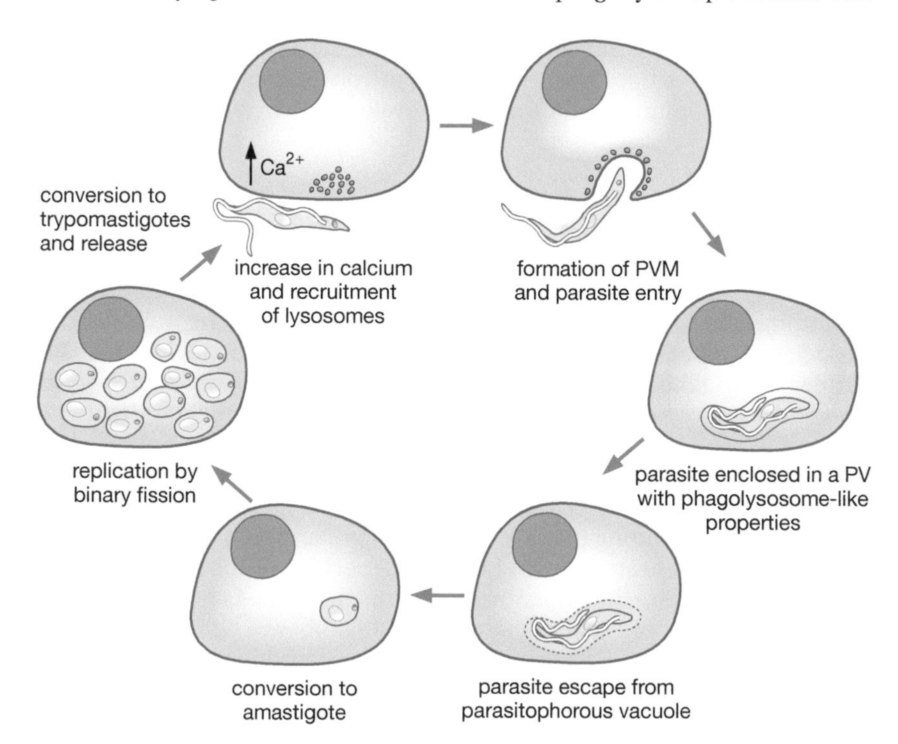

Figure 9.3 Host cell entry and parasite replication. The trypomastigote makes contact with the host cell and lysosomes are recruited to this location on the cell surface. Fusion of these lysosomes with the plasma membrane forms the nascent parasitophorous vacuolar membrane (PVM) into which the parasite enters and is eventually enclosed within a parasitophorous vacuole (PV). The parasite escapes from the parasitophorous vacuole by disrupting the PVM and the trypomastigote develops into an amastigote which replicates within the host cell cytoplasm by binary fission. The amastigotes convert back into trypomastigotes and are released from the infected host cell.

This intracellular cycle lasts an average of 4 days and after several rounds of division the amastigotes will differentiate back into trypomastigotes and be released from the infected cell and circulate throughout the body as non-dividing trypomastigotes. Some of these trypomastigotes invade other host cells, transform back into amastigotes, and repeat the replication cycle. These cycles of asexual replication will continue until the vertebrate host is cured of the infection or dies. Without treatment the infection may persist for the entire life of the human host.

The parasite replicates within the gut of the vector

Trypomastigotes in the circulatory system can also be taken up by a triatomine bug during feeding (Figure 9.2). Within the midgut of the vector the trypomastigotes differentiate into **epimastigotes** which undergo multiple rounds of binary fission. The triatomine bugs remain infected for their entire life (1–2 years). Some of the epimastigotes quit dividing and migrate to the hindgut where they differentiate into metacyclic trypomastigotes which will be excreted with the feces. The complete cycle within the vector generally takes 10–15 days. The epimastigote stage is also the form that is most readily cultivated *in vitro*.

The process and regulation of metacyclogenesis (i.e., the formation of metacyclic forms) is not well understood. There are morphological changes in which the parasite becomes more slender and smaller and the relative positions of the nucleus and kinetoplast change as the parasite converts from an epimastigote to a trypomastigote. In conjunction with these morphological changes the metacyclic trypomastigotes become more motile than epimastigotes. Changes in protein expression between epimastigotes and metacyclic trypomastigotes have also been noted. In particular, changes in the proteins expressed on the parasite surface have been speculated to be involved in cell invasion, complement resistance, and other adaptations to the mammalian host. For example, only the mature metacyclic trypomastigotes are infective for the vertebrate host. This noninfectivity of epimastigotes is due in part to the lysis of epimastigotes by host complement.

A variation of this life cycle has been observed in opossums. Epimastigotes and metacyclic trypomastigotes, which are normally only found in the vector, are found within the anal glands of opossums and these anal gland derived parasites are infective to experimental mice. Infected anal glands are not common in nature and the significance of this observation in regard to transmission is not clear. However, transmission via anal gland secretions may account in part for the high prevalence of *T. cruzi* in opossums.

Transmission and Control

The inefficiency of transmission from vector to human necessitates a lot of contact between vector and human host. High levels of vector–host contact are generally associated with houses becoming infested with triatomines or triatomines living near the house. Thatched roofs and adobe walls provide ideal habitats for the vector and some vector species have a high propensity to adapt to human habitations (Figure 9.4). In some situations the infestation of the house and peridomiciliary locations (i.e., immediate vicinity) is due to the proximity of houses to **sylvatic** cycles. In addition, animal stalls adjacent to the house may provide a source of reservoir hosts and contribute to disease transmission. Under such conditions the triatomine bugs will come out of their hiding places during the night and feed upon the inhabitants. It is this repeated feeding over long periods of time that allow for transmission despite the rather inefficient mechanism. As further evidence of the inefficiency, only some members of a family will be infected with *T. cruzi* despite the fact that all members live in a triatomine infested house.

Risk factors for the transmission of *T. cruzi* are usually associated with rural poverty and the presence of the vector (Table 9.1). However, poverty

Figure 9.4 Examples of conditions promoting the transmission of *T. cruzi*. Adobe walls and thatched roofs provide habitats for the triatomines. Proximity to the sylvatic cycle provides a source of infected triatomines and domestic animals near the house provide reservoirs. (Photographs from Antonio D'Alessandro, Tulane University, Department of Tropical Medicine.)

per se is not a risk factor, but the conditions associated with poverty are conducive toward the transmission of *T. cruzi*. For example, houses with dirt floors, adobe walls, thatched roofs, and little or no plaster provide habitats for the triatomine bugs and increase the chances for the triatomines to infest the house. Similarly, the tendency to pile construction or agricultural materials near the house provides peridomiciliary habitats for the triatomines. In addition, having domestic animals living in or near the house provides a reservoir for the parasite. But these conditions alone will not promote the transmission of Chagas' disease if there are no vectors present. In this regard, a good surrogate marker for the risk of Chagas' disease is the ability to recognize triatomines. Generally people who can correctly identify triatomine bugs are from endemic areas and are familiar with the bugs. Quite often people from nonendemic areas will identify a triatomine bug as a cockroach. However, the ability to identify triatomines usually does not translate into knowledge about the role the triatomine plays in the transmission of Chagas' disease.

Another major factor in regard to the transmission of Chagas' disease is the destruction of the natural habitats of the triatomine bugs and natural reservoirs. *T. cruzi* can infect a wide range of animals and sylvatic cycles are prevalent in many areas. In fact, the history of infection of humans with *T. cruzi* has been quite short compared with the evolution of *T. cruzi* in other animals (Box 9.1). The encroachment of humans into the forest and then the subsequent destruction of the forest to build houses and farm land lead to a loss of habitat for the triatomine bugs and the natural reservoirs (Figure 9.5). Some of these animals and triatomine bugs will then occupy habitats around the houses and start a peridomiciliary cycle. In cases where the human habitats are adjacent to the forest there can be exchange between the sylvatic and peridomiciliary cycles. Some of the triatomines, such as *T. infestans* and *R. prolixus*, will permanently occupy houses and establish a domiciliary cycle involving humans and domestic animals as well as natural reservoirs such as rodents.

Table 9.1 Risk factors associated with *T. cruzi* infection
Lower socioeconomic status
Having lived in a rural area
Being able to recognize triatomines
Received blood transfusion(s)

Control of domestic triatomine populations

The Southern Cone Initiative against Chagas' disease was formalized between Argentina, Bolivia, Brazil, Chile, Paraguay, and Uruguay in 1991. The goal of this initiative was to interrupt the transmission of *T. cruzi* by eliminating *T. infestans* from domestic settings. The control efforts consisted of spraying all houses and peridomiciliary areas with pyrethroid-based insecticides. Spraying uninfested houses prevents the triatomines from simply moving house to house during the campaign. Community support was also enlisted to monitor for the reappearance of triatomines and houses were resprayed if triatomines reappeared. In addition, blood donors were also screened to prevent transmission via blood transfusions. These

Box 9.1 Molecular epidemiology and evolution of *Trypanosoma cruzi*

T. cruzi exhibits a high degree of polymorphism when analyzed by various biochemical and molecular techniques [1, 2]. Despite this high level of polymorphism, two major phylogenetic groups are recognized regardless of the method used. However, there are a few parasite isolates that do not fit into either of these groups or may be hybrids. The two major groups have been designated as *T. cruzi* I (TcI) and II (TcII) in an effort to consolidate nomenclatures. There is even some discussion that TcI and TcII should be considered as distinct subspecies or species in that the genetic differences between them are greater than the differences between some *Leishmania* species.

In regard to its evolutionary history, *T. cruzi* is estimated to have emerged as a species more than 150 million years ago (mya). Phylogenetic analysis suggests that TcI and TcII diverged between 37 and 88 mya [3]. This is a similar time period as the geographical isolation of North and South America. In other words, the split of Gondwanaland occurring approximately 100 mya led to the divergence of the two *T. cruzi* groups. The TcI type parasites originated in what would become North America and were primarily associated with marsupials, whereas the TcII type parasites were associated with placental mammals in South America. The continuous land connection at the Panama Isthmus, which formed 2–5 mya, led to an exchange of mammals and thus a geographical mixing of the two types. Although, both types are found throughout North and South America, TcI appears to be more predominant in the northern part of South America and northward, whereas TcII tends to be more predominant in the southern half of South America.

TcI and TcII also exhibit different biological and epidemiological features (Figure 9A). For example, there is a preferential association of TcII with placental mammals, whereas TcI exhibits a preference for marsupials. TcI is a major agent for human disease north of the Amazon Basin, but is also ubiquitous in sylvan transmission cycles throughout the Americas. In the southern cone region of South America, TcII causes most human infection and overall is more prevalent. Furthermore, the TcI infections are more likely to be diagnosed during the acute stage, suggesting that TcI parasites may be cleared by the human host and are less likely to develop into long-term chronic infections. This higher prevalence of TcII infections and disease in humans cannot be attributed to a preference for particular triatomine species since the vectors appear capable of transmitting TcI and TcII equally well.

Human contact with *T. cruzi* and the establishment of the domestic cycle are much more recent than the divergence of these two phylogenetic groups. Humans have been present in South and Central America for at least 15 000 years and perhaps for as long as 20 000–25 000 years. Interestingly, lesions consistent with Chagas' disease and both TcI and TcII kinetoplastid DNA has been identified in mummies up to 9000 years old from Chile, Peru, and Brazil [4] indicating the presence of Chagas' disease in prehistoric hunter-gatherer populations. As humans domesticated plants and animals and became more sedentary, the triatomine vectors adapted to human habitations and this adaptation allowed for the establishment of domiciliary and peridomiciliary transmission cycles that exist today.

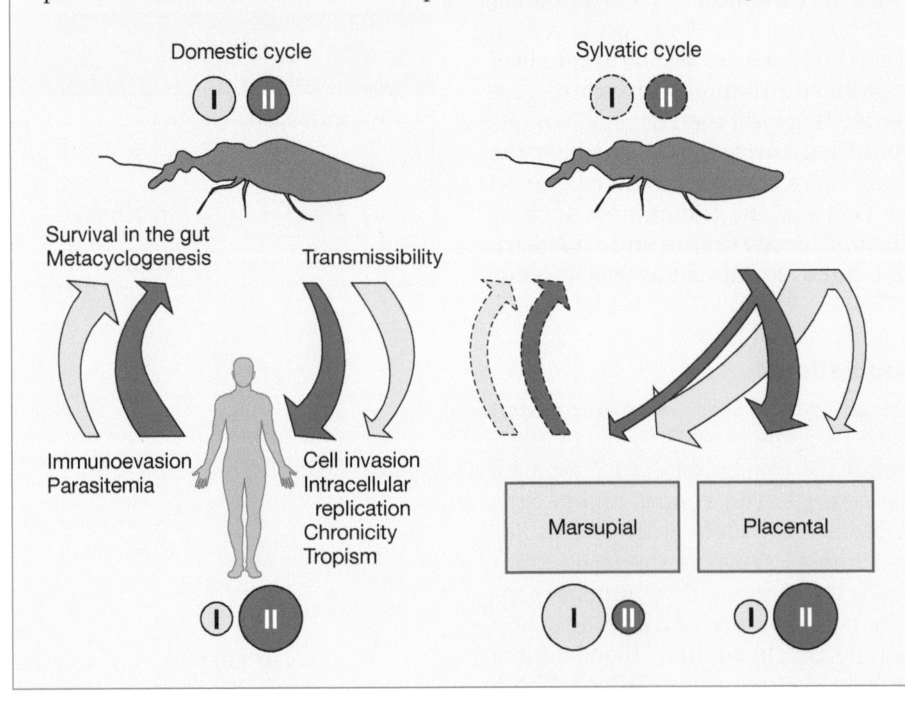

Figure 9A Comparison of TcI and TcII. Schematic drawing comparing ecological (domestic and sylvatic cycles), epidemiological (scaled circles), and biological features (width of arrows) of TcI (light blue) and TcII (gray). Dashed lines indicate that insufficient experimental data are available. (Redrawn from *Buscaglia and Di Noia [1]*).

1. Buscaglia, C.A. and Di Noia, J.M. (2003) *Trypanosoma cruzi* clonal diversity and the epidemiology of Chagas disease. *Microb. Infect.* 5: 419–427.

2. Macedo, A.M., Machado, C.R., Oliveira, R.P. and Pena, S.D.J. (2004) *Trypanosoma cruzi*: genetic structure of populations and relevance of genetic variability to the pathogenesis of Chagas' disease. *Mem. Inst. Oswaldo Cruz* 99: 1–12.

3. Briones, M.R., Souto ,R.P., Stolf, B.S. and Zingales, B. (1999) The evolution of the two *Trypanosoma cruzi* subgroups inferred from rRNA genes can be correlated with the interchange of American mammalian faunas in the Cenozoic and has implications to pathogenicity and host specificity. *Mol. Biochem. Parasitol.* 104: 219–232.

4. Lima, V.S., Iñiguez, A.M., Otsuki, K., Ferreira, L.F., Araújo, A., Vicente, A.C.P. and Cansen, A.M. (2008) Chagas' disease in ancient hunter-gatherer population, Brazil. *Emerg. Infect. Dis.* 14: 1001–1002.

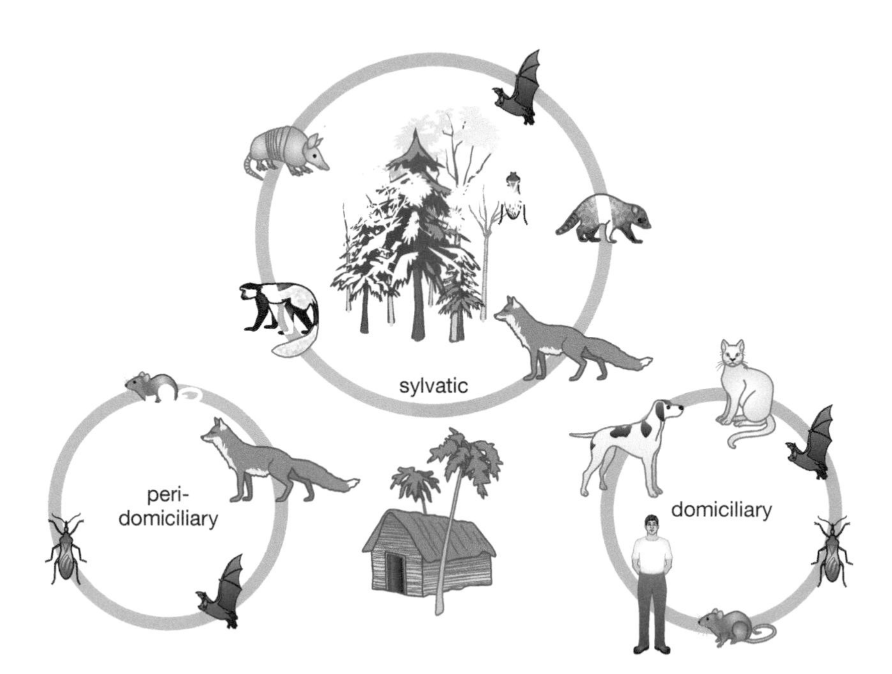

Figure 9.5 Transmission cycles of *T. cruzi*. The intersection of human habitats with the sylvatic transmission cycle allows for the introduction of the triatomines and natural reservoirs into the peridomiciliary cycle and then to domiciliary cycles.

efforts have resulted in the interruption of *T. cruzi* transmission in large areas of these countries. Disease prevalence through South and Central America has been reduced from an estimated 16–18 million people in 1990 to less than 10 million in 2006. There has also been a progressive decline in Chagas' disease associated hospitalizations.

The success of this initiative is largely due to *T. infestans* being almost exclusively domestic and being responsible for an estimated 80% of the transmission in these countries. Furthermore, it is believed that *T. infestans* had been accidentally imported throughout the southern cone countries from Bolivia in relatively recent times—particularly during the last 100 years—through human migration. Therefore, there are no sylvatic and minimal peridomiciliary cycles of *T. infestans* that can lead to reinfestation of the houses. The relatively slow reproduction of the triatomine bugs and their low population variability also results in a low level of selection for insecticide resistance. Thus, the biological characteristics that make *T. infestans* a good domestic vector also render it a vulnerable target for elimination. In addition, the active involvement of the community and education about the role of the triatomine in the transmission of Chagas' disease contributed significantly to the success of these programs.

Two other initiatives have been formalized as a result of the success of the Southern Cone Initiative: the Andean Pact Initiative between Venezuela, Colombia, Ecuador, and Peru and the Central American Initiative (including Mexico). In many of these areas *Rhodnius prolixus* is the principal vector. Like *T. infestans*, *R. prolixus* is well adapted to houses and peridomestic areas, is found in relatively few sylvan habitats, and is believed to have been artificially introduced into many regions. Although sylvan forms of *R. prolixus* are relatively uncommon, there is some controversy as to whether sylvan forms of *R. prolixus* will invade houses. If this is possible, then control efforts will need to include measures to evaluate and remedy post-control reinvasion. There are also foci of transmission due to other vectors found throughout South and Central America. *Triatoma dimidiata*, found in many locations throughout South and Central America, and *Panstrongylus megistus*, found in parts of Brazil, are particularly noteworthy. Both appear to be in the process of adapting to houses, and at the same time are still found in many sylvan habitats.

Another effective control strategy involves the improvement of human dwellings so that they no longer provide resting and breeding sites for the

Table 9.2 Effective control activities for eliminating domestic vector populations

Pyrethroid insecticides	Spray internal and external walls, inside roof structures, and peridomiciliary animal stalls
	Treat all houses in community
House improvements	Replace thatched roofs
	Plaster internal walls
	Install ceilings
Health education and community empowerment	Recognition of triatomine vector and its importance
	Vigilance against reinfestation

triatomines (Table 9.2). For example, thatched roofs can be replaced with metal, interior walls can be plastered, and dirt floors covered with cement. These efforts reduce the contact between humans and triatomines. In addition, the separation of animal stalls from the house reduces the role domestic animals play as reservoirs. These efforts to improve human dwellings and outbuildings also need to be combined with education. Quite often the connection between triatomines and Chagas' disease is not realized because of the long latent period between infection and the development of disease. Treating houses with insecticides, improving houses, and education are all targeted at eliminating the domestic vector populations. Because the distribution of Chagas' disease is coincident with triatomines in the household it is possible to break the transmission cycle through a combination of these methods.

Blood transfusions are a common mode of transmission

As with other blood-borne pathogens, it is also possible to acquire the disease through blood transfusion, which is the second most common mode of transmission, or organ transplants. In contrast to natural vector transmission, blood transfusions are often associated with transmission in urban areas. The long asymptomatic latent period associated with Chagas' disease results in transfusion-associated transmission being a problem in endemic areas since blood donors may be unaware of their infections. Generally it is estimated that 10–15% of *T. cruzi* infections are due to blood transfusions. Blood donors from endemic areas should be screened for *T. cruzi* infections and rejected if positive. If it is not feasible to screen blood donors, then donated blood should be treated with gentian violet to kill the parasite.

In addition to natural vector transmission and blood transfusions other modes of transmission are congenital, accidents, and ingestion. The parasite can be congenitally transmitted from mother to fetus during either the acute or chronic phases. Congenital infections are often associated with more severe outcomes. The ability of the metacyclic trypomastigotes to invade mucous membranes poses problems for laboratory workers. In addition to needle sticks, transmission can occur through rubbing the eyes with contaminated hands. It is also possible to acquire the infection by ingesting material contaminated with the parasite because the metacyclic trypomastigote can penetrate the mucous membranes around the gums and in the oral cavity. In regard to numbers of persons infected, these are all relatively minor modes of transmission.

Clinical Symptoms and Manifestations

Acute infections usually exhibit mild or no symptoms

Chagas' disease is a complex process with a poorly understood pathophysiology. The disease exhibits three phases: acute, indeterminate (or latent), and chronic. The acute stage is characterized by an active infection that is often asymptomatic or only exhibits mild symptoms. Children are more likely to exhibit symptoms than adults. A hallmark of acute Chagas' disease is the **Romaña sign** (Figure 9.6), a swelling of the eye characterized by periorbital edema and conjunctivitis (i.e., inflammation of the conjunctiva). The Romaña sign is only observed when the parasite enters through the mucosa membranes around the eye and is not a generalized symptom. Another distinctive symptom that can occur is a localized erythematous skin lesion known as a chagoma. In both cases, these lesions represent the entry site for the parasite—either the mucous membranes around the eye or the bite wound—and a localized replication of the parasite as intracellular amastigotes. Lysis of the infected host cells and release of the parasites result in the localized inflammation. Generally these inflammatory lesions appear a few days after infection and gradually subside over 2–3 months.

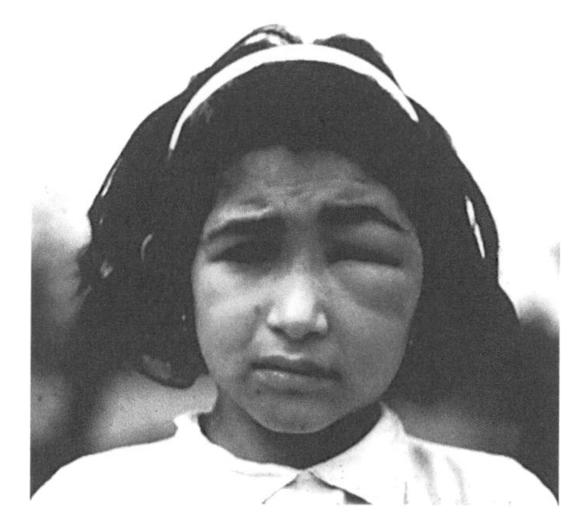

Figure 9.6 Romaña sign. Swelling of the eye and conjunctivitis are sometimes seen during the acute stage of Chagas' disease. (Photograph from Antonio D'Alessandro, Tulane University, Department of Tropical Medicine.)

Symptoms of a systemic infection may also appear during the acute stage and can include: fever, malaise, lymphadenopathy, hepatosplenomegaly, vomiting, and diarrhea. Generally, these symptoms spontaneously resolve within 3–4 months and most infected individuals are unaware of being infected. Trypomastigotes can also be detected in the blood during the acute stage of the infection, but drop to undetectable levels after a few months. Host immune responses are able to control the infection, but not eliminate it. Thus, reactivation of the acute stage can occur during immunosuppression (e.g., AIDS or chemotherapy). In the case of AIDS, this reactivation has a high mortality rate with the majority of the patients presenting with meningoencephalitis and/or acute myocarditis. Among nonimmunocompromised patients an extremely small number of individuals will develop and die of complications associated with an acute myocarditis or meningoencephalitis. Necropsy reveals extensive tissue pathology correlated with numerous amastigotes in cardiac, skeletal, and smooth muscle, as well as glial cells.

Chagas' disease is characterized by a long latent period

Following the acute phase is an extended period of seropositivity, but an absence of detectable parasitemia and overt clinical symptoms. Many of the patients in the latent phase do exhibit a subclinical level of cardiac involvement when monitored by ultrasonic cardiography. This indeterminate or latent phase can last for decades and possibly persists for the life of the infected person. The average time between the acute infection and development of severe Chagas' disease is estimated to be 20–35 years. Approximately one-third of the infected individuals will eventually develop cardiac or gastrointestinal sequelae characteristic of chronic Chagas' disease.

Chronic disease is often associated with cardiac symptoms

Myocardial involvement in chronic Chagas' disease includes arrhythmias, conduction defects, cardiomegaly, congestive heart failure, and thromboembolic events (Table 9.3). All parts of the heart (i.e., endocardium, myocardium, pericardium) can be affected. Autonomic ganglia and cardiac nerves are also frequently damaged in chagasic hearts. Electrocardiogram (ECG) abnormalities are approximately two-fold higher in seropositive persons as compared to seronegative persons. Heart disease is frequently associated with specific types of conduction disturbances, such as a complete

Table 9.3 Pathophysiology of chronic Chagas' disease

Pathology	Clinical symptoms or possible outcomes
Heart	
Arrhythmias and conduction defects	Abnormal ECG, palpitations, lightheadedness, fainting
Enlargement of the heart	Aneurysm
Inflammation and fibrosis	Congestive heart failure and thromboembolic events
Gastrointestinal tract	
Megaesophagus	Difficulty swallowing, regurgitation, heartburn, hiccups, increased salivation and speech impairments
Megacolon	Prolonged constipation

atrioventricular block or a right bundle branch block. Symptoms associated with these cardiac arrhythmias may include: rapid heart rate (i.e., palpitations), lightheadedness, dizziness, and fainting.

Progressive enlargement of the heart (i.e., cardiomegaly), congestive heart failure, and blood clots are other manifestations associated with Chagas' disease. The cardiomegaly is due to a combination of hypertrophy and dilation. Thinning of the heart wall, particularly at the apex, is also commonly seen in hearts from Chagas' patients (Figure 9.7). Aneurysms at the apex of the left ventricle have been observed in 20–30% of the autopsies. Intracardiac thrombi (i.e., blood clots) are also commonly observed in chagasic hearts and these thrombi can be carried to other parts of the body to form embolisms in other organs. These various cardiomyopathies are probably due to a progressive destruction of the myocardium and conduction system as suggested by the defects in the electrocardiography. Furthermore, these abnormalities are cumulative over time and the patient progressively worsens during the latent and chronic stages of the disease. Death is usually the result of rhythm disturbances, congestive heart failure or thromboembolism. Although chronic Chagas' heart disease tends to be insidious and slowly progressing, on occasion the presentation of arrhythmias and congestive heart failure can occur abruptly and lead to sudden death.

The micropathology of the heart reveals areas of cellular infiltration and inflammation (Figure 9.7). The infiltration is primarily lymphocytes

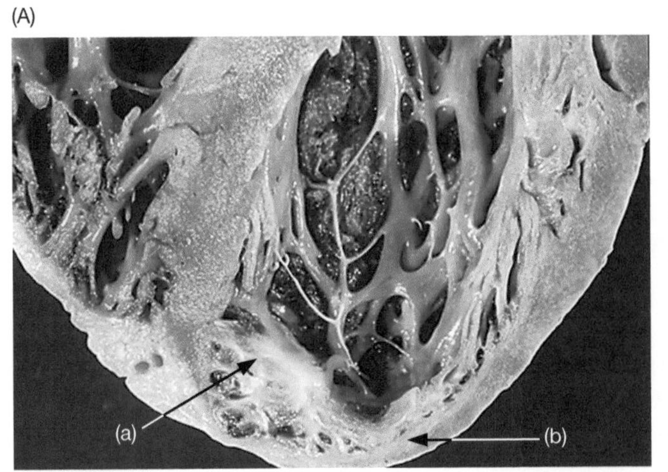

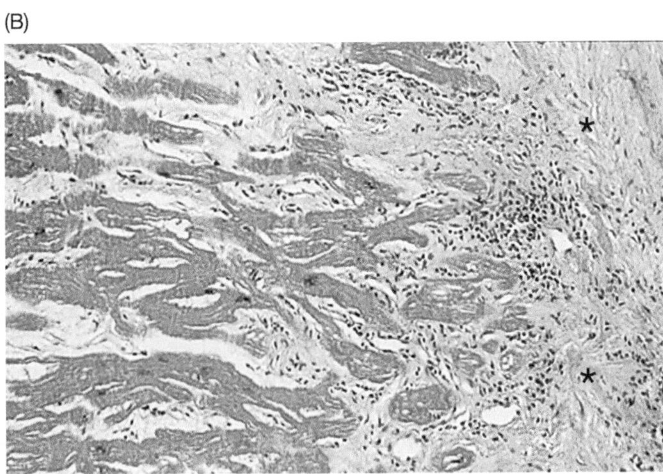

Figure 9.7 Cardiomyopathy associated with Chagas' disease. (A) Section of heart showing thinning apex (b) and fibrosis (a). (B) Micrograph of heart section showing fibrosing lymphocytic myocarditis. Note the numerous lymphocyte nuclei (dark dots) and replacement of myofibrils by fibrotic material on right side (*). (Photographs from Maria de Lourdes Higuchi, University of São Paulo Medical School, São Paulo, Brazil.)

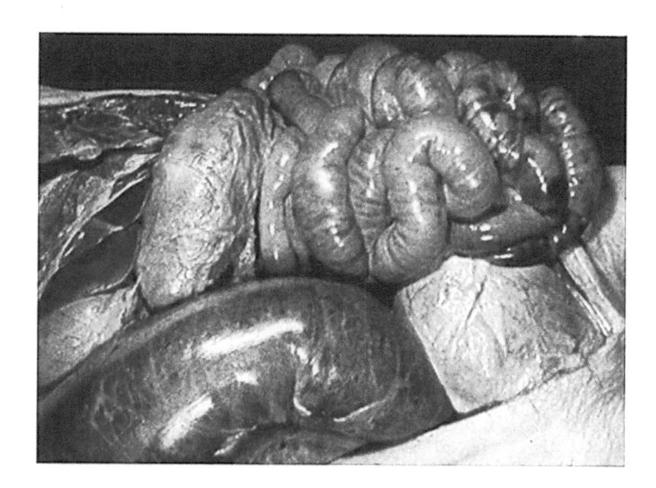

Figure 9.8 Megacolon. Dilation of hollow organs is sometimes seen in chronic Chagas' disease. (Figure reprinted from *A Colour Atlas of Tropical Medicine and Parasitology*, 3rd ed.,with kind permission from Elsevier Ltd.)

and macrophages, whereas polymorphonuclear cells (PMNs), neutrophils, and eosinophils are rarely seen. CD8 lymphocytes predominate over CD4 lymphocytes in these inflammatory lesions. This focal accumulation of infiltrating cells is probably correlated with the release of parasites from the infected cells and is often accompanied by damage to the myocardial fibers. Extensive fibrosis associated with the replacement of previously damaged myocardial tissue is also observed. In addition, focal and diffuse areas of myocellular hypertrophy may also be seen with or without inflammatory infiltrates. The cardiac symptoms correlate best with the fibrosis and cellular hypertrophy suggesting that this damage at the cellular level leads to the cardiomegaly, thromboembolic events, and arrhythmias.

Megaviscera are clinical manifestations of chronic disease

Gastrointestinal manifestations characterized by the dilation of the digestive tract are also frequently observed in chronic Chagas' disease (Table 9.3). The most commonly affected organs are the esophagus and the colon. Megaesophagus and megacolon (Figure 9.8) can occur independently of each other, in association with each other, or in conjunction with heart disease. The dilation of these organs is possibly due to autonomic neuronal dysfunction. Upon autopsy a reduction in the number of ganglia associated with the colon and the esophagus has been noted in patients exhibiting megacolon and megaesophagus, respectively. In many patients megaesophagus precedes heart disease. Clinical symptoms associated with megaesophagus are difficult or painful swallowing and regurgitation of food. Other manifestations include heartburn, hiccups, increased salivation, and speech impairments.

In general, megacolon occurs later in the course of Chagas' disease than megaesophagus. The primary symptom associated with megacolon is prolonged constipation. This will lead to dilation of the remaining intestines, pain, and constant physical distress. Complications associated with megacolon typically include intestinal obstruction and rupture. Possible secondary effects due to the long-term use of laxatives can include ulcers, septicemia, perforation of the intestines, and peritonitis.

Basis of Pathogenesis

The pathogenesis in chronic Chagas' disease is associated with inflammatory damage and/or loss of parasympathetic nerves in the heart or digestive system. The parasympathetic nervous system is part of the autonomic nervous system which regulates cardiac muscle, smooth muscle, and glandular tissue. Loss of function of the parasympathetic nerves could account for the dilation seen in the megasyndromes and the heart, and denervation of these tissues has been noted—especially in the esophagus and colon. Denervation could be due to parasite infection of the parasympathetic nerves or immunological destruction of the nerves. The causes and mechanisms of the pathogenesis are not known. The slow development of chronic stage and the lack of good animal models impede experimental studies.

Inflammation is a key component of chagasic cardiomyopathy

The lesions associated with Chagas' disease are consistent with an immunological attack on host tissues. Histological analysis of tissues from chronically infected hosts reveals inflammatory cells or fibrosis with few or no parasites (Figure 9.7B). However, parasites are routinely detected when more sensitive techniques, such as immunohistochemistry or PCR, are used. Furthermore, the foci of tissue damage are correlated with the presence of parasites as determined by immunohistochemical or molecular methods. Therefore, the inflammatory response associated with chronic Chagas' disease appears to be due to the presence of parasites. In addition, some studies demonstrate that treatment of chronic Chagas' disease lessens the clinical disease. Thus, an emerging picture is that *T. cruzi* persists in the host at a low level of parasitemia and results in a progressive disease with foci of inflammation and damage. Eventually the accumulating foci of damage reach a clinically relevant level and produce the symptoms of cardiomyopathy or megasyndromes.

The observation that chronic chagasic cardiomyopathy is generally accompanied by few, if any, parasites led to a prevailing dogma that the pathogenesis has an autoimmune or indirect etiology. The delayed onset of the disease and the organ specificity are consistent with an autoimmune etiology. In addition, anti-self antibodies and lymphocytes have been observed during the course of infection. Mechanisms for generating autoimmunity could include molecular mimicry between parasite and host antigens (e.g., similar antigenic epitopes) or bystander activation (e.g., released autoantigens due to tissue destruction). However, autoreactive antibodies (e.g., anti-myosin) and immune effector cells are seen in other non-chagasic cardiomyopathies. Therefore it is not clear whether the autoimmunity is the cause of the disease or a result of the infection. In addition, immunosuppression tends to exacerbate rather than alleviate the disease, indicating that autoimmunity does not completely explain the pathogenesis.

Others have proposed that an altered immune response and/or toxic factors may play a role in the pathology. For example, the balance of Th1 and Th2 cells correlates with cardiomyopathy and disease severity. In addition, inflammatory cytokines and chemokines have been associated with pathogenesis. These various hypotheses are not mutually exclusive in that the disease may result from several factors. For example, the persistence of parasites may stimulate an immune response that indirectly leads to autoimmune responses which contribute to the pathogenesis of the disease. The fact that approximately 30% of infected individuals eventually develop severe clinical disease emphasizes the need to better understand the basis of pathogenesis of Chagas' disease.

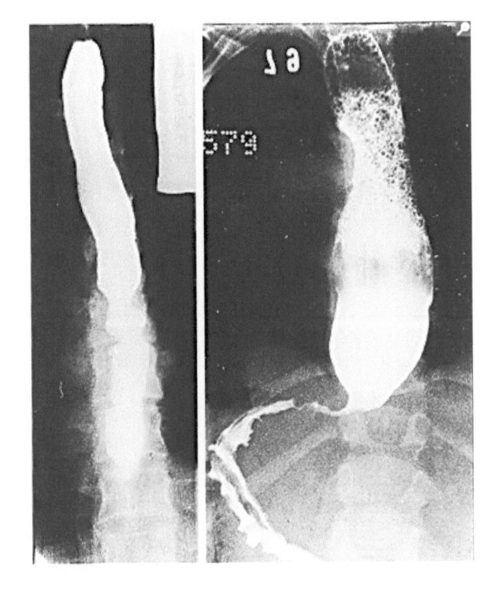

Figure 9.9 Demonstration of megasyndrome. Radiograph following a barium swallow showing normal esophagus (left) compared with megaesophagus (right). (Photographs from Giovanni Cerri, University of São Paulo Medical School, São Paulo, Brazil.)

Diagnosis and Treatment

Chagas' disease is suspected in persons exhibiting symptoms compatible with Chagas' disease and having an epidemiological history of possible exposure (i.e., residence in substandard housing or receipt of unscreened blood products from endemic areas). X-rays and other imaging techniques can also be used to detect the enlargement of the affected organs during the chronic stage (Figure 9.9).

Low parasitemias make parasite detection difficult

Confirmation of Chagas' disease is through laboratory diagnosis. The simplest methods are to examine fresh blood for motile trypomastigotes or Giemsa-stained thin or thick blood smears. However, direct detection of the parasite is generally only possible during the first month or two of the acute stage in that the levels of parasitemia drop to undetectable levels as the disease enters the latent phase. Concentration techniques such as examining the buffy coat following centrifugation of the blood in a microhematocrit tube may increase sensitivity. Other methods to increase the sensitivity are based on parasite replication and include inoculation of mice, *in vitro* culture, and xenodiagnosis. However, detection of parasites will take several weeks by these methods. Xenodiagnosis is performed by allowing uninfected laboratory-raised triatomine bugs to take a blood meal from the patient and examining the bugs for infection. *In vitro* culture or xenodiagnosis will be positive in most acute cases and some of the chronically infected patients. Recently, PCR (polymerase chain reaction) has been used as a sensitive method for the detection of *T. cruzi*. However, PCR is currently only available in research laboratories.

Serology is important in the diagnosis of chronic Chagas' disease

Various serological tests are available for the diagnosis of Chagas' disease. These include assays based on indirect hemagglutination, immunofluorescence and ELISA. During suspected acute-stage infections blood is examined for specific IgM or rising titers of IgG in paired samples taken at different times. However, methods that are based on parasite detection or replication are generally preferred during the acute stage. Because direct detection of parasites is not possible during the chronic stage and even methods based on parasite replication exhibit a low sensitivity, diagnosis of the chronic stage relies on serological tests for *T. cruzi*-specific IgG. Even these tests are not 100% sensitive and it is recommended to carry out two or more distinct serological tests. In addition, a related trypanosome, *T. rangeli*, can be confused with *T. cruzi* due to antibody cross-reactivity (Box 9.2).

Box 9.2 *Trypanosoma rangeli*

T. rangeli is another trypanosome with a geographical range that overlaps *T. cruzi*. It is also transmitted by triatomine bugs and infects a wide variety of mammals including humans. *T. rangeli* differs from *T. cruzi* in that metacyclic trypomastigotes are found in the salivary glands of the triatomine bug and transmission is probably not via fecal contamination. *T. rangeli* is not pathogenic for the mammalian host and parasitemias tend to be quite low and transient. Interestingly, *T. rangeli* is pathogenic for the triatomine vector. The life cycle of *T. rangeli* in the mammal and vector is not completely known. Dividing trypomastigotes have been observed in mammalian blood and amastigotes have been observed in *in vitro* cultures. However, the significance of these potential replicative forms is not clear. Similarly, the role of the various morphological forms found in the vector and how the parasite reaches the salivary glands is not clear. Because of the similar geographical ranges, vectors, hosts, and some serological cross-reactivity, it is possible to confuse *T. rangeli* and *T. cruzi*. However, the blood stage trypomastigotes are morphologically distinct. *T. rangeli* is a longer parasite with a smaller kinetoplast, similar to the African trypanosomes (Figure 9B).

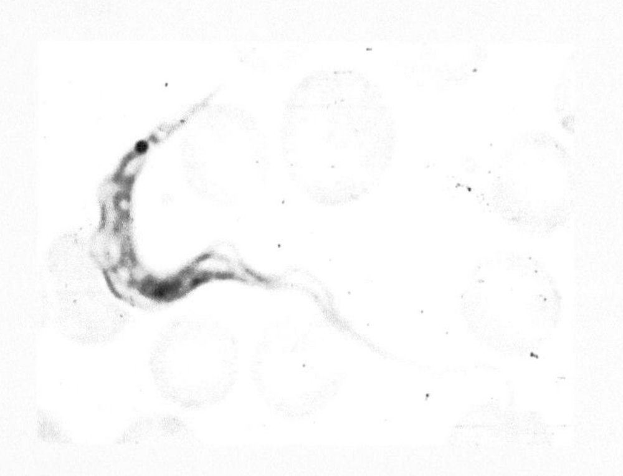

Figure 9B Blood smear of *Trypanosoma rangeli*. A nonpathogenic trypanosome found in the Americas. Can be distinguished from *T. cruzi* by the small kinetoplast and long slender morphology (Photograph from Antonio D'Alessandro, Tulane University, Department of Tropical Medicine.)

Treatment includes chemotherapy and supportive care

The two approved drugs for the treatment of Chagas' disease are nifurtimox (Lampit®, Bayer) and benznidazole (Rochagan® or Radanil®, Roche). Both drugs are believed to act primarily on the bloodstream trypomastigotes and suppress parasitemia. The mechanism of action of nifurtimox is believed to involve the generation of reactive oxygen intermediates (i.e., superoxide and hydrogen peroxide), whereas benznidazole seems to act via a reductive stress mechanism. Parasitological cures of up to 80% have been obtained with treatment by either drug during the acute phase of the disease. However, the effectiveness of these drugs during the chronic stage is controversial. Some studies have shown a benefit to treating indeterminate or chronic stage patients in that clinical symptoms were lessened and parasitological cures could be demonstrated. But other studies have not found evidence of parasitological cures or amelioration of the clinical outcomes following treatment during the chronic stage. In addition, both drugs cause relatively severe and frequent side effects, probably as a consequence of oxidative or reductive damage to host tissues. Over two-thirds of the patients experience side effects. Thus many patients are unable to complete the long treatment schedules and many physicians have strong reservations about the use of nifurtimox and benznidazole during the chronic stage due to the ambiguous risk-to-benefit profile of these drugs. Although controversial, a set of guidelines for the treatment of Chagas' disease patients has been recommended by a panel of experts (Table 9.4). Alternative drugs are clearly needed, but few are currently under development.

Supportive treatment directed at relieving the symptoms may also be necessary in Chagas' disease. Bed rest, hydration, and antipyretics may be prescribed during symptomatic acute infections. In the case of heart failure during the chronic stage the supportive care is the same as for other forms of heart failure. Such care can include bed rest, diuretics, small doses of digoxin, and possibly anticoagulants. Patients can also benefit from supervised physical activity. Anti-arrhythmic agents can be used to manage symptomatic or life-threatening arrhythmias. Pacemakers are also indicated in some situations. Mild symptoms associated with megaesophagus may respond to dietary changes or drugs to relax the lower esophageal sphincter. Surgical interventions may be prescribed in more advanced cases. Similarly, cases of megacolon which do not respond to diet, laxatives and enemas may require surgical intervention.

Table 9.4 Treatment guidelines*

Drug	Condition	Dose (mg/kg/d)	Duration (days)
Nifurtimox	Acute phase	10–12	30–60
	Chronic phase	8–10	60–90
Benznidazole	Acute, children	7.5	60
	Acute, adults	5	30–60
	Chronic phase	5	30

*Meeting held April 23–25, 1998 at Fundación Oswaldo Cruz, Rio de Janeiro, Brazil (PAHO/WHO. Etiological Treatment of Chagas' Disease. *Rev. Pat. Trop.* 28: 247–249, 1998).

Summary and Key Concepts

- *Trypanosoma cruzi* is transmitted by triatomine bugs.
- Transmission is primarily associated with living in houses infected with the vector and control efforts to eliminate the domestic populations of triatomine bugs are successful at reducing the prevalence of Chagas' disease.
- The parasite replicates as an intracellular amastigote form within the mammalian host.
- Chagas' disease is usually characterized by a mild or asymptomatic acute phase followed by a long latent period before the appearance of symptoms associated with the chronic disease.
- The most common manifestations of chronic Chagas' disease are cardiomyopathy, megaesophagus, and megacolon.
- Disease pathogenesis may involve parasite-mediated destruction of host cells and an immunological attack on host tissues.
- Diagnosis relies on the detection of parasites or serological methods in conjunction with disease symptoms.
- Treatment includes chemotherapy to eliminate the parasites and supportive care to alleviate the symptoms.

Further Reading

Barrett, M.P., Burchmore, J.S., Stich, A., Lazzari, J.O., Frasch, A.C., Cazzula, J.J. and Krishna, S. (2003) The trypanosomiases. *Lancet* 362: 1469–1480.

Girones, N. and Fresno, M. (2003) Etiology of Chagas' disease myocarditis: autoimmunity, parasite persistence, or both? *Tr. Parasitol.* 19: 19–22.

Higuchi, M.L., Benvenuti, L.A., Reis, M.M. and Metzger, M. (2003) Pathophysiology of the heart in Chagas' disease: current status and new developments. *Cardiovasc. Res.* 60: 96–107.

Mott, G.A. and Burleigh, B.A. (2008) Chapter 13. The role of host cell lysosomes in *Trypanosoma cruzi* invasion. In *Molecular Mechanisms of Parasite Invasion* (ed. B.A. Burleigh and D. Soldati-Favre), pp. 165–173. Landes Bioscience and Springer Science + Business Media, Austin, Texas.

Schofield, C.J. and Dias, J.C.P. (1999) The Southern Cone Initiative against Chagas' disease. *Adv. Parasitol.* 42: 2–27.

Schofield, C.J., Jannin, J. and Salvatella, R. (2006) The future of Chagas' disease control. *Tr. Parasitol.* 22: 583–588.

Tanowitz, H.B., Machado, F.S., Jelicks, L.A., Shirani, J., Campos de Carvalho, A.C., Spray, D.C., Factor, S.M., Kirchhoff, L.V. and Weiss, L.M. (2009) Perspectives on *Trypanosoma cruzi*-induced heart disease (Chagas' disease). *Prog. Cardiovasc. Dis.* 51: 524–539.

Tarleton, R.L. (2003) Chagas' disease: a role for autoimmunity? *Tr. Parasitol.* 19: 447–451.

Teixeira, A.R.L., Nascimento, R.J. and Sturm, N.R. (2006) Evolution and pathology in Chagas' disease – a review. *Mem. Inst. Oswaldo Cruz* 101: 463–491.

Urbino, J.A. and Docampo, R. (2003) Specific chemotherapy of Chagas' disease: controversies and advances. *Tr. Parasitol.* 19: 495–501.

10 Leishmaniasis

Disease(s)	Cutaneous or visceral leishmaniasis
Etiological agent(s)	*Leishmania* species
Major organ(s) affected	Skin and reticuloendothelial system
Transmission mode or vector	Sand fly
Geographical distribution	Foci in tropical and subtropical areas throughout the world
Morbidity and mortality	Cutaneous forms of the disease are generally self-limiting while visceral leishmaniasis is usually fatal
Diagnosis	Detection of parasite in cutaneous lesions or bone marrow and serological tests
Treatment	Pentavalent antimonials
Control and prevention	Minimize human/sand fly contact

Several species of *Leishmania* are capable of infecting humans and cause a myriad of disease manifestations ranging from self-limiting cutaneous lesions to more serious disseminated diffuse and mucocutaneous disease as well as visceral disease. This wide range of disease manifestations depends somewhat on the specific parasite species. However, host responses and poorly understood host–parasite–vector interactions also contribute to the disease. In addition, transmission of the various species, and thus the various clinical outcomes, are also dependent on the geographic area and ecological setting. For example, many of the species are also capable of infecting other mammals and in many situations leishmaniasis is a zoonotic disease, whereas anthroponotic transmission predominates in other areas.

Leishmaniasis is endemic in tropical and subtropical areas throughout the world. Transmission occurs in settings ranging from rain forests to deserts and can be found in rural or peri-urban areas. Leishmaniasis is a focal disease in both space and time. Distinct species of parasites, vectors, and reservoir hosts maintain the transmission cycle in any given area. The number of reported cases and geographical range of endemic leishmaniasis is increasing, and thus leishmaniasis can be considered as an emerging disease.

Life Cycle

Leishmania are transmitted by a small biting fly known as a **sand fly** (Figure 10.1). Two different genera of phlebotomine sand flies transmit *Leishmania*: *Phlebotomus* in the eastern hemisphere (i.e., Old World including Africa, Asia, and Europe) and *Lutzomyia* in the western hemisphere (i.e., New World including South and Central America). Only the females take blood meals and thus the males play no direct role in transmission. Sand flies

Figure 10.1 Sand fly, *Lutzomyia longipalpis*. (Photo kindly provided by Jose Ribeiro of the National Institute of Allergy and Infectious Disease, Bethesda, Maryland.)

are pool feeders in that they drink blood and lymph from the bite wound. **Metacyclic promastigotes** are transferred to the vertebrate host during feeding. However, in contrast to many other vector-transmitted protozoa, the promastigotes are not in the salivary glands, but are found in the anterior portion of the gut and pharynx. The transmission mechanism is not entirely understood and there is some debate about the exact nature of this mechanism. Several factors are known to influence transmission though. For example, infection of the sand fly leads to a dysfunction in the valve at the entrance to the midgut. When the sand fly takes a blood meal the valve cannot close and blood moves back into the mouthparts and the bite wound. This reflux results in an expulsion of promastigotes from the anterior portions of the sand fly gut into the bite wound of the mammalian host. In addition, a gel-like substance is secreted by the promastigotes forming a physical obstruction packed with promastigotes in the foregut. This blockage interferes with the ability of the sand fly to take a blood meal and therefore the sand fly needs to expel, or regurgitate, this promastigote-filled gel. The secreted gel may also contain factors that exacerbate the disease. In addition, the saliva is also expelled into the bite wound and there are also factors in the saliva of the sand fly which potentiate the infectivity of *Leishmania* for the vertebrate host. Some of these salivary gland compounds have immunosuppressive activities against lymphocytes and macrophages which may explain this potentiation.

Leishmania amastigotes replicate within host macrophages

Within the mammalian host the promastigotes are phagocytosed by macrophages or other professional phagocytes (Figure 10.2). The parasite cannot actively invade host cells and is dependent upon the phagocytic activity of the host cell. Therefore, *Leishmania* amastigotes are only found in phagocytic cells. Attachment of promastigotes to macrophages involves various receptors that are utilized during phagocytosis. In particular, complement receptor (CR) 1 and CR3 play major roles in the uptake of the parasite. Possible ligands include the C3 component of complement bound to the surface of promastigotes; a parasite surface protein known as gp63; and a lipophosphoglycan (LPG) found on the promastigote surface. Other macrophage receptors such as CR4, the fibronectin receptor, and the mannose receptor have been implicated in parasite binding and subsequent phagocytosis. As expected, the phagocytosed promastigotes initially enter into the endocytic pathway and the parasite-containing phagosome fuses with a lysosome. In contrast to the normal situation in which a pathogen is destroyed within the phagolysosome, *Leishmania* is resistant to the acidic pH and hydrolytic enzymes present in the phagolysosome. In addition, the parasite decreases the generation of reactive oxygen intermediates by the macrophage.

The promastigote converts to an **amastigote** within the phagolysosome by losing its flagellum and becoming more spherical. These amastigotes replicate by binary fission and fill up the infected macrophage (Figure 10.3). Amastigotes are released from the host cell and taken up by another macrophage leading to another round of replication. The release of the amastigotes is usually assumed to be due to the infected macrophage bursting open, but it has also been suggested that amastigote release may involve a fusion of the parasitophorous vacuolar membrane (i.e., the phagolysosomal membrane) with the host cell plasma membrane, and thus resembles an exocytosis-like process. These repeated rounds of replication within the macrophages will continue until the host spontaneously resolves the infection, is cured via chemotherapy, or dies. In some situations the macrophages remain at the initial site of inoculation resulting in a simple cutaneous lesions, whereas in other situations the infected macrophages metastasize to other locations in the body resulting in more serious disease (see Clinical Manifestations).

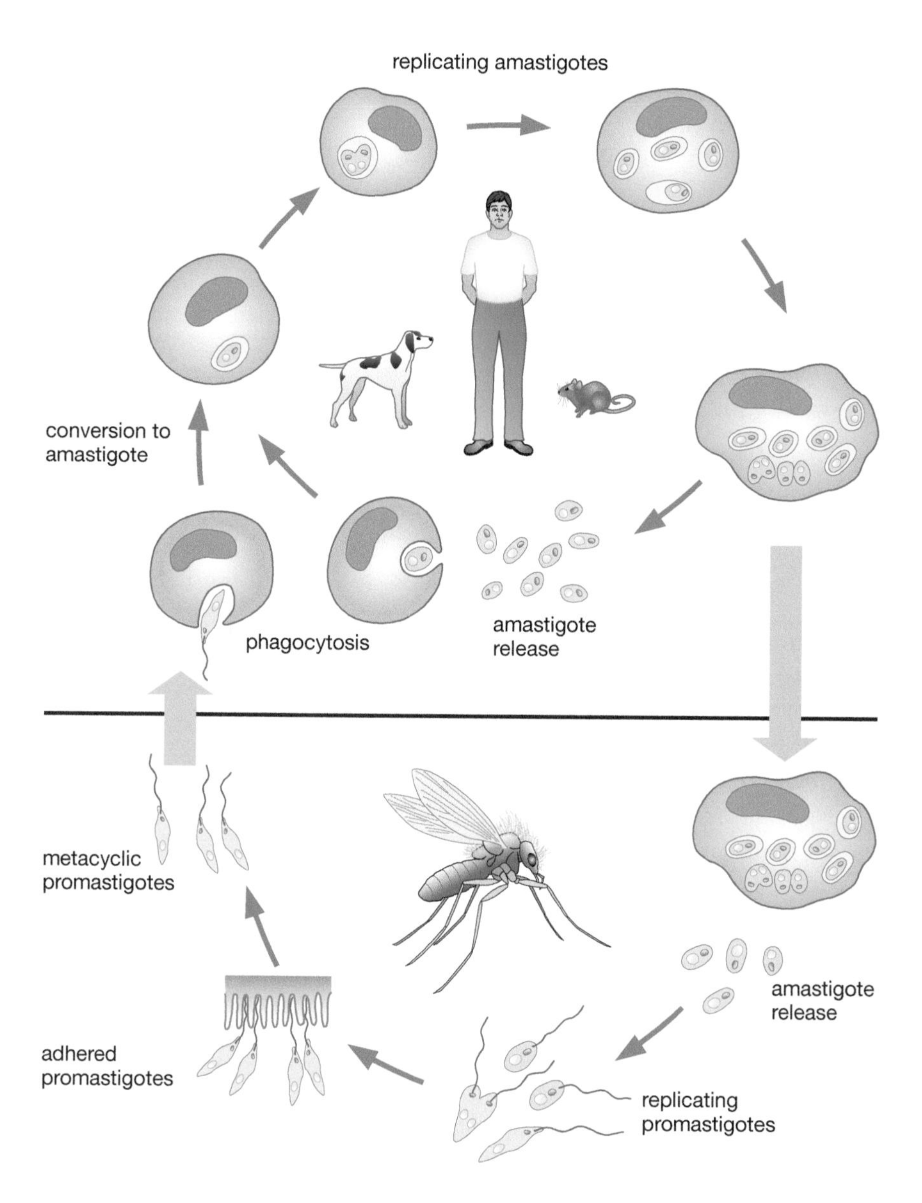

Figure 10.2 Life cycle of *Leishmania*.
During its life cycle *Leishmania* alternates between amastigotes which replicate within macrophages in the vertebrate host and promastigotes which replicate in the gut of the sand fly vector. Transmission between host and vector (block arrows) occurs while the sand fly is feeding on the vertebrate.

Leishmania promastigotes replicate and develop in the sand fly gut

Ingestion of infected macrophages by sand flies results in the release of amastigotes which convert to promastigotes (Figure 10.2). These extracellular promastigotes replicate by binary fission in the gut of the fly. In addition, the promastigotes differentiate through several morphological forms including forms that attach to the midgut epithelium of the fly. Attachment prevents excretion of the parasite with the feces and is mediated by

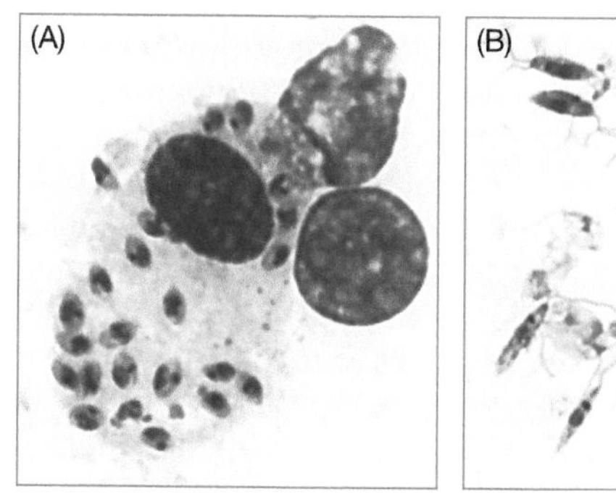

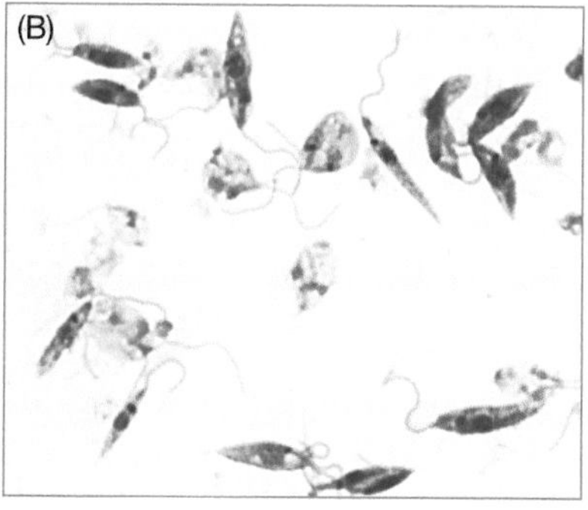

Figure 10.3 Morphological forms of *Leishmania* stained with Giemsa.
(A) Amastigotes within an infected macrophage. (B) Promastigotes from *in vitro* culture.

Box 10.1 The lipophosphoglycan of *Leishmania*

Lipophosphoglycan (LPG) is an abundant cell surface associated glycoconjugate that is a major component of the dense glycocalyx surrounding the parasite [1]. The LPG molecule is composed of four major domains: a glycolipid membrane anchor, a core glycan, phosphodisaccharide repeats, and a small oligosaccharide cap (Figure 10A). However, the precise structure of LPG varies according to species. The lipid anchor and the glycan core are identical in all species of *Leishmania*, whereas polymorphisms in regard to the specific carbohydrate groups occur in the cap and the repeating units. For example, the repeating units can be unbranched or contain a variety of oligosaccharide side chains substituted to various extents. These polymorphisms play a role in determining the range of vector species that can be infected by any particular *Leishmania* species because LPG interacts with specific lectins found on the epithelial cells of the sand fly gut.

LPG has several important functions throughout the life cycle of the parasite and in particular helps the parasite survive the relatively hostile environments of the sand fly gut, the mammalian bloodstream, and the phagolysosome of the macrophage. For example, LPG protects the promastigote from the digestive activities found in the sand fly gut as well as playing a role in adherence to the gut epithelium that prevents excretion of the parasite. During the process of metacyclogenesis, in which the promastigote converts from a noninfectious state to a highly infectious metacyclic form, LPG is remodeled. This remodeling can include an increase in the number of disaccharide repeats, changes in the repeating subunits, as well as changes to the side chains, or changes in the cap. Substitutions in the cap and side-chain carbohydrates can result in a detachment of the promastigote due to the specificity of the lectin activity associated with the gut epithelium.

The near doubling in the number of disaccharide repeat units leads to an elongation of the LPG which is associated with a thickening of the glycocalyx on the surface of the promastigote from 7 nm to 17 nm. Corresponding to this glycocalyx thickening is an increased complement (i.e., serum) resistance in metacyclic promastigotes as compared with other promastigote forms. LPG can also interact directly or indirectly with receptors on the macrophage and thus play a role in the phagocytosis of the parasite. But LPG is not absolutely required for phagocytosis and other promastigote surface molecules can substitute as the parasite ligand. However, within the phagolysosome LPG is known to be important in the ability of the parasite to withstand or shut down the antimicrobial activities of the macrophage [2]. Thus, LPG plays a role in the infectivity to the vertebrate host and contributes to the virulence of the parasite.

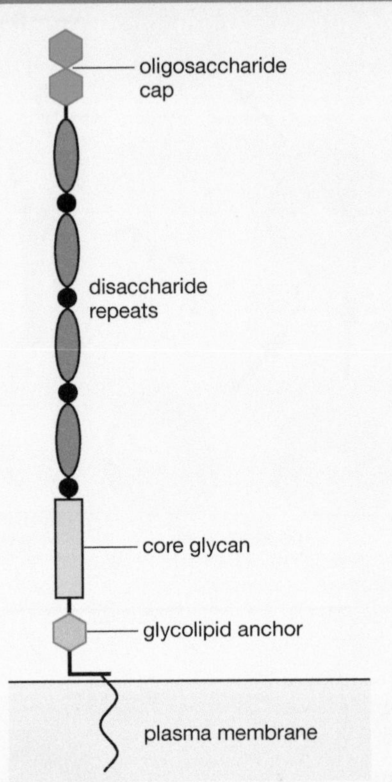

Figure 10A Structure of LPG. Schematic diagram illustrating the major domains of LPG.

1. Turco, S.J. and Descoteaux, A. (1992) The lipophosphoglycan of *Leishmania* parasites. *Ann. Rev. Microbiol.* 46: 65–94.

2. Descoteaux, A. and Turco, S.J. (2002) Functional aspects of *Leishmania donovani* lipophosphoglycan during macrophage function. *Microbes Infect.* 4: 975–981.

lipophosphoglycan (LPG) and other glycoproteins that cover the surface of the promastigote. The replicating and adhered promastigotes are noninfective to the mammalian host and need to differentiate to the nondividing metacyclic promastigotes. One step in this differentiation is the detachment of the promastigotes from the gut epithelium and migration to the anterior part of the gut and pharynx. Morphological and biochemical changes are also noted during this differentiation. For example, the metacyclic promastigotes are more slender and have longer flagella than the other forms. These changes result in a more motile parasite and are probably important for infection of the mammalian host. Changes in the promastigote surface, and especially in the LPG (Box 10.1), that lead to detachment of the promastigotes from the sand fly gut epithelium and prepare the parasite for life within the vertebrate host, have also been noted.

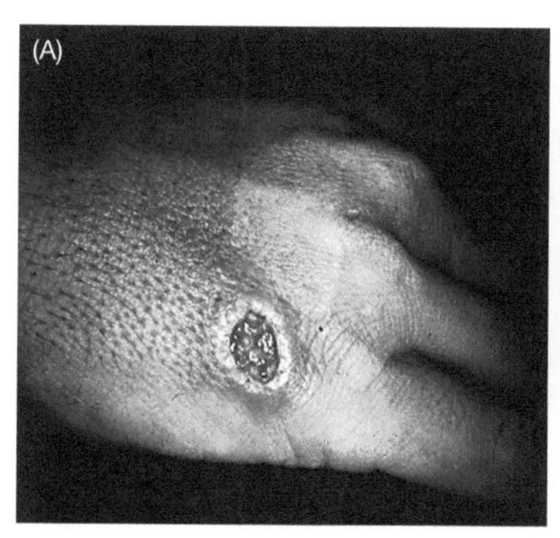

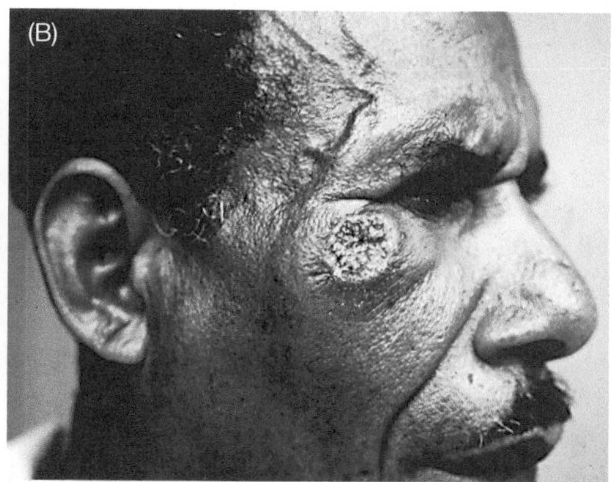

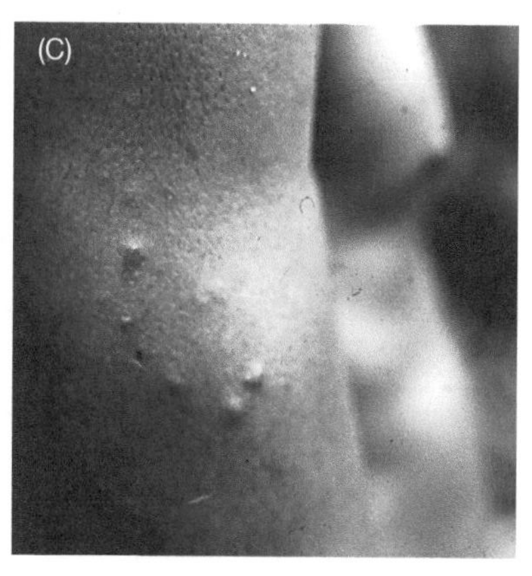

Figure 10.4 Examples of cutaneous leishmaniasis. (A) An ulcerated wet lesion, (B) a dry lesion, and (C) satellite papular lesions surrounding the scar of a healed lesion. (Photos kindly provided by Antonio D'Alessandro, Tulane University, Department of Tropical Medicine.)

Clinical Manifestations of Leishmaniasis

Infection with *Leishmania* can result in a wide range of clinical manifestations ranging from benign skin lesions to a fatal systemic disease. **Cutaneous leishmaniasis** (CL) is the most common manifestation of the disease and is characterized by benign self-healing lesions that are generally painless and nonpruritic. The lesions can be ulcerated with a raised border or exhibit a papular or nodular nature (Figure 10.4). Quite often there is a progression from papules, to nodules, and finally to ulcers. The ulcerated lesions can be classified as wet or dry depending on whether a crust forms over the lesion. Generally there is no pus associated with the wet lesions unless they become secondarily infected with bacteria. The lesion represents a localized infection at the site of the sand fly bite. The infection spreads outward from this point due to the continuing cycles of replication within the macrophages. The raised border of the lesion is where the active infection is occurring. Satellite lesions in the vicinity of the original lesion are sometimes observed. At some point the lesion will quit expanding and healing will start to take place. These lesions generally take weeks to months to heal. Scarring or pigmentation can be associated with the healed lesions.

Leishmaniasis **recidivans** is a rare form of relapsing cutaneous leishmaniasis associated with *L. tropica* infections. Recidiva is a chronic recurrence of nodular lesions or a rash that is generally found at the margins of a nearly healed lesion. Few amastigotes are detected and this condition is considered to be a hypersensitivity in which most parasites are eliminated but the infection is not completely cured.

Infected macrophages can metastasize and cause disseminated disease

Generally cutaneous leishmaniasis is characterized by localized lesions near the site of the original infection. However, infected macrophages can on occasion spread to other locations. Two examples of disseminated disease are **diffuse cutaneous leishmaniasis** (DCL) and **mucocutaneous leishmaniasis** (MCL). DCL is characterized by disseminated nodular lesions that resemble lepromatous leprosy (Figure 10.5). These lesions tend to be scaly and not ulcerated and can metastasize distally to cover large areas of the body. This condition is generally associated with anergy and amastigotes are abundant in the nodules.

MCL is primarily due to members of the *L. braziliensis* complex. A small proportion (<5%) of patients with simple cutaneous leishmaniasis will develop MCL. Mucocutaneous disease begins as simple skin lesions that metastasize via the bloodstream or lymphatics, particularly to the mucosae of the nose and mouth. The expression of this form of the disease can occur while the primary lesion is still active or can occur several years after the primary lesion has healed. Two distinct forms of MCL are recognized (Figure

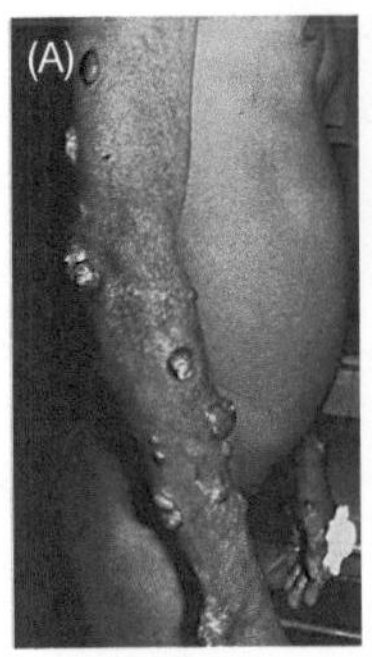

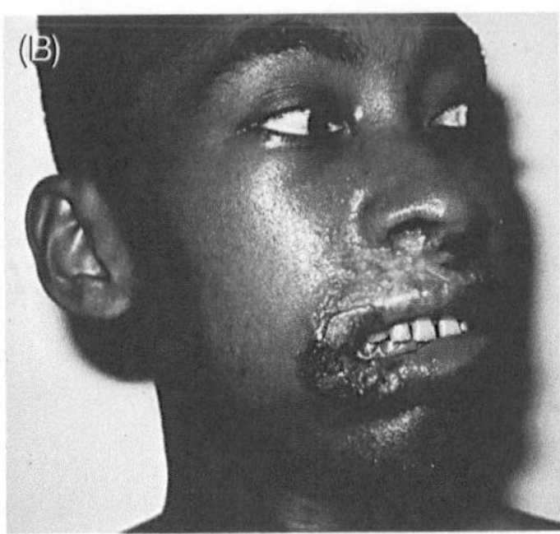

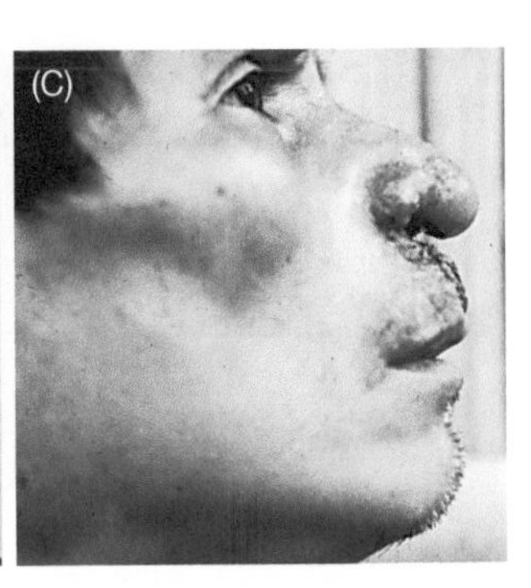

Figure 10.5 Disseminated forms of cutaneous leishmaniasis. (A) DCL is characterized by nodules all over the body. MCL is characterized by lesions around the nose and mouth that can be ulcerative with extensive mutilation (B), destroy the cartilage of the nose (C), or characterized by edema of the upper lip (C). (Photos kindly provided by Antonio D'Alessandro, Tulane University, Department of Tropical Medicine. Photo B reprinted from *Clinical Parasitology* with kind permission from Wolters Kluwer Publishers.)

10.5). One is an ulcerative disease with a rapid and extensive mutilation of soft tissue and cartilage. The other is nonulcerative and characterized by local edema and hypertrophy, particularly of the upper lip. This disease will generally continue to progress and can lead to severe pathology and deformity if not treated.

Visceral leishmaniasis (VL) is the most serious form of the disease and is usually fatal if not treated. Kala-azar, meaning black sickness, is a widely used Hindi name for the disease in reference to a darkening of the forehead and hands that is sometimes observed. However, this clinical presentation is rarely observed now and is perhaps a feature of the disease in an era of prolonged illness in the absence of effective treatment. Visceral leishmaniasis is a generalized infection of the reticuloendothelial system (RES) involving the spleen, liver, bone marrow, and lymph nodes. The disease is characterized by fever, spleno- and hepatomegaly, and enlarged lymph nodes and tonsils. Fever is a prominent symptom associated with VL and the fever may be sudden with rigors or gradually rising. Continuous, intermittent, or remittent fevers lasting for days to weeks or even months have been observed. Other possible clinical manifestations include edema, anemia, leucopenia, monocytosis, lymphocytosis, and thrombocytopenia. As the disease progresses, malaise, tiredness, lassitude, and weakness become more prominent and patients often exhibit a wasting syndrome despite good appetite. A classic picture associated with a protracted disease is a protuberant abdomen and muscle wasting of the limbs and thorax (Figure 10.6). Symptoms and signs of bacterial co-infections, such as pneumonia, diarrhea, or tuberculosis, can also confound the clinical presentation.

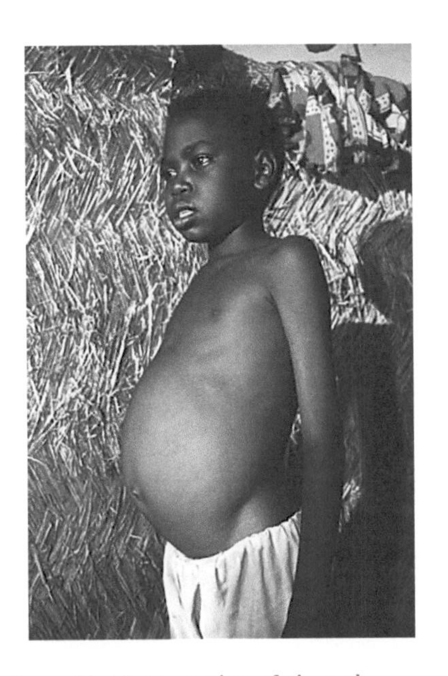

Figure 10.6 Presentation of visceral leishmaniasis. Profile view of a teenage boy suffering from visceral leishmaniasis exhibiting splenomegaly, distended abdomen, and muscle wasting. (Photo kindly provided by WHO/TDR/Crump—TDR Image Library.)

Visceral leishmaniasis is usually considered to have a high case fatality rate of up to 90% in untreated cases. However, some outbreaks of *L. infantum* have been associated with relatively low numbers of fatalities and serious disease, but high levels of positive skin tests (see Diagnosis). The positive skin test indicates exposure to the parasite but no development of disease. Similarly, high levels of skin test positivity with little or no human disease have been observed in foci of canine leishmaniasis. The situation may be similar for *L. donovani* in that skin-test-positive villagers without apparent disease have been detected in villages of Kenya and Ethiopia in both endemic and epidemic situations.

Disease manifestations depend partly on the species of parasite

More than 20 different *Leishmania* species have been described as infecting humans and many are considered to be zoonotic. The species have defined geographical ranges and tend to exhibit certain types of disease (Table 10.1). The defined geographical ranges are determined for the most part by the particular species of sand fly transmitting the parasite. There is also a tendency for each parasite species to exhibit a particular disease manifestation or range of manifestations. For example, in the New World *L. mexicana* infection tends to only cause simple CL, whereas *L. braziliensis* is capable of metastasis and causing the more serious MCL in addition to CL.

Table 10.1 Major types of Leishmania species infecting humans

Species	Clinical manifestations	Geographical distribution
L. mexicana complex	CL and rare cases of DCL	Central America, northern and central S. America, very rare in southern USA
L. braziliensis complex	CL with some cases developing MCL later	Central and South America
L. major	CL (wet type)	Northern and central Africa, Middle East, southern Asia
L. tropica	CL (dry type) and rare cases of recidiva	Middle East and southern Asia
L. aethiopica	CL and some cases of DCL	Highlands of Ethiopia and Kenya
L. donovani	VL with rare cases of post-kala-azar CL	East Africa, sub-Sahara, southern Asia including India and Iran
*L. infantum**	VL or CL according to strain	North Africa and southern Europe
*L. chagasi**	VL or some CL	Foci in Brazil, Venezuela, and Colombia. Isolated cases throughout S. America

CL, cutaneous leishmaniasis; DCL, diffuse cutaneous leishmaniasis; MCL, mucocutaneous leishmaniasis; VL, visceral leishmaniasis. *Possibly the same species introduced to the New World by the European colonists (Box 10.2).

L. infantum causes both cutaneous and visceral disease. Analysis of zymodemes (i.e., isoenzymes), however, indicates the presence of viscerotropic and dermotropic strains. Interestingly, dermotropic strains will frequently cause visceral disease in AIDS patients. Because the various *Leishmania* species cannot be distinguished based on morphological criteria with the light microscope, geographical and clinical criteria are often used to identify *Leishmania* species. However, positive species identification often requires immunological, molecular, or biochemical criteria.

Host responses also contribute to the disease manifestations

Both host and parasite factors are involved in disease outcome. Leishmaniasis can be viewed as a polar disease ranging from a lethal visceral disease to a self-curing cutaneous disease. Studies in rodents indicate that the host immune response plays an important role in determining which of these extremes is expressed. Mouse strains resistant to *Leishmania* mount a strong Th1, or cellular, response while susceptible mouse strains mount a predominant Th2, or humoral, response. The immunological aspects of human disease outcomes are more complex and less polarized than those in mice. However, there does tend to be a correlation with the differential expression of Th1 or Th2 immune responses (Figure 10.7). For example, localized self-healing cutaneous lesions have few parasites and the patient exhibits a positive leishmanin skin test (see Diagnosis) and usually has low antibody levels. The skin test reflects a delayed hypersensitivity (i.e., Th1 response). At the other extreme are DCL and visceral leishmaniasis, which are characterized by high parasitemias, a negative skin test, and high antibody titers (i.e., Th2 response). MCL and recidivans exhibit both Th1 and Th2 responses. These lesions are slow to heal or fail to heal even though parasites are difficult to detect.

New World Cutaneous Leishmaniasis

Foci of new world cutaneous leishmaniasis are found from southern Texas in the United States to northern Argentina. These various foci represent unique combinations of vectors, reservoir hosts, and humans. In general,

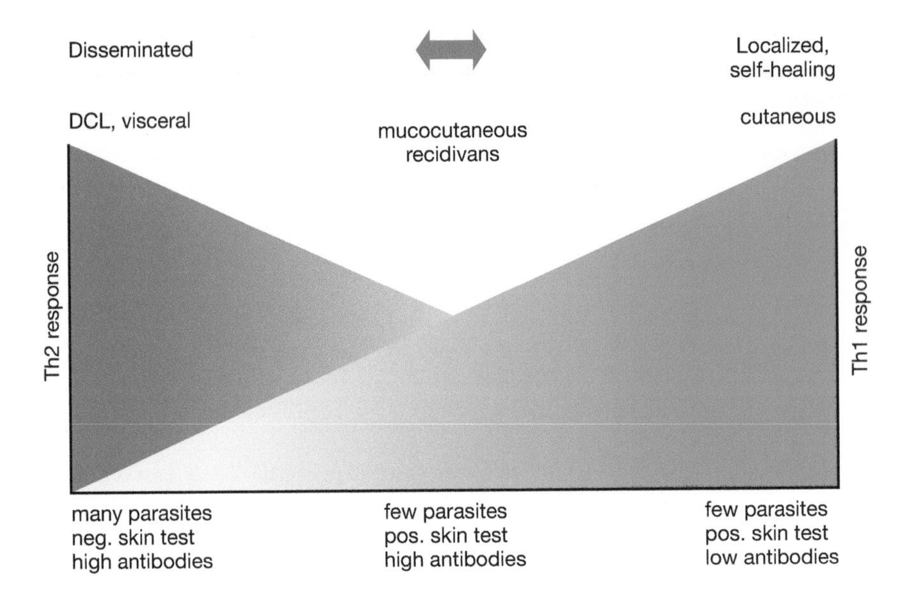

Figure 10.7 Correlation of immunological responses and disease outcome in leishmaniasis. Th1 immune responses are associated with localized self-healing lesions whereas Th2 responses are associated with disseminated disease.

leishmaniases in the Americas are sylvatic zoonoses due to humans entering the forest or other areas where the natural life cycle involving the reservoir host and vector is maintained. Two species complexes are recognized: *L. mexicana* and *L. braziliensis.* These two species complexes are also assigned to different subgenera. Members of the *L. braziliensis* complex are in the subgenus *Viannia* and *L. mexicana* subspecies, along with all other *Leishmania* species infecting humans, are in the subgenus *Leishmania.* These two subgenera are defined by differences in their vector stages and other biological differences (Table 10.2). The species within the subgenus *Leishmania* are confined to the mid- and foregut of the vector; whereas members of the *L. braziliensis* complex exhibit replicative stages in the hindgut of the vector which subsequently migrate through the midgut to the foregut. Species within the *Leishmania* subgenera also generally grow rapidly in culture and produce extensive lesions when injected into hamsters. This is in contrast to the *Viannia* subgenus which grows slowly in culture and results in slowly developing lesions in hamsters.

The two most common species within the mexicana complex are *L. mexicana* and *L. amazonensis. L. mexicana* is found in Mexico, especially the Yucatan peninsula, Belize, and Guatemala. It is generally associated with a dry scrub forest and rodents are the usual reservoir. It primarily causes a single self-healing cutaneous lesion and rarely leads to DCL. Lesions on the ear leading to a localized destruction of the cartilage, or the chiclero ulcer, are common. The name chiclero ulcer is in reference to the common occurrence of the disease among the gum (i.e., chicle) harvesters who work in the forests. *L. amazonensis* is found primarily in the Amazon basin and

Table 10.2 Subgenera of *Leishmania*

Subgenus *Leishmania* (*L. mexicana* complex + others)	Subgenus *Viannia* (*L. braziliensis* complex)
Species include: *L. mexicana*, *L. amazonensis*	Species include: *L. braziliensis*, *L. guyanensis*, *L. panamensis*, *L. peruviana*
Rapid *in vitro* growth	Slow *in vitro* growth
Large lesions in hamsters	Slowly developing lesions in hamsters
Development confined to foregut of vector	Development in hindgut and foregut of vector

primarily causes simple CL. However, it results in multiple lesions and DCL more frequently than *L. mexicana*. Reservoirs are arboreal rodents and marsupials.

Most of the species in the *L. braziliensis* species complex are associated with lowland rain forests of South and Central America. Arboreal rodents, sloths, primates, and marsupials serve as reservoirs. *L. braziliensis* is the most likely species to involve the disfiguring mucocutaneous form of the disease which has been called espundia. This species is generally associated with the rain forests of Brazil and adjacent areas but has been found throughout South and Central America. *L. guyanensis* is found in Guyana and surrounding areas and causes pian bois (forest yaws). This disease is characterized by a high frequency of multiple ulcers due to metastasis. *L. panamensis* is found primarily in Panama, Colombia, and Costa Rica. It is quite similar to *L. guyanensis* in regards to the disease it produces as well as its biology and biochemical and molecular markers. In contrast to the other *L. braziliensis* complex species, *L. peruviana* is associated with dry highlands on the western slopes of the Andes in Peru. It produces a relatively mild disease with self-healing lesions called uta. Transmission is frequently found in villages and dogs are often the reservoir host. This disease has been known for a long time in that depictions of the cutaneous lesions can be found on artifacts dating to the Peruvian Nazca period (200 BC–600 AD).

Old World Cutaneous Leishmaniasis

Foci of old world cutaneous leishmaniasis are found throughout Asia, Africa, and southern Europe. The four species causing cutaneous leishmaniasis are distinguished by geographical distribution, reservoir hosts, and the types of lesions. *L. major* exhibits a patchy distribution in parts of India, central Asia, the Middle East, and northern and central Africa. Transmission is generally in dry or desert rural areas and rodents are the primary reservoirs. The lesions are described as being the "wet type" and are characterized by an acute rapidly ulcerating lesion with a lot of inflammation. Healing generally occurs in 6–12 months.

In contrast, *L. tropica* causes a chronic "dry type" of lesion and is sometimes referred to as oriental sore. It begins as a small papule and progresses to a nodular lesion. The lesion may ulcerate with a crust covering the lesion and upon healing a pigmented scar may form. It is found primarily in urban areas in the Middle East and southern Asia with major foci in Afghanistan, Iran, and Syria. *L. tropica* typically exhibits an anthroponotic transmission. In some settings, though, dogs might be important reservoirs. Generally the lesions take 1–2 years to heal and on rare occasions the infection can develop into the chronic nonhealing recidivans type of disease. *L. tropica* infections tend to occur in outbreaks lasting a relatively short time because infection is self-curing and induces immunity. Therefore, only children and immigrants will be susceptible to infection and thus an anthroponotic transmission cycle will be difficult to maintain. Seasonal weather changes resulting in larger than normal sand fly populations or human migration are factors associated with epidemics.

L. aethiopica is restricted to the highlands of Ethiopia and Kenya. The reservoir host is the rock hyrax, a guinea-pig-sized mammal found in dry savannas of eastern Africa. It generally causes a simple cutaneous disease that on occasion progresses to DCL characterized by nodular lesions that resemble leprosy.

L. infantum is found in southern Europe and northern Africa. Analysis of zymodemes (i.e., isoenzymes) indicates the presence of viscerotropic and dermotropic strains. In addition the immune status of the host affects disease outcome. For example, visceral disease is found more often in children whereas adults tend to exhibit a cutaneous disease. Similarly, AIDS patients will exhibit a visceral disease.

Box 10.2 *L. chagasi = L. infantum?*

The three *Leishmania* species causing visceral leishmaniasis are all quite similar and generally considered part of the *L. donovani* complex. Various types of genetic and biochemical analyses indicate *L. chagasi* and *L. infantum* are closely related and do not form separate clades [1]. Therefore, *L. chagasi* is now considered to be *L. infantum* by many investigators. However, the exact taxonomic status of *L. chagasi* is being debated [2]. Some data suggest that the parasite was introduced to the Americas via infected dogs or rats with the arrival of the Spanish and Portuguese colonists. Furthermore, *L. infantum* infects *Lutzomyia* as effectively as *Phlebotomus* indicating that a rapid transition to New World vectors is possible.

1. Maurício, I.L., Stothard, J.R. and Miles, M.A. (2000) The strange case of *Leishmania chagasi*. *Parasitol. Today* 16: 188–189.

2. Dantas-Torres, F. (2006) Final comments on an interesting taxonomic dilemma: *Leishmania infantum* versus *Leishmania infantum chagasi*. *Mem. Inst. Oswaldo Cruz* 101: 929–930.

Visceral Leishmaniasis

Infection of macrophages in the **reticuloendothelial system**, especially the spleen, liver, bone marrow, and lymph nodes, results in a slowly progressing systemic infection that is often fatal. The species most often associated with visceral disease are *L. donovani*, *L. infantum*, and *L. chagasi*. *L. donovani* is found in India and foci in East Africa. In India and East Africa the disease is primarily anthroponotic, whereas in Sudan it might be a zoonosis with dogs or wild animals as the reservoir hosts. *L. infantum* occurs primarily around the Mediterranean region and the Middle East. The domestic dog is the principal reservoir in most areas. However, rodents may also serve as reservoir hosts. *L. chagasi* is found in foci throughout South and Central America and especially in Brazil. Domestic dogs have been implicated as reservoirs as well as wild foxes. *L. chagasi* is probably the same as *L. infantum* and was introduced to the New World by European colonists (see Box 10.2).

Recently, a significant number of cases of cutaneous leishmaniasis due to *L. donovani* have been described in India and Sri Lanka. In addition, *L. donovani* can also cause a cutaneous disease known as post-kala-azar dermal leishmaniasis (PKDL). Nodular lesions are often first localized around the mouth and become more generalized. In Sudan approximately half of the patients diagnosed with visceral leishmaniasis will manifest PKDL. These lesions will generally self-cure within a year without additional treatment and a lengthy re-treatment is generally needed to clear the dermal lesions. In India, 5–10% of patients will develop PKDL several years after an apparently successful treatment of visceral leishmaniasis and it is believed that inadequate treatment may contribute to the PKDL. Although small in numbers, PKDL patients are of potential epidemiological importance in that the presence of numerous parasites in the skin makes them readily accessible to sand flies, and thus facilitates the anthroponotic transmission cycle.

Diagnosis

Diagnosis in endemic areas can be based on clinical features. However, definitive diagnosis depends on detecting amastigotes in clinical specimens or culturing the promastigotes (Table 10.3). The ability to detect parasites will depend on the abundance and distribution of amastigotes and whether the lesion is active or healing. In the case of cutaneous leishmaniasis it is critical to collect the specimen from the border of the lesion where the infection is most active (Figure 10.8). An incision is made in the raised border of the ulcer and material scraped from the incision is examined microscopically. Alternatively, a biopsy can be taken from the lesion. The biopsy can be used to prepare impression or smears or sectioned. The lesion can also be aspirated by injecting sterile saline under the skin and withdrawing this fluid. The aspirate is then used to inoculate *in vitro* cultures. *In vitro* culture generally takes a minimum of 1 week. Inoculation of hamsters with the clinical specimens can also be carried out, but may take 2–3 months before the animals become positive.

Examination of an aspirate or biopsy from bone marrow is the method of choice for diagnosis of visceral leishmaniasis. Aspirates or biopsies from lymph nodes or the spleen can also be used. Detection of amastigotes from visceral cases is sometimes easier than cutaneous cases since the parasitemias are generally higher in visceral leishmaniasis. The aspirates from microscopy-negative samples can also be used to inoculate *in vitro* cultures to increase sensitivity.

Serology or intradermal skin tests are also used in diagnosis

Serological tests include direct agglutination, immunofluorescence, and ELISA using whole promastigotes or promastigote extracts. Among these,

Table 10.3 Diagnosis of leishmaniasis

	Dermal	Visceral
Suspected	Geographical presence of parasite	Geographical presence of parasite
	History of sand fly bite	History of sand fly bite
Clinical symptoms	Chronic, painless, skin lesion (CL), nasopharyngeal lesions (MCL), or nodular lesions (DL)	Prolonged fever, splenomegaly, hepatomegaly, anemia, leucopenia, thrombocytopenia
Parasite detection	Amastigotes in scrapings or aspirates	Amastigotes in bone marrow aspirate or biopsy
	Culture of promastigotes	Culture of promastigotes
	Inoculation of hamsters	
Immuno-diagnosis	Leishmanin (Montenegro) skin test	Serological tests (direct agglutination, K39 dipstick)

the direct agglutination is the most readily adapted to field conditions. One limitation of serological tests is the persistence of antibody long after the infection has been cured and the resulting inability to distinguish past infections from current infections. In addition, serology is most useful for the diagnosis of visceral leishmaniasis because cutaneous leishmaniasis is often characterized by low antibody responses. More recently an abundant amastigote antigen referred to as K39 has been developed into an immunochromatographic strip test for the diagnosis of *L. donovani* and *L. infantum/chagasi.* K39 contains a 39 amino acid repeat and infection with members of the *L. donovani* complex leads to extraordinarily high antibody titers against this repeat. This allows for a rapid, sensitive, and specific test for visceral leishmaniasis from finger-stick blood. Furthermore, the antibody titers drop in subclinical and healing infections indicating that the dipstick test can distinguish active infections from past infections.

An intradermal skin test, often called the **leishmanin skin test**, can also be used in the diagnosis of leishmaniasis. In Latin America this test is often called the Montenegro skin test in reference to the developer. Classically the

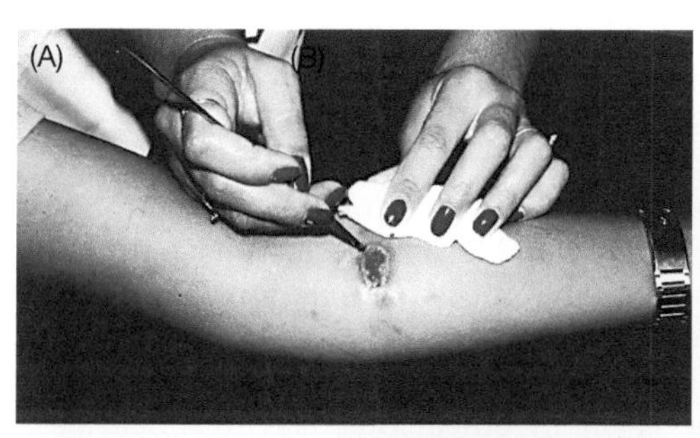

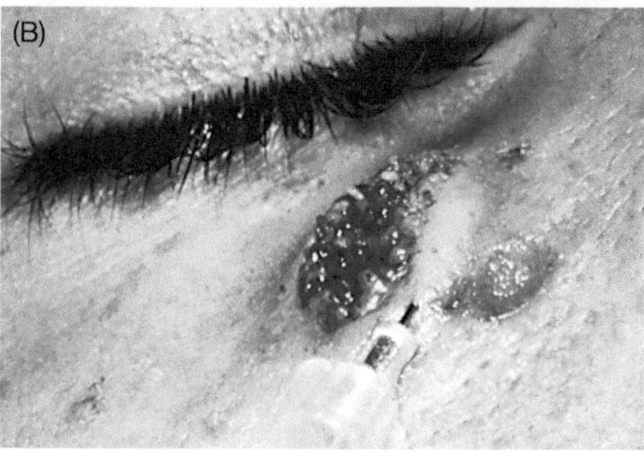

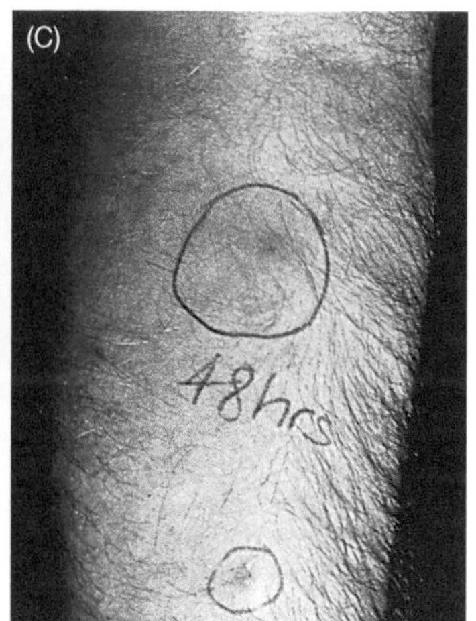

Figure 10.8 Diagnosis of cutaneous leishmaniasis. (A) An incision along the raised border is made and dermal scrapings are examined by Giemsa staining and microscopy. (B) Sterile saline is injected into the raised border and fluid is aspirated. The aspirate can be examined microscopically or preferably used to inoculate *in vitro* cultures or hamsters. Note the scar of the poorly positioned biopsy punch. Biopsies should be taken at the raised border of the lesion. (C) Montenegro delayed-type hypersensitivity reaction. (Photos kindly provided by Antonio D'Alessandro, Tulane University, Department of Tropical Medicine.)

test has been used in epidemiological studies for detection of exposure to *Leishmania.* However, the skin test can be useful in the diagnosis of cutaneous and mucocutaneous disease. The procedure involves the intradermal injection of whole killed promastigotes or a promastigote extract and monitoring for a local delayed-type hypersensitivity reaction. Within 48–72 hours an induration with some erythema will develop at the inoculation site if the patient has been exposed to *Leishmania.* A positive or negative skin test can be used to confirm or rule out *Leishmania* as the etiological agent of a cutaneous lesion. As with many serological tests, the skin test cannot distinguish current infections from past infections. The skin test can also be used as a prognostic indicator following chemotherapy of visceral leishmaniasis. During acute visceral leishmaniasis patients are skin test negative, but convert to positive if therapy is successful. This reflects a conversion from a predominantly Th2 response to a Th1 response as discussed above.

Treatment

A limited number of drugs are available for treatment of leishmaniasis and many of these have toxicity problems and require parenteral administration and a long duration of therapy. Most simple cutaneous lesions will self-heal over a few months and lesions in the process of healing can simply be observed. Treatment of cutaneous leishmaniasis is used to accelerate healing, reduce scar formation, prevent relapses, or to prevent dissemination such as mucocutaneous, diffuse or visceral diseases. Especially large or persistent (>6 months) lesions, multiple lesions, or lesions over joints should be treated, as well as disseminated forms of the disease (i.e., visceral, mucocutaneous, diffuse). Mucocutaneous and diffuse leishmaniases are notably unresponsive to treatment.

Pentavalent antimonials—either sodium stibogluconate (Penostam®) or meglumine antimoniate (Glucantime®)—are usually the first line of treatment. These drugs have been used for more than 60 years and are generally effective. However, they require parenteral administration and a long duration of therapy. Toxic side effects, such as pancreatitis, hepatitis, myalgias, fatigue, headache, rash, and nausea, are also common. In addition, treatment failures have been observed over recent years suggesting the emergence of drug resistance. This is particularly true in certain areas of India where widespread resistance to antimonials has resulted in these drugs no longer being effective. Pentamidine (Pentacarinat®) can be used in cases of treatment failures with pentavalent antimonials.

Amphotericin B (Fungizone®) is an alternative to the pentavalent antimonials. It is effective, but does exhibit toxicity and also requires parenteral administration and a long duration of treatment. It is also more expensive than the pentavalent antimonials. Lipid formulations of amphotericin B have proven to be effective against visceral leishmaniasis with fewer toxic effects due to the lower doses. However, these formulations are prohibitively expensive for use in developing countries. An oral drug, miltefosine (Impavido®), for the treatment of visceral leishmaniasis has recently been registered in India. Miltefosine is highly efficacious, less toxic, easier to administer, and will probably cost less than current therapies since hospitalization will not be required.

Control

Control activities are important in that the geographical range and incidence of leishmaniasis appears to be increasing (Box 10.3). Specific control activities will depend on the local transmission conditions. In general, the exposure to sand fly bites needs to be reduced. Personal protection activities can include: protective clothing, insect repellents, bed nets, and insecticide spraying in domiciliary and peridomiciliary transmission settings.

Box 10.3 Emerging and re-emerging disease

Leishmaniasis can be considered an emerging or re-emerging disease in that its geographical range is expanding and there appears to be an increasing incidence of disease. Factors underlying these changes are difficult to identify but may be related to global warming or changes in the behaviors of humans, vectors, and reservoirs. Climatic changes may allow an extension of breeding seasons for existing vectors or the establishment of new vector species. For example, before 1990 northern Italy was almost free of sand flies and canine leishmaniasis. Since then the incidence of canine leishmaniasis has increased and several species of *Phlebotomus* are now found in northern Italy [1]. In addition, there appears to be a northward expansion of *L. infantum* in other parts of Europe.

Human factors affecting the endemicity of leishmaniasis include migration, deforestation, urbanization, and an increase in susceptibility to infection due to immunosuppression (e.g., AIDS) and malnutrition. Urbanization in South America is leading to an increase in leishmaniasis [2]. Suburbs constructed on the outskirts of primary forests bring people into contact with the sylvatic cycle. In addition, these suburbs are often shanty towns where sand flies and dogs are common, sanitation is poor, and malnutrition is everywhere. Furthermore, the deforestation associated with this urbanization often leads to a domestication of the transmission cycle through adaptation of the vector and parasite to new reservoir hosts such as rodents and dogs. Accordingly, an increase in the peri- and intradomiciliary transmission is being observed throughout Latin America, whereas previously working in the forest was the primary risk factor. Similar increases in the suburban transmission of leishmaniasis in the Old World have also been noted.

1. Gramiccia, M. and Gradoni, L. (2005) The current status of zoonotic leishmaniases and approaches to disease control. *Int. J. Parasitol.* 35: 1169–1180.

2. Desjeux, P. (2001) The increase in risk factors for leishmaniasis worldwide. *Trans. R. Soc. Trop. Med. Hyg.* 95: 239–243.

For example, bed nets have proven effective in preventing anthroponotic transmission in situations where sand flies are biting within the houses. *L. tropica* has been eradicated in Azerbaijan in the former Soviet Union through a campaign involving insecticide treatment of houses.

In a few situations destruction of stray dogs and decreasing this potential reservoir has been effective in controlling leishmaniasis. However, in most other situations elimination of infected dogs is not effective at controlling leishmaniasis. Some studies suggest that deltamethrin impregnated dog collars or topical treatment with permethrin reduces the incidence of canine leishmaniasis and leads to a subsequent drop in human leishmaniasis. *L. major* infections have been controlled in Uzbekistan and Turkmenistan through poisoning the rodent reservoirs (*Meriones* sp.) and destroying their burrows. However, rodent control in other areas involving other species may not be as effective or economically feasible

Currently no antileishmanial vaccine is available. Historically vaccination by inducing the disease in a controlled manner has been practiced for the prevention of *L. tropica* in the Middle East and eastern Europe. The procedure involves either the encouragement of sand fly bites or to inoculate with living material obtained from an active lesion on a normally unexposed area of the skin such as the buttocks. The resulting lesion would heal spontaneously and prevent cutaneous leishmaniasis on the face or other exposed areas. Although capable of providing lasting immunity, it is not recommended because of the risk of developing recidiva or visceral leishmaniasis. Efforts to develop a vaccine based on killed promastigotes or proteins are underway. Most studies indicate that vaccination does lead to conversion to leishmanin skin test positive, which is generally associated with a lower incidence of disease.

Summary and Key Concepts

- Human leishmaniasis exhibits a focal distribution in South and Central America, north and central Africa, the Mediterranean and southern Europe, the Middle East, and Asia.
- A wide range of disease manifestations are associated with infection.

- Humans are infected by several distinct species of *Leishmania* with defined geographical ranges, specific vector species and reservoir hosts, and a propensity to cause a particular type of disease.
- The most common disease manifestation is a chronic skin lesion that generally heals in a few months.
- Simple cutaneous leishmaniasis can develop into a disfiguring mucocutaneous leishmaniasis or a diffuse cutaneous leishmaniasis.
- The most lethal form of the disease is a systemic infection of the reticuloendothelial system known as visceral leishmaniasis.
- The infection is transmitted by sand flies of the genus *Lutzoymia* in the New World and *Phlebotomus* in the Old World.
- The parasite replicates as amastigotes within macrophages within the vertebrate host and as promastigotes within the gut of the sand fly.
- The epidemiology of transmission and control measures depend on the complex dynamics of the local *Leishmania* species, vector species, and reservoir.
- Diagnosis is made based on clinical symptoms, detection of the parasite, and/or immunological methods.
- A limited number of drugs are currently available for treatment with pentavalent antimonials generally being the first line of treatment.

Further Reading

Bates, P.A. (2007) Transmission of *Leishmania* metacyclic promastigotes by phlebotomine sand flies. *Int. J. Parasitol.* 37: 1097–1106.

Chappuis, F., Sundar, S., Hailu, A., Ghalib, H., Rijal, S., Peeling, R.W., Alvar, J. and Boelaert, M. (2007) Visceral leishmaniasis: what are the needs for diagnosis, treatment and control? *Nat. Rev. Microbiol.* 5: S7–S16.

Handman, E. and Bullen, D.V.R. (2002) Interaction of *Leishmania* with the host macrophage. *Tr. Parasitol* 18: 332–334.

Herwaldt, B.L. (1999) Leishmaniasis. *Lancet* 354: 1191–1199.

Murray, H.W., Berman, J.D., Davies, C.R. and Saravia, N.G. (2005) Advances in leishmaniasis. *Lancet* 366: 1561–1577.

Reithinger, R., Dujardin, J.C., Louzir, H., Pirmez, C., Alexander, B. and Brooker, S. (2007) Cutaneous leishmaniasis. *Lancet Infect. Dis.* 7: 581–596.

General Apicomplexan Biology

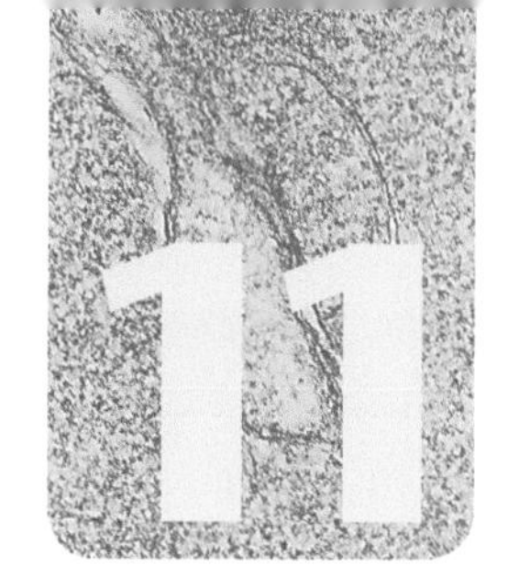

The apicomplexa are an extremely large and diverse group (>5000 named species). Seven genera infect humans (Table 11.1) and several other apicomplexan species are important pathogens of livestock. Most notable among the apicomplexan parasites that are important in terms of veterinary medicine and agriculture are *Babesia* and *Theileria* in cattle and *Eimeria* in poultry. *Plasmodium* species, as the causative agent of malaria, probably have the greatest impact on human health of any protozoan pathogen. Most of the medically important Apicomplexa are transmitted via an oral route and some cause intestinal disease. Many of these parasites, most notably *Cryptosporidium* and *Toxoplasma*, generally cause benign self-limiting diseases if the host has an intact immune system, but can cause serious and even life-threatening disease in immunocompromised patients. The AIDS epidemic has resulted in many of these organisms, which were previously considered to be rare and exotic infections, becoming prominent opportunistic infections of humans.

Apicomplexa, along with ciliates and dinoflagellates, form a higher order group known as Alveolata. The apicomplexa probably evolved from predatory flagellates and like many dinoflagellates have retained a chloroplast remnant called the apicoplast (Box 11.1). A major defining characteristic of the **alveolates** are the cortical alveoli, which are membrane-bound vesicles that are found just underneath the plasma membrane. In the apicomplexa these vesicles are flattened and give the appearance of a tri-layered pellicle and the alveolar membranes are often called the **inner membrane complex**. The Apicomplexa are a monophyletic group composed almost entirely of parasitic species. At some point during their life cycle, members of the Apicomplexa either invade or attach to host cells. This interaction between the parasite and host cell is mediated by unique organelles localized to one end of the parasite. These "apical organelles" are the defining characteristic of the Apicomplexa. The apical organelles are only evident by electron microscopy, and thus this group was not defined until after the application of electron microscopy to the study of protozoa during the 1960s. Formerly the apicomplexa were part of a group called sporozoa and this name is sometimes still used.

Table 11.1 Apicomplexan genera infecting humans

Genera	Transmission	Disease or comment
Plasmodium	Mosquito	Malaria
Cryptosporidium	Fecal–oral	Watery diarrhea
Isospora	Soil	Watery diarrhea
Cyclospora	Soil	Watery diarrhea
Toxoplasma	Felines are definitive host	Neurological manifestations
Sarcocystis	Predator–prey	Extremely rare infection
Babesia	Tick	Rare zoonotic disease

Box 11.1 Algal origins of the Apicomplexa

Historically the apicomplexa have been described as a group with only parasitic forms. This and their unique apical organelles raise questions in regard to the origin of the group. Phylogenetic analysis indicates that members of the genus *Colpodella* form a sister group with the apicomplexa [1]. The colpodellids are predatory flagellates that feed on unicellular algae by a process called myzocytosis. Myzocytosis involves the predator (or parasite) attaching to the prey (or host) and literally sucking out the cytoplasm of the prey cell via specialized structures. In these predatory apicomplexans, this attachment and interaction with the prey cell are mediated by organelles similar to those that are utilized by the parasitic apicomplexans for attachment to or invasion of host cells. Thus the evolution of the apicomplexa likely evolved from this myzocytotic predation to myzocytotic parasitism, as exhibited by gregarines and *Cryptosporidium* (see Box 12.1), to intracellular parasitism.

Other myzocytotic organisms with apicomplexan-like apical organelles include *Perkinsus*, parasites of oysters and clams, and *Parvilucifera*, a predator of dinoflagellates. These perkinsids, however, form a sister group with the dinoflagellates and not the apicomplexa (Figure 11A). This suggests that the progenitor of dinoflagellate and apicomplexan clades may have been a predatory flagellate and that the apical organelles were retained in the apicomplexan clade, but lost in most members of the dinoflagellate clade.

The other connection between algae and the apicomplexa is a chloroplast remnant, called the apicoplast, found in most apicomplexans [2]. The apicoplast is likely the result of a secondary endosymbiosis of a red alga and is likely the same endosymbiotic event giving rise to the plastids of dinoflagellates. The apicoplast is nonphotosynthetic but exhibits activities associated with type II fatty acid biosynthesis, isoprenoid biosynthesis, and possibly heme synthesis. These pathways are essentially prokaryotic and represent excellent drug targets. A photosynthetic alveolate, *Chromera velia*, that appears to be the earliest branching apicomplexan has also been identified [3].

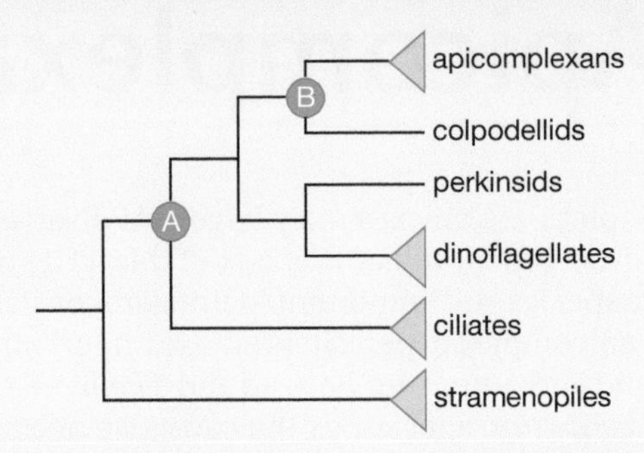

Figure 11A Cladogram showing relationships between apicomplexans and other Alveolata. Node A is the Alveolata branch and node B is the Apicomplexa branch.

1. Kuvardina, O.N., Leander, B.S., Aleshin, V.V., Myl'nikov, A.P., Keeling, P.J. and Simdyanov, T.G. (2002) The phylogeny of colpodellids (Alveolata) using small subunit rRNA gene sequences suggest they are the free-living sister group to Apicomplexans. *J. Eukaryot. Microbiol.* 49: 498–504.

2. Wilson, R.J.M. (2002) Progress with parasite plastids. *J. Mol. Biol.* 319: 257–274.

3. Moore, R.B., Obornik, M., Janouskovec, J., Chrudimsky, T., Vancova, M., Green, D.H., Wright, S.W., Davies, N.W., Bolch, C.J.S., Heimann, K., Slapeta, J., Hoegh-Guldberg, O., Logsdon, J.M. and Carter, D.A. (2008). A photosynthetic alveolate closely related to apicomplexan parasites. *Nature* 451: 959–963.

General Apicomplexan Structure and Life Cycle

The apicomplexa generally exhibit complex life cycles involving specialized invasive stages with apical organelles and other ultrastructural features (Figure 11.1). At the most anterior end of the invasive stages are rings of microtubules known as the polar ring. In some species, a hollow cone-shaped structure, called a conoid, is part of the polar ring. The polar ring is a microtubule organizing center and microtubules which run the length of the organism emanate from the polar ring. These longitudinal microtubules are also associated with the inner membrane complex. Three distinct types of secretory organelles are also found at the apical ends of the invasive stages: rhoptries, micronemes, and dense granules.

Rhoptries are described as being tear-drop or club shaped membrane-bound organelles due to a duct connecting with the anterior end of the organism. **Micronemes** are small elliptical shaped vesicles found in close proximity to the polar ring. **Dense granules** are secretory vesicles found throughout the organism. However, some of the dense granules are concentrated at the apical end and appear to play a role in invasion.

Life cycle phases

The apicomplexa have complex life cycles that are characterized by three distinct processes: **sporogony**, **merogony**, and **gametogony** (Figure 11.2).

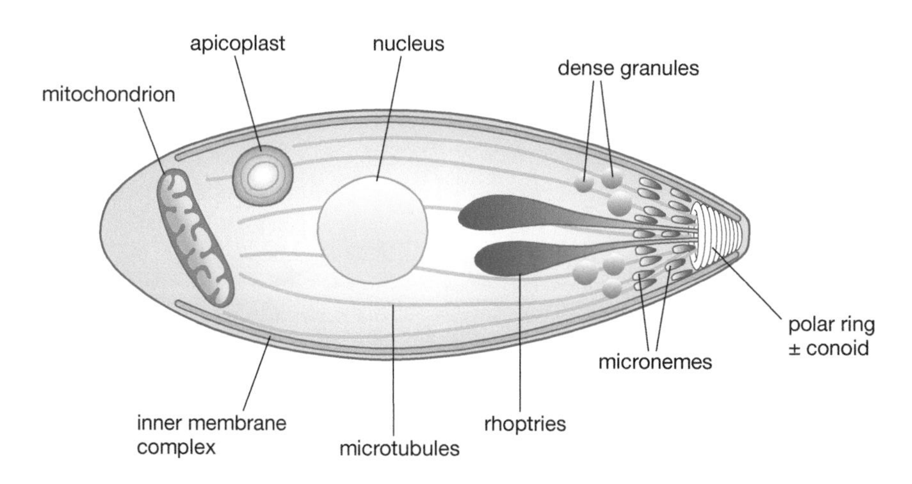

Figure 11.1 Schematic representation of generalized apicomplexan. Invasive and motile forms of apicomplexa exhibit distinctive apical organelles and ultrastructural features in addition to the typical eukaryotic organelles. (Reprinted from *Parasitology: The Biology of Animal Parasites* by Noble with kind permission from Wolters Kluwer Publishers.)

Although most apicomplexa exhibit this overall general life cycle the details vary between species. Furthermore, the terminology used to describe these various life cycle stages varies between the species. The life cycle consists of both asexually reproducing forms and sexual stages. In monoxenous species all three of these processes will be carried out in a single host and often in a single cell type or tissue, whereas in heteroxenous species the various processes will be carried out in different hosts and generally involve different tissues.

Sporogony occurs immediately after a sexual phase and consists of an asexual reproduction that culminates in the production of sporozoites. Sporozoites are an invasive form that will invade cells and develop into forms that undergo another asexual replication known as merogony. Merogony and the resulting merozoites are known by different names depending on the species. In contrast to sporogony, in which there is generally only one round of replication, quite often there are multiple rounds of merogony. In other words, the merozoites, which are also invasive forms, can reinvade cells and initiate another round of merogony. Sometimes these multiple rounds of merogony will involve a switch in the host organism or a switch in the type of cell invaded by the parasite resulting in distinct stages of merogony. As an alternative to asexual replication merozoites can develop into gametes through a process variously called gametogony, gamogony, or gametogenesis. As in other types of sexual reproduction, the gametes fuse to form a zygote which will undergo sporogony.

Many apicomplexa exhibit unique modes of cell division

Apicomplexans generally do not exhibit a typical binary fission as is often associated with the replication of most other protozoa. This is in part due to replication in apicomplexans generally resulting in the formation of specialized invasive forms, or **zoites**. Furthermore, cellular replication in several of the apicomplexa is a process by which the organism undergoes numerous rounds of nuclear division without cytoplasmic division resulting in a multinucleated form. During merogony these multinucleated forms are generally called either schizonts or meronts. The progeny are then formed by a budding process in which a single nucleus will associate with the nascent apical organelles resulting in the formation of the invasive stages. Microtubule organizing centers (i.e., polar rings) are established at several positions on the plasma membrane of the meront. The nascent apical organelles and nuclei associated with these microtubule organizing centers and the formation of the budding zoites is driven by the emergence of the subpellicular microtubules from the forming apical complex. The inner membrane complex is also formed along the subpellicular microtubules during the budding process. The mitochondria and apicoplasts also divide and are incorporated into the newly formed zoite.

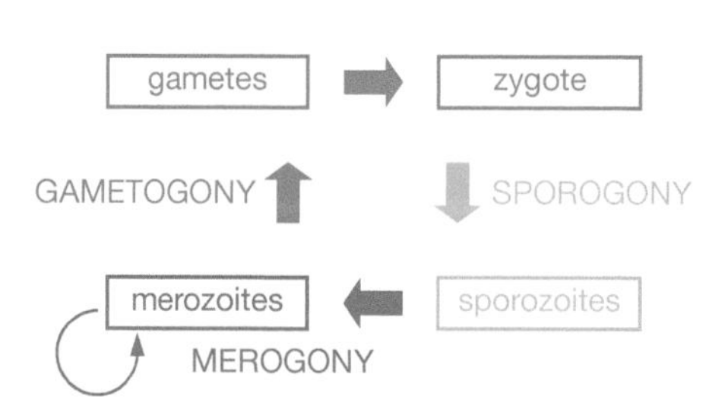

Figure 11.2 Generalized life cycle of apicomplexan parasites. Apicomplexans have a complicated life cycle involving sexual reproduction and the formation of gametes, and two types of asexual reproduction (sporogony and merogony). Merozoites and sporozoites are invasive forms that result from merogony and sporogony, respectively.

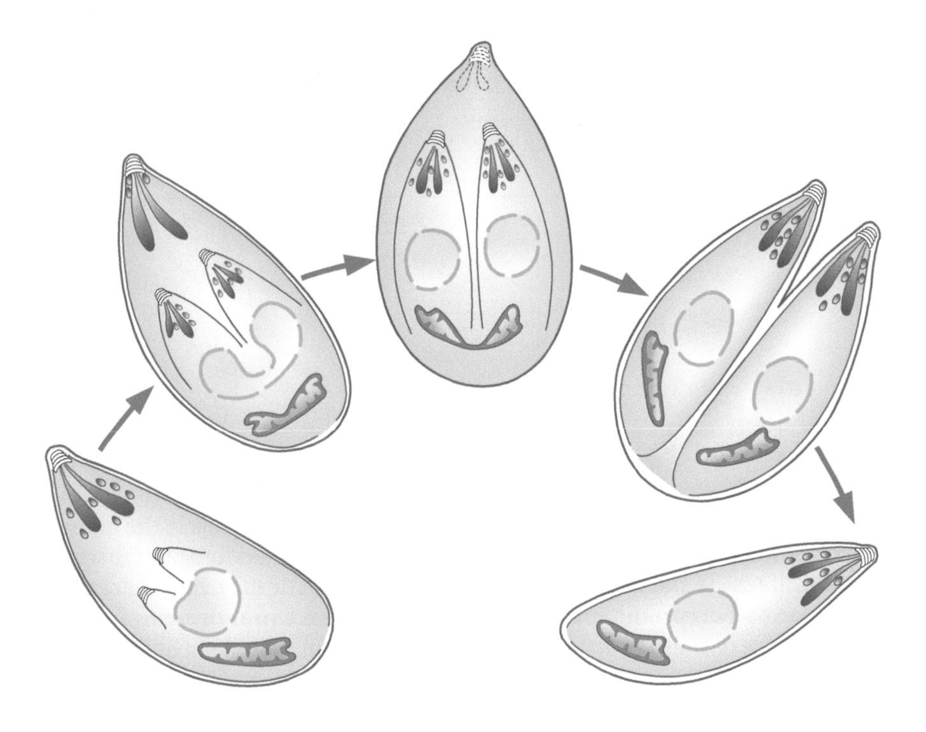

Figure 11.3 Schematic representation of endodyogeny. The zoites of *Toxoplasma* are formed by internal budding of the daughter cells. (Reprinted from *Modern Parasite Biology* with kind permission from W. H. Freeman.)

Some life cycle stages of *Toxoplasma* exhibit a unique form of binary fission called **endodyogeny** (Figure 11.3). During endodyogeny the nascent apical organelles and inner pellicular membranes of the daughter cells start to form within the cytoplasm of the cell instead of at the plasma membrane. Other organelles (i.e., nuclei, mitochondria, and apicoplasts) divide and also associate with the newly forming daughter cells within the mother cell. The inner pellicular membranes of the mother cell disappear and are replaced by the inner pellicular membranes of the daughter cells. The outer plasma membrane of the mother cell is reused to form the outer plasma membrane of the daughter cells. Other stages of *Toxoplasma* and related parasites replicate by **endopolygeny**. This is similar to both schizogony and endodyogeny in that the organism undergoes multiple rounds of nuclear division without cytoplasmic division and formation of the progeny is initiated within the cytoplasm of the parent cell instead of at the plasma membrane.

Mechanism of Host Cell Invasion

The vast majority of the apicomplexa exhibit an intracellular stage during part of their life cycle. The exceptions to this are some gregarines and *Cryptosporidium*. In these cases the parasite attaches to the host cell and derives its nutrients from the host cell by myzocytosis. The attachment and invasion processes are related in that the apical organelles are involved in both processes. The mechanism of invasion is best characterized in the more medically important genera of *Plasmodium* and *Toxoplasma* and has been especially well characterized in erythrocyte invasion by *Plasmodium* merozoites. Although the details of invasion differ between species and life cycle stages, the overall mechanism is presumably conserved throughout the Apicomplexa. During the invasion process the invasive forms (e.g., merozoites and sporozoites) of most apicomplexan species orientate themselves so that the apical end is juxtaposed to the host cell. The contents of the rhoptries, micronemes, and dense granules are expelled as the parasite invades. Experiments in *Toxoplasma gondii* indicate that the micronemes are expelled first and the expulsion occurs with initial contact between the parasite and host. The rhoptries are discharged immediately after the micronemes and dense granule contents are released last. Many dense granules are extruded after the parasite has completed its entry, and thus probably play a role in modifying the host cell.

Adhesion proteins are found in the micronemes

Many proteins found within the micronemes are characterized by having domains that are homologous to known adhesive domains. For example, the thrombospondin-related anonymous protein (TRAP) family of micronemal proteins is characterized by having multiple copies of thrombospondin type 1 domains and von Willebrand A domains (Figure 11.4). Thrombospondin is a multifunctional adhesive protein that binds to a wide range of cellular and extracellular matrix molecules. Other adhesive domains found in other micronemal proteins include epidermal growth factor (EGF)-like domains, apple domains, and lectin domains. These "adhesins" play a role in the adherence of the motile and invasive zoites to either host cells or the substratum. In this sense, the micronemal adhesins can be viewed as ligands that bind to host cell receptors. Some micronemal proteins are specific and only bind to particular molecules on the host, whereas as others exhibit a wider range of binding activity. The specificity of the adhesins will reflect the host and cell specificity exhibit by the various species and stages of apicomplexan zoites.

The Duffy binding protein of *P. vivax* and erythrocyte binding antigen (EBA)-175 of *P. falciparum* are two rather well characterized proteins of *Plasmodium* merozoites which are not related to the TRAP family and other micronemal adhesins. The Duffy binding protein binds the Duffy antigen on the surface of the erythrocyte, whereas EBA-175 binds to sialic acid moieties on erythrocyte glycophorins. Thus, *P. vivax* and *P. falciparum* utilize different receptors on the host erythrocyte. Persons lacking the Duffy antigen on their erythrocytes are refractory to *P. vivax* infection. In particular, *P. vivax* is not very prevalent in western Africa where a large portion of the population is Duffy negative. This recognition of specific receptors on the host cell partly accounts for the host species specificity exhibited by *Plasmodium* species as well as the specificity *Plasmodium* merozoites have for erythrocytes.

Despite the different binding specificities the Duffy binding protein and erythrocyte binding antigen exhibit similar structures (Figure 11.4) and are members of a gene family of erythrocyte binding proteins. In particular, the receptor-binding activity has been mapped to a domain in which the cysteine and aromatic amino acid residues are conserved between species. This sequence motif is referred as a Duffy-binding domain and is found in other adhesive proteins expressed by the malaria parasite (see Chapter 15). The Duffy-binding domain is duplicated within the EBA-175 protein. Signal sequences and transmembrane domains are found on the N- and

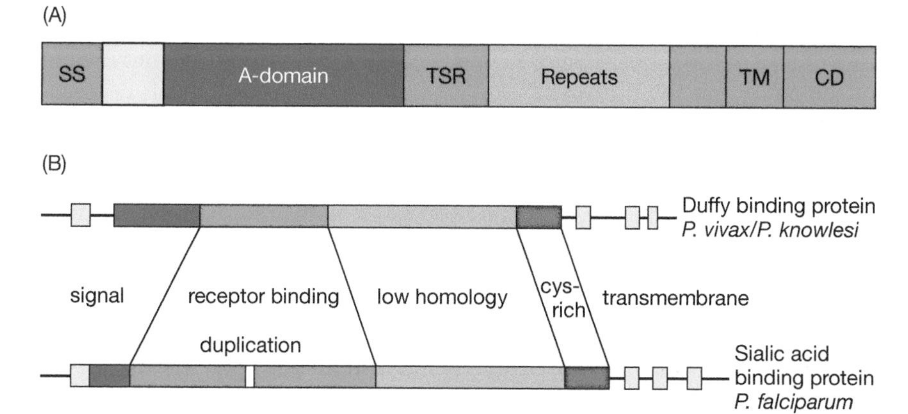

Figure 11.4 Schematic representation of micronemal adhesins. (A) Diagram of TRAP showing signal sequence (SS); von Willebrand A domain; thrombospondin type 1 repeat (TSR); transmembrane domain (TM); and cytoplasmic domain (CD). (B) Comparison of the gene structures of *Plasmodium* erythrocyte binding proteins illustrating similarity in overall domain structure. Thin line represents introns and other nontranslated regions.

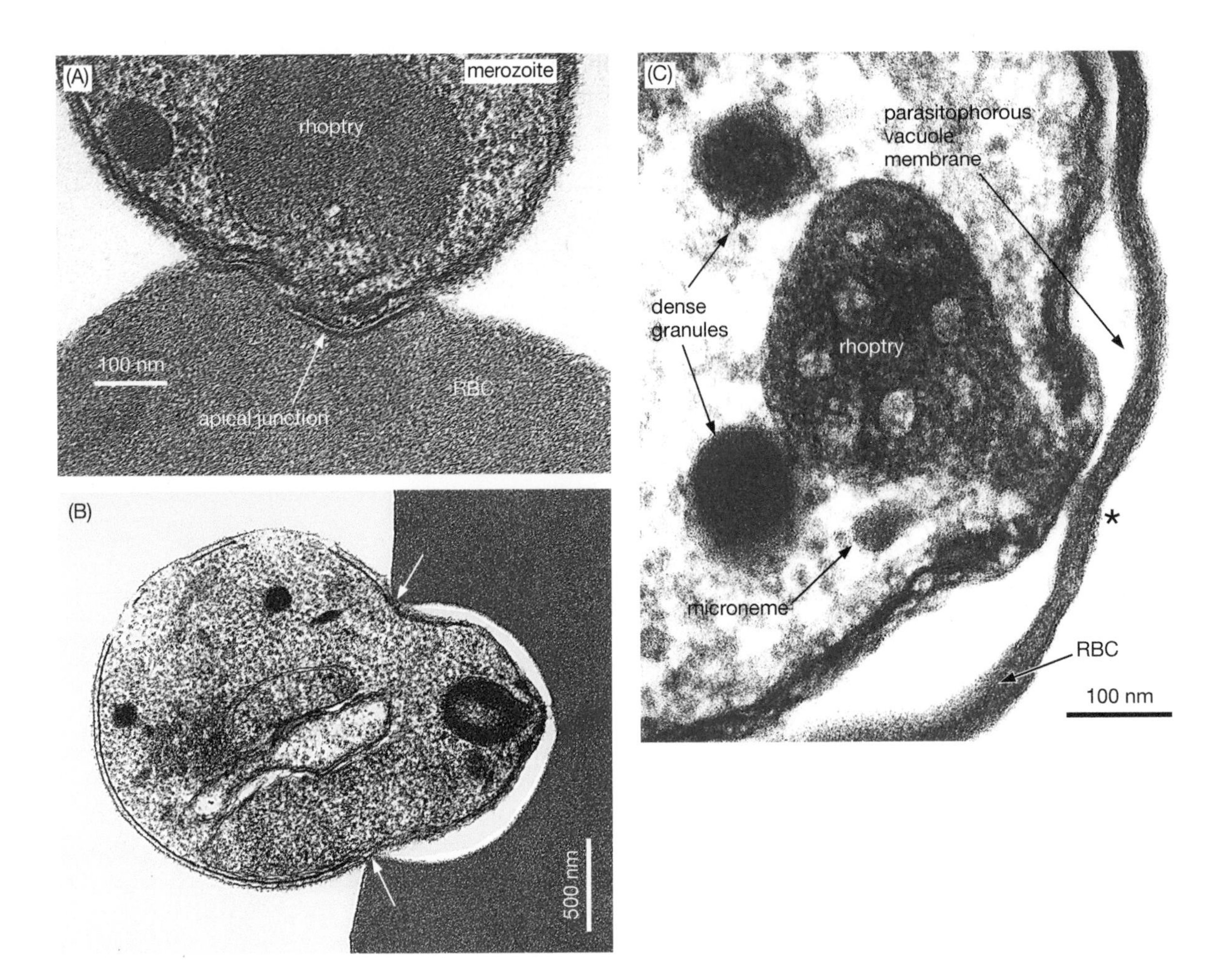

Figure 11.5 Electron micrographs showing merozoite attachment and entry into an erythrocyte. (A) An electron dense junction (arrow) is observed between the erythrocyte (RBC) and the apical end of the merozoite. (Photo kindly provided by Lawrence Bannister. Reprinted from *Critical Reviews in Oncology/Hematology*, Vol. 8, No. 4, with kind permission from Elsevier.) (B) As the parasite enters the host cell the junction becomes annular in shape (arrows) and is pulled back along the parasite. (Photo kindly provided by Lawrence Bannister. Reprinted from *Parasitology*, Vol. 92, with kind permission from Cambridge University Press.) (C) A connection (*) is seen between the parasitophorous membrane and the ends of the rhoptry ducts. (Photo kindly provided by Lawrence Bannister. Reprinted from *Journal of Eukaryotic Microbiology*, 2007, with kind permission from John Wiley and Sons.)

C-termini, respectively. The topography of the transmembrane domain is consistent with the parasite ligands being integral membrane proteins with the receptor-binding domain exposed within the lumen of the microneme. This is also true for members of the TRAP family and other micronemal proteins. Thus, the fusion of the microneme membrane with the plasma membrane during microneme discharge will result in the exposure of the receptor-binding domain on the merozoite surface at the apical end.

Coincident with the release of the microneme contents is the formation of an electron dense junction between the host and parasite (Figure 11.5A). This and the receptor–ligand interactions imply that the contents of the micronemal adhesins play a role in the formation of this junction and the attachment of the parasite to the host cell. This junction converts from its initial disk shape to a ring, or band, surrounding the zoite and is pulled back along the length of the invading zoite (Figure 11.5B). The translocation of this "moving junction" toward the back of the parasite results in a forward movement of the parasite into the host cell. Coincident with the movement of the zoite into the host cell is the formation of a parasitophorous vacuole. The parasitophorous vacuolar membrane (PVM) is likely derived from both the host membrane and parasite components and expands as the parasite enters the erythrocyte. Formation of the parasitophorous vacuole is coincident with the release of the rhoptries and the contents of the rhoptries are believed to be important in the formation of the parasitophorous vacuolar membrane (Figure 11.5C). The parasitophorous vacuole continues to expand as the zoite enters the host cell and when the moving junction reaches the posterior end of the parasite there is a closure resulting in a separation of the PVM and erythrocyte membrane and the enclosure of the parasite within the parasitophorous vacuole.

Invasion and zoite motility are powered by a molecular motor complex

Apicomplexan parasites actively invade host cells and, in contrast to some intracellular pathogens like *Leishmania*, entry is not due to uptake or phagocytosis by the host cell. In addition, the zoites are often motile forms that crawl along the substratum by a type of motility referred to as "gliding motility." Gliding motility, like invasion, also involves the release of micronemal adhesins, attachment to the substratum, and a capping of the adhesins at the posterior end of the zoite. One difference between gliding motility and invasion is that the micronemes must be continuously released as the organism is moving along a surface. Thus, gliding motility does not involve this relatively small moving junction, but a continuous formation of new junctions between the zoite and the substratum. In addition, the adhesins are cleaved from the surface of the zoite as the adhesions reach the posterior of the zoite and a trail of the adhesive molecules are left behind the moving zoite on the substratum. However, the overall mechanism of motility and invasion are quite similar, and thus during invasion the parasite literally crawls into the host cell through the moving junction. In addition, some apicomplexans use this type of motility to escape from cells and can traverse biological barriers, such as epithelial cell layers, by entering and exiting cells.

Cytochalasins inhibit zoite entry and this inhibition suggests that the force required for parasite invasion is based upon actin–myosin cytoskeletal elements. The ability of myosin to generate movement and force is well characterized (e.g., muscle contraction). A myosin unique to the Apicomplexa has been identified and localized to the inner membrane complex which lies just beneath the plasma membrane of the zoites. This myosin is part of a motor complex which is linked to the adhesins (Figure 11.6). Members of

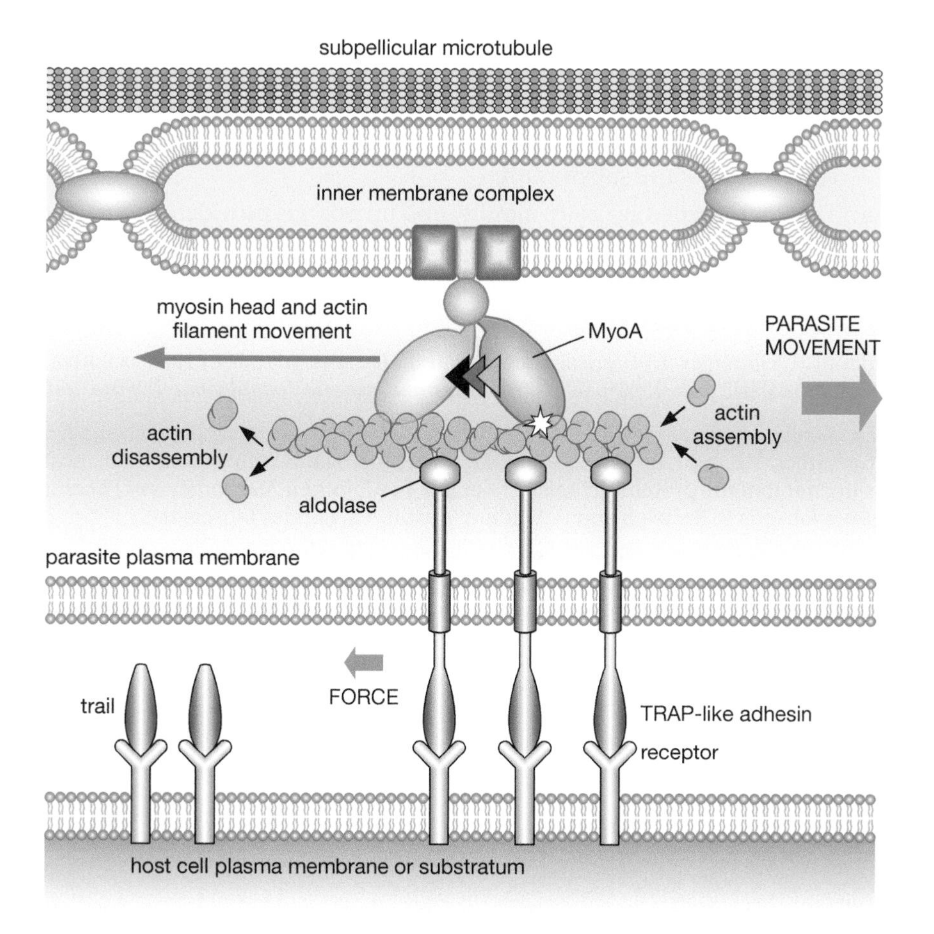

Figure 11.6 Current model of the motor protein complex driving gliding motility. Transmembrane adhesins released from the micronemes bind to receptors on the host cell or substratum. The intracellular C-termini of the adhesins are linked to short actin filaments. These actin filaments and the adhesin cargo are propelled towards the posterior of the parasite by a myosin attached to the inner membrane complex. Associated with the movement of the actin filament is a polarized disassembly and assembly of the actin filaments. The force generated through the ATPase activity of the myosin and the resulting conformation change results in a net forward movement of the parasite. When the adhesins reach the posterior end of the parasite they are cleaved and remain associated with substratum. (Modified from Baum J. et al., (2008) *Tr. Parasitol.* 24: 557–563.)

the TRAP family and other adhesins have a conserved cytoplasmic domain. This cytoplasmic domain is linked to short actin filaments via aldolase. The actin filaments and myosin are oriented in the space between the inner membrane complex and plasma membrane so that the myosin propels the actin filaments toward the posterior of the zoite. The myosin is anchored into the inner membrane complex and does not move. Therefore, the transmembrane adhesins are pulled through the fluid plasma membrane due to their association with the actin filaments. Thus the complex of adhesins and actin filaments is transported toward the posterior of the cell. Because the adhesins are either complexed with receptors on the host cell or bound to the substratum, the net result is a forward motion of the zoite. When the adhesins reach the posterior end of the parasite they are proteolytically cleaved and shed from the zoite surface.

A trophic period characterized by the parasite taking up nutrients from the host and increasing in size generally follows the completion of parasite entry. This is then followed by a replicative stage which leads to the production of more invasive forms. In some Apicomplexa the mechanism of zoite egress involves the apical organelles and motility in a fashion similar to invasion. In other cases the egress of the invasive forms involves a destruction of the host cell and the release of the zoites.

Summary and Key Concepts

- The Apicomplexa are a monophyletic group of almost exclusively parasitic protozoa characterized by specialized invasive stages.
- The invasive stages exhibit a polarity and contain unique organelles and subcellular structures at the apical end (i.e., apical complex).
- During part of their life cycles most apicomplexans invade and replicate within the host cells.
- Micronemes and rhoptries are specialized membrane-bound apical organelles that play a major role in interactions with and invasion of host cells.
- Micronemes contain adhesive proteins that are important in attaching to host cells or the substratum.
- The force needed for zoite motility and invasion is provided by a motor protein complex including a myosin unique to apicomplexans.

Further Reading

Baum, J., Gilberger, T.W., Frischknecht, F. and Meissner, M. (2008) Host-cell invasion by malaria parasites: insights from *Plasmodium* and *Toxoplasma*. *Tr. Parasitol.* 24: 557–563.

Carruther, V.B. and Tomley, F.M. (2007) Receptor–ligand interactions and invasion: microneme proteins in apicomplexans. In *Molecular Mechanisms of Parasite Invasion* (eds B. Burleigh and D. Soldati). Austin, TX, Landes Bioscience.

Iyer, J., Gruner, A.C., Renia, L., Snounou, G. and Preiser, P.R. (2007) Invasion of host cells by malaria parasites: A tale of two protein families. *Mol. Microbiol.* 65: 231–249.

Striepen, B., Jordan, C.N., Reiff, S. and van Dooren, G.G. (2007) Building the perfect parasite: Cell division in Apicomplexa. *PLoS Pathog.* 3(6): e78.

Cryptosporidium

Disease(s)	Cryptosporidiosis
Etiological agent(s)	*Cryptosporidium hominis* and *C. parvum*
Major organ(s) affected	Intestinal tract
Transmission mode or vector	Fecal–oral, waterborne, contact with cattle
Geographical distribution	Worldwide
Morbidity and mortality	Generally a self-limiting diarrhea, but much more severe in AIDS patients and can be fatal
Diagnosis	Detection of oocysts in feces by acid-fast staining
Treatment	Paromomycin and nitazoxanide have some clinical effect, but neither drug completely clears parasitemia. Treatment of severe cryptosporidiosis should include supportive care such as rehydration therapy and antimotility agents
Control and prevention	Avoid fecal contamination of food or water from both humans and cattle

Members of the genus *Cryptosporidium* are apicomplexan parasites that infect intestinal cells and exhibit a fecal–oral type of life cycle. Since its initial identification in 1907 several *Cryptosporidium* species have been identified in a wide variety of animals ranging from fish to humans. Most *Cryptosporidium* species are host adapted and only infect a narrow range of natural hosts. Two species are responsible for the majority (>90%) of infections in humans: *C. hominis*, which is generally restricted to humans and *C. parvum*, which is found primarily in calves and humans. The first human cases of cryptosporidiosis were reported in 1976 and were characterized as a diarrheal disease associated with immune suppression. Initially cryptosporidiosis was believed to be a rare and exotic disease. During the 1980s *Cryptosporidium* was recognized as a major cause of diarrhea in AIDS patients and often resulted in death. Thus, *Cryptosporidium* was viewed as an opportunistic infection. However, it is now recognized that *Cryptosporidium* is a common cause of diarrhea in immunocompetent persons, as well as the immunocompromised, and has probably been a human pathogen since the beginning of humanity.

Life Cycle

Cryptosporidium is often classified as a coccidian and exhibits a life cycle similar to other intestinal **coccidia** (Chapter 13). However, *Cryptosporidium* is more closely related to the gregarines (see Box 12.1) and this is reflected in some aspects of its life cycle. The infection is acquired through the ingestion of sporulated **oocysts**. After passing through the stomach, sporozoites emerge from the oocyst and attach to intestinal epithelial cells. In contrast to coccidian parasites, and more similar to the gregarines, *Cryptosporidium* **sporozoites** do not invade the enterocytes. Instead they attach to the intestinal epithelial cells and induce the fusion of the microvilli resulting in the parasite becoming encapsulated by a double membrane of host origin (Figure 12.1). The interaction of parasite with the host enterocytes likely involves the apical organelles (i.e., rhoptries and micronemes) which function in host cell invasion in other apicomplexa (Chapter 11). This unique localization in regard to the host cell is sometimes designated as being "extracytoplasmic" since the parasite is not within the host cell, but at the same time the parasite is surrounded by membranes derived from the host. A junction, called the "**feeder organelle**" or the "adhesion zone," forms between the parasite and the host enterocyte. The parasite likely derives nutrients from the host cell via this junction.

After association with the enterocyte the parasite converts to a trophozoite and then will exhibit a typical apicomplexan development by undergoing an asexual replication usually referred to as merogony. Following a trophic period in which the parasite increases in size, the parasite will undergo an asexual replication consisting of two to three rounds of nuclear division without cytoplasmic division followed by sequential rounds of cytoplasmic division. Typically four to eight **merozoites** are produced and are subsequently released into the intestinal lumen. These merozoites are functionally the same as the sporozoites and infect new intestinal epithelial cells followed by another round of merogony (Figure 12.2). In immunocompetent persons the number of rounds of merogony is limited by an

Box 12.1 Gregarines

Gregarines are a diverse group of apicomplexan parasites that infect invertebrates—especially arthropods, annelids (segmented worms), and mollusks. The gregarines are thought to be the earliest lineage of apicomplexans. A typical gregarine life cycle generally only consists of gametogony and sporogony and only a few species exhibit merogony. Instead of undergoing merogony, the sporozoites will generally develop into large extracellular trophozoites. In addition, many gregarines do not exhibit intracellular stages and the trophozoites attach to the host cell via a structure called either the mucron or epimerite (depending on species). These specialized structures are derived from the conoid at the apical end and function to anchor the parasite to the host (Figure 12A). This attachment to the host cell also functions in feeding in that the cytoplasm of the host is taken up by the attached parasite (i.e., myzocytosis). Gametogony begins when two trophozoites pair, referred to as a syzygy, and develop into a gametocyst. Gametes are formed, conjugate, and develop into sporulated oocysts within the gametocyst.

Analysis of small subunit ribosomal RNA gene sequences indicates that *Cryptosporidium* is more closely related to the gregarines than to coccidians [1]. In addition, the electron-dense junction observed between the host cell and *Cryptosporidium* is quite similar to the feeding mechanism exhibited by most gregarines. This suggests that the feeding organelle of *Cryptosporidium* is an evolutionary modification of the myzocytotic feeding exhibited by other alveolata. Furthermore, trophozoite and gamont-like stages similar to the gregarines have been observed in *in vitro* cultures. These observations suggest that the gregarines and *Cryptosporidium* share a common ancestor.

1. Barta, J.R. and Thompson, R.C.A. (2006) What is *Cryptosporidium*? Reappraising its biology and phylogenetic affinities. *Tr. Parasitol.* 22: 463–468.

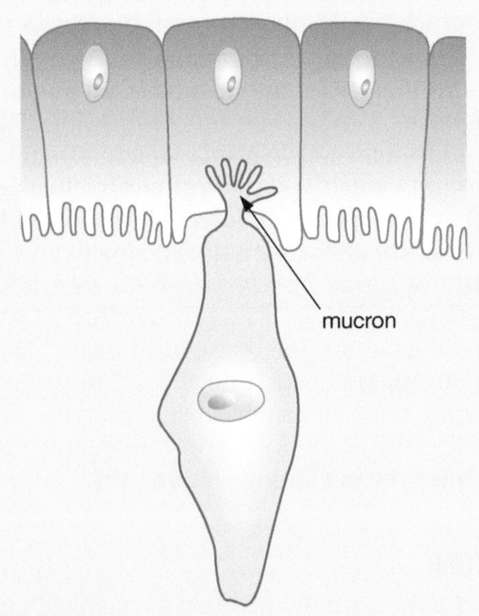

Figure 12A Attachment of a gregarine trophozoite to intestinal epithelium. A schematic representation demonstrating the interaction of the mucron with host cell. The mucron attaches to the cell and functions in parasite feeding via myzocytosis.

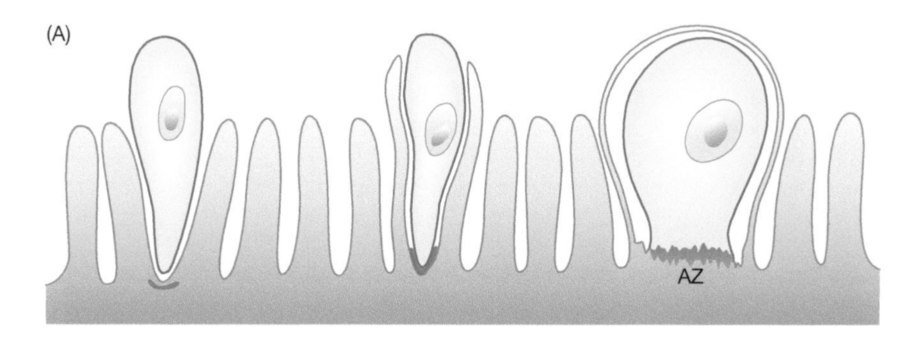

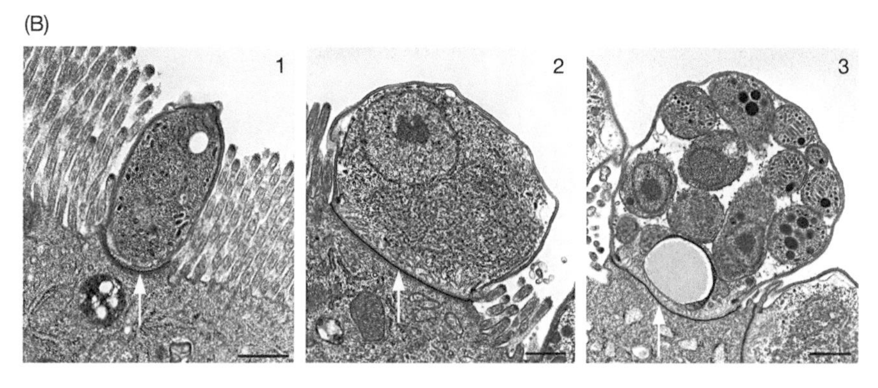

Figure 12.1 Extracytoplasmic association of *Cryptosporidium* with intestinal epithelial cells. (A) Schematic representation showing the microvilli of the intestinal epithelial cells becoming longer and flatter and then fusing to surround the parasite. A junction called either the adhesive zone (AZ) or the feeder organelle also forms between the parasite and host intestinal epithelial cell. (B) Transmission electron micrographs showing (1) a trophozoite shortly after internalization; (2) a fully developed trophozoite before division into merozoites; and (3) mature meront with eight merozoites ready to be released into the gut lumen to infect new cells. Note also the electron-dense adhesive zone between the parasite and intestinal epithelial cell (arrowed). Bar = 500 nm. (Photos kindly provided by Saul Tzipori, School of Veterinary Medicine, Tufts University, North Grafton, MA. Reprinted from *Microbes and Infection*, Vol. 4, Issue 10, with kind permission from Elsevier.)

adequate immune response. The increased severity of the disease in AIDS and immunocompromised patients is due in part to their inability to limit these additional rounds of merogony and to control the infection.

Sexual reproduction results in the production of infectious forms

As an alternative to merogony, the merozoites can undergo **gametogony** and develop into either macro- or microgametes. Microgametogenesis involves several rounds of replication followed by the release of numerous microgametes (analogous to sperm) into the intestinal lumen. The **microgametes** fertilize **macrogametes** (analogous to ova) still attached to the intestinal epithelial cells. The resulting zygote undergoes sporogony which consists of two rounds of replication resulting in the production of four sporozoites.

The mature sporulated oocysts are excreted with the feces and are immediately infectious if ingested by an appropriate host. An autoinfection in which excystation occurs within the same host is also possible. Thick-walled and thin-walled oocysts have been observed and it is believed that the thin-walled oocysts are responsible for the autoinfection and the thick-walled oocysts are more likely to be involved in person-to-person transmission. The possibility of autoinfection may also contribute to the increased disease severity observed in immunocompromised patients.

Transmission and Molecular Epidemiology

Cryptosporidium exhibits a fecal–oral type of transmission and the risk factors for transmission are similar to other fecal–oral diseases (Chapter 2). These include direct contact with infected persons or animals or ingestion of contaminated food or water. However, waterborne cryptosporidiosis outbreaks have been especially notable. This includes contamination of drinking water sources and recreational water. Outbreaks associated with recreational waters most often involve fecal accidents in swimming pools by diapered infants and toddlers. Contamination of drinking water is usually presumed to be due to contaminated surface water and the failure of water treatment to remove the oocysts. The durability of the oocysts and their small size are major factors in the waterborne transmission of

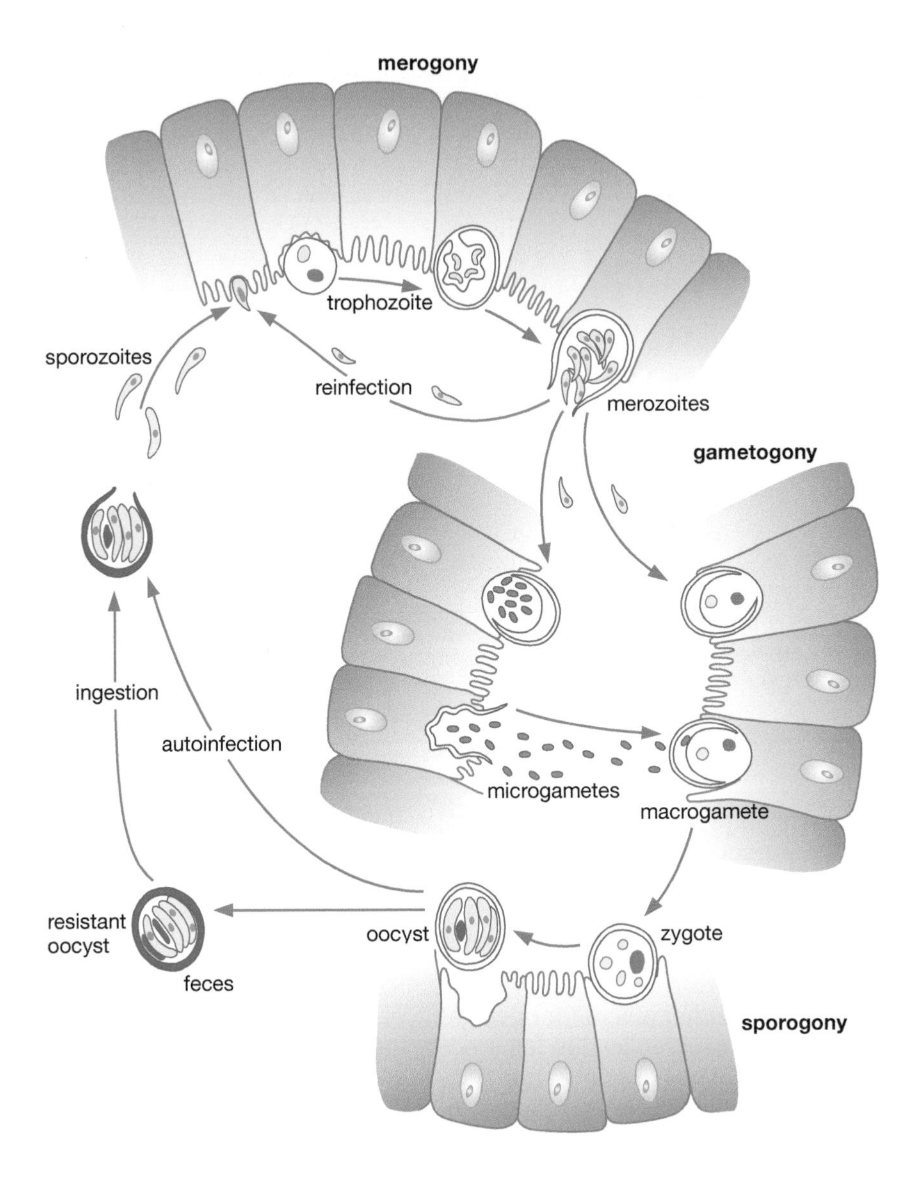

Figure 12.2 Life cycle of *Cryptosporidium*. Following ingestion, sporozoites are released from the oocyst and associated with an intestinal epithelial cell. The resulting trophozoite will undergo an asexual replication producing merozoites. The merozoites can either undergo additional rounds of asexual replication (i.e., merogony) or undergo a sexual phase (gametogony) which consists of macrogametes or microgametes. The motile microgametes are released and fertilize the macrogamete. The resulting zygote undergoes sporogony which involves two rounds of asexual reproduction producing four sporozoites within the oocyst. Mature sporulated oocysts are released in the feces or can cause an autoinfection within the same host.

Cryptosporidium (Table 12.1). The small size of the oocysts and their resistance to standard chlorine treatment makes them difficult to remove from the water supply. In addition, the zoonotic potential of *C. parvum* contributes an additional risk of outbreaks from animal sources and in particular cows.

Table 12.1 Factors that contribute to *Cryptosporidium* waterborne outbreaks
Small size of oocysts
Low infective dose (1–10 oocysts)
Robust oocysts (resistant to chlorine)
Infectious sporulated oocysts excreted
Monoxenous development
Shared host specificity (*C. parvum*)
Large number of oocysts excreted (up to 100 billion per calf)
Proximity of domestic animals to surface water sources

One of the most infamous *Cryptosporidium* outbreaks occurred in Milwaukee during the spring of 1993 in which an estimated 400 000 people developed symptomatic cryptosporidiosis. It is believed that Lake Michigan was the source of the *Cryptosporidium*. However, the origin of the oocysts and when they entered Lake Michigan is not known. Subsequent molecular characterization indicated that the human type, or *C. hominis*, was the cause of the outbreak suggesting human sewage as a source. Regardless of the source, coagulation and filtration processes apparently did not adequately remove the oocysts. Higher levels of turbidity, but within acceptable limits, were noted in the treated water during the weeks preceding the outbreak suggesting that deficiencies in the water treatment were responsible for the outbreak.

In addition to waterborne outbreaks, food-borne and human-to-human transmission also occur quite frequently. For example, secondary cases in households range from 5 to 20%. Furthermore, outbreaks in hospitals, institutions, and day care centers have also been noted. This person-to-person type of transmission is also common in fecal–oral transmission of other intestinal protozoa. Thus despite the notoriety of waterborne outbreaks the cycle in humans is probably maintained primarily through direct transmission between persons. There is also potential for a zoonotic transmission of *C. parvum* and other *Cryptosporidium* species from livestock, companion animals, and wildlife. Zoonotic transmission can be through direct contact with animals or waterborne transmission.

Molecular studies indicate that two distinct species infect humans

Genotyping studies using molecular markers revealed that humans are primarily infected with two distinct genotypes. Originally these were designated as genotype 1 and genotype 2 of *C. parvum*. Subsequently genotype 1 was elevated to the status of species and named *C. hominis*. *C. hominis* has only been isolated from human sources and is noninfective for mice and calves, whereas *C. parvum* (i.e., genotype 2) has been isolated from both animal (primarily bovine) and human sources and is infective for mice and calves. Further subtyping indicates that some isolates of *C. parvum* appear to be adapted to humans and thus exhibit an anthroponotic transmission. Other species and genotypes of *Cryptosporidium*, such as *C. felis* (cat), *C. canis* (dog), *C. suis* (pig), and *C. meleagridis* (turkey), are rarely isolated from humans and such infections are almost always associated with AIDS patients or other immunocompromised patients.

These genetic data imply that there are primarily two distinct transmission cycles in humans (Figure 12.3) involving distinct populations of

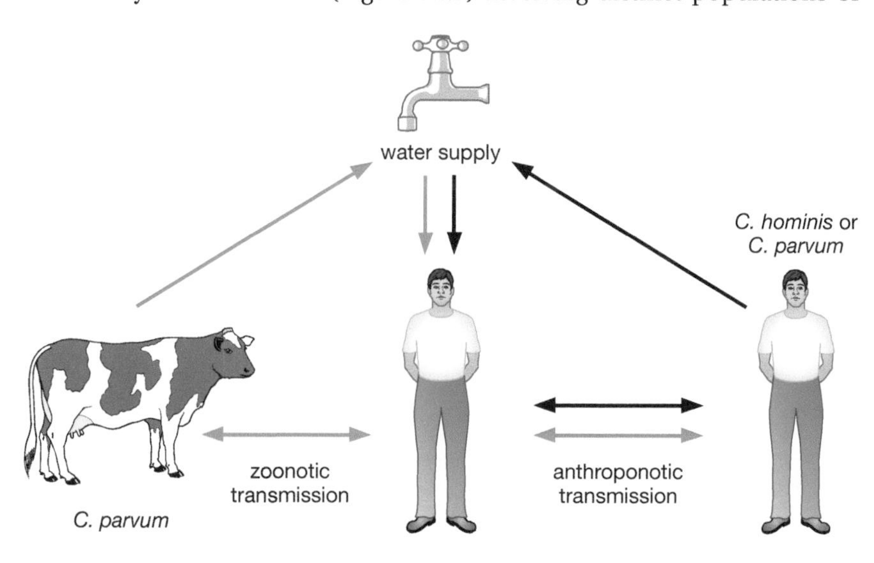

Figure 12.3 Transmission cycles of *Cryptosporidium*. A schematic representation of the primary modes of transmission of *Cryptosporidium*. The black arrows represent *C. hominis* and the blue arrows *C. parvum*.

Cryptosporidium. There is an exclusively **anthroponotic** (i.e., human-to-human) cycle caused by *C. hominis* and sometimes *C. parvum,* and a **zoonotic** cycle caused by *C. parvum.* Both species have been demonstrated to be the etiological agent in waterborne outbreaks. Waterborne outbreaks linked to *C. hominis* are likely due to contamination of water with human sewage, whereas waterborne outbreaks associated with *C. parvum* are likely due to contamination of surface water with cow manure. Consistent with this is a higher prevalence of *C. parvum* in rural areas with extensive animal husbandry and generally a higher prevalence of *C. hominis* in urban areas. It is still not clear the relative importance of the various types of transmission routes, but molecular epidemiological studies may help determine the relative importance of direct contact with humans or animals versus food-borne or waterborne transmission.

Clinical Aspects and Pathogenesis

The most common clinical manifestation of cryptosporidiosis is a mild to profuse watery diarrhea. This diarrhea is generally self-resolving and persists from several days up to 1 month. Recrudescences are common with intermittent episodes of diarrhea and symptom-free periods. Abdominal cramps, anorexia, nausea, weight loss, and vomiting are additional manifestations which may occur during the acute stage. Data from the Milwaukee outbreak indicate that watery diarrhea, cramps, and weight loss are the most common symptoms (Table 12.2). Cryptosporidiosis in children in both developing and developed countries is common even though adults and children appear to be equally susceptible to infection. In malnourished children in developing countries the infection can result in severe dehydration and lead to weight loss and failure to thrive. This can have devastating and long-term effects on physical and cognitive development.

The disease is generally much more severe and potentially life threatening in patients with AIDS or other immunological deficits. In the earlier stages of the AIDS epidemic cryptosporidiosis was a relatively common opportunistic infection in patients with advanced HIV infection and in particular in patients with low levels of $CD4^+$ T cells. This type of T cell is destroyed by the HIV and indicates that cell-mediated immunity is important in recovering from cryptosporidiosis. Although AIDS patients can exhibit a transient or an intermittent diarrhea, most often the diarrhea is persistent and can last for months or years until the immunodeficiency is removed or the patient dies. The diarrhea associated with AIDS is profuse

Table 12.2 Symptoms of patients with confirmed cryptosporidiosis during the Milwaukee outbreak

Symptoms	%
Watery diarrhea Median = 9 days (1–55) Median = 12 stools/day (1–90) 39% recrudescence rate	93
Abdominal cramps	84
Weight loss Median = 10 lb (1–40 lb)	75
Fever Median = 38.3 (37.2–40.5)	57
Vomiting	48

Modified from Mac Kenzie et al., 1994.

and voluminous and is often described as being cholera-like. Intravenous rehydration therapy is often necessary and the fatality rate can be quite high in these fulminant cases. In more advanced AIDS cases extraintestinal infection, particularly of the bile duct (i.e., cholangitis), can be observed.

Cellular basis of pathogenesis

The basis of the pathogenesis caused by *Cryptosporidium* is not well understood. Diarrhea can have osmotic, inflammatory, or secretory components (Chapter 2). Experimental evidence does suggest that glucose-coupled sodium absorption is decreased and chloride secretion is increased. Therefore, the diarrhea associated with *Cryptosporidium* appears to be primarily osmotic in nature and involving the small intestine. Associated with this disruption of enterocyte (i.e., intestinal epithelial cells) function is some blunting of the villi and crypt cell hyperplasia. Thus one possible mechanism of pathogenesis is the damage to the enterocytes caused by the infection of intestinal epithelial cells with *Cryptosporidium*. This eventually leads to enterocyte death and triggers cell division in the crypt region (i.e., hyperplasia) to replace the damaged cells. The combination of destruction of absorptive cells at the tips of the villi and the increase in the chloride-secreting cells in the crypts leads to more water in the intestinal lumen and thus diarrhea.

In addition, a minimal amount of local inflammation in the submucosal layer (i.e., lamina propria) is associated with *Cryptosporidium* infection. Associated with this inflammation is the production of various cytokines, in particular interferon-γ, prostaglandins, and other products such as reactive oxygen intermediates. These factors could also contribute to the secretory process and increased intercellular permeability and thus contribute to the diarrhea. A more extensive disruption of the epithelial barrier and a greater level of infiltration of the lamina propria with inflammatory cells are observed in immunodeficient patients and may account for the increased severity of the disease.

The parasite exhibits a tropism for the jejunum and ileum in immunocompetent persons, whereas the infection is more widespread in AIDS patients and can include the stomach, duodenum, colon, and biliary tract. This more extensive anatomical range in AIDS patients is presumably due to the inability of the immune system to control and limit the infection. Cell-mediated immunity appears to be the major component of the immune response in eliminating the infection as evidenced by the correlation between lower $CD4^+$ T cells and the risk and severity of cryptosporidiosis. Interferon-γ, interleukin-12, and tumor necrosis factor-α are involved in protection against *Cryptosporidium* infection.

Diagnosis, treatment, and prevention

The most widely used method for diagnosis is the detection of oocysts in acid-fast stained stool samples. The small size of *Cryptosporidium* oocysts as compared with oocysts of *Cyclospora* (see Chapter 13) allows these two organisms to be distinguished. Kits utilizing antibodies to detect the oocysts via immunofluorescence are also available. Molecular techniques are needed to distinguish *C. hominis* from *C. parvum* and other *Cryptosporidium* species.

There is no completely satisfactory treatment for *Cryptosporidium*. It is hypothesized that the extracytoplasmic location of *Cryptosporidium* shelters it from drugs. The parasite acquires its nutrients directly from the host epithelial cell and therefore nonabsorbed drugs (i.e., luminal agents) do not move into the parasitophorous space created by the fused microvilli. Paromomycin (Humatin®), and more recently nitazoxanide (Alinia®), are reported to have some clinical efficacy. However, nitazoxanide does not appear to benefit AIDS patients and neither drug completely clears parasitemia. Treatment of severe cryptosporidiosis should include supportive care (e.g., rehydration and nutritional support) and antimotility agents.

In general preventive measures will be similar to other diseases transmitted by the fecal–oral route (Chapter 2). Such preventive activities include hand-washing, avoiding contact with stool from humans or animals, and avoiding ingestion of recreational water. Protection of the water supply and proper water treatment are also important elements in the control of waterborne cryptosporidiosis. Persons with compromised immune systems need to be especially diligent in avoiding infection with *Cryptosporidium* since there is no completely effective treatment. Frequent hand-washing and drinking water that has been adequately filtered or boiled are highly recommended.

Summary and Key Concepts

- *Cryptosporidium* species are monoxenous apicomplexan parasites that infect a wide range of animals including humans.
- The majority of human infections are associated with two species: *C. hominis* (found almost exclusively in humans) and *C. parvum* (found in both humans and animals).
- Transmission is via the fecal–oral route involving a sporulated oocyst and exhibits either anthroponotic or zoonotic cycles.
- *Cryptosporidium* exhibits a typical apicomplexan life cycle involving merogony, gametogony, and sporogony.
- The small and robust oocyst is immediately infectious when excreted in the feces and is ideally suited for waterborne transmission.
- The most common clinical symptom of cryptosporidiosis in immunocompetent persons is a watery diarrhea that can range from mild to profuse.
- Immunocompetent persons exhibit a self-limiting and transient disease, whereas the disease can be long lasting and severe in AIDS patients and other immunocompromised persons.
- Damage of intestinal epithelial cells by the parasite probably results in an osmotic type of diarrhea.
- Currently there is no totally effective therapy for eliminating *Cryptosporidium* other than an intact immune system.

Further Reading

Davies, A.P. and Chalmers, R.M. (2009) Cryptosporidiosis. *Brit. Med. J.* 339: 963–967.

Mac Kenzie, W.R., Hoxie, N.J., Proctor, M.E., Gradus, M.S., Blair, K.A., Peterson, D.E., Kazmierczak, J.J., Addiss, D.G., Fox, K.R., Rose, J.B. and Davis, J.P. (1994) A massive outbreak in Milwaukee of *Cryptosporidium* infection transmitted through the public water supply. *N. Engl. J. Med.* 331: 161–167.

Rose, J.B., Huffman, D.E. and Gennaccaro, A. (2002) Risk and control of waterborne cryptosporidiosis. *FEMS Microbiol. Rev.* 26: 113–123.

Sunnotel, O., Lowery, C.J., Moore, J.E., Dooley, J.S.G., Xiao, L., Millar, B.C., Rooney, P.J. and Snelling, W.J. (2006) *Cryptosporidium*. *Lett. Appl. Microbiol.* 43: 7–16.

Xiao, L. and Feng, Y. (2008) Zoonotic cryptosporidiosis. *FEMS Immunol. Med. Microbiol.* 52: 309–323.

Xiao, L. and Ryan, U.M. (2004) Cryptosporidiosis: an update in molecular epidemiology. *Curr. Opin. Infect. Dis.* 17: 483–490.

Monoxenous Intestinal Coccidia

Disease(s)	Isosporiasis, cyclosporiasis, coccidiosis
Etiological agent(s)	*Isospora belli* and *Cyclospora cayetanensis*
Major organ(s) affected	Small intestine
Transmission mode or vector	Soil transmission, food-borne, and waterborne
Geographical distribution	Worldwide, but more prevalent in the tropics
Morbidity and mortality	Generally a self-limiting diarrhea, but much more severe in AIDS patients
Diagnosis	Detection of oocysts in feces
Treatment	Trimethoprim-sulfamethoxazole (TMP-SMX)
Control and prevention	Avoid ingestion of contaminated food or water

The coccidia are characterized by a thick-walled oocyst stage that is typically excreted with the feces. The oocysts of many species are quite resistant to environmental factors and can survive for months, even years, if kept cool and moist. Some coccidia carry out their entire life cycle within the intestinal epithelial cells of a single host (i.e., **monoxenous**) and are transmitted in a fashion similar to the fecal–oral route (Chapter 2). Other coccidia, such as *Toxoplasma* and *Sarcocystis*, are **heteroxenous** and have more complicated life cycles involving a predator–prey type of life cycle in multiple hosts (Chapter 14). The monoxenous coccidia that infect humans are *Isospora belli* and *Cyclospora cayetanensis*. *Cryptosporidium* (Chapter 12) is often grouped with these intestinal coccidian. However, *Cryptosporidium* is more closely related to the gregarines and exhibits some differences in the life cycle. Despite being monoxenous and exhibiting a fecal–oral type of life cycle the intestinal coccidia exhibit a typical apicomplexan life cycle which includes the processes of merogony, sporogony, and gametogony and produce invasive motile forms called merozoites and sporozoites (Chapter 11).

Isosporiasis

Isospora belli is believed to only infect humans and is probably the only species of *Isospora* that routinely infects humans. Thus, it is generally accepted that there are no nonhuman reservoirs of *Isospora*. Molecular studies also confirm that probably a single species of *Isospora* infects humans and that transmission is probably confined to an anthroponotic cycle. *Isospora* has a worldwide distribution but it is more common in tropical regions and areas with poor sanitation. Infections are often asymptomatic and those with symptoms tend to exhibit a self-limiting diarrhea lasting a few weeks. Therefore, the incidence of isosporiasis is probably underestimated. Infections are more common and the symptoms more severe in AIDS patients than in immunocompetent persons.

Life cycle

The infection is acquired through the ingestion of food or water contaminated with sporulated **oocysts** (Figure 13.1). **Sporozoites** are released from the oocysts in the intestinal lumen and invade intestinal epithelial cells. As is typical in apicomplexan parasites, the intracellular form will initially go through a trophic period followed by asexual reproduction. The intracellular parasite will initially divide by endodyogeny to form two **merozoites.** Endodyogeny is a type of cell division in which the daughter cells form within the mother cell (Chapter 11). The merozoites can undergo additional rounds of endodyogeny or may eventually form multinucleated meronts. In this case, the merozoites form by budding from the multinucleated meront as is observed in other apicomplexa such as *Cryptosporidium* or *Plasmodium.* Merozoites are released from the enterocyte, reinvade intestinal epithelial cells, and can undergo additional rounds of merogony. The parasites are predominantly found in the enterocytes lining the villi of the small intestine and are rarely found in the large intestine.

As an alternative to this asexual reproduction, the merozoites can develop into either micro- or macrogamonts which respectively develop into **microgametes** and **macrogametes**. The microgamont is multinucleated and produces numerous motile biflagellated microgametes. These microgametes are released into the lumen of the intestines and fertilize the macrogamete. The resulting zygote develops into the oocyst and is passed in the feces. The oocysts found in the feces are immature and therefore not infectious. Maturation into infectious sporulated oocysts occurs in the environment at ambient temperature and in the presence of oxygen. Studies in *Isospora* species from animals indicate that sporogony is inhibited by temperatures greater than 40°C and less than 20°C and completion of sporulation can occur in less than 24 hours and is generally completed within 2 to 3 days. This type of transmission is often referred to as **soil transmission** because the infection cannot be acquired through direct contact and requires an incubation period outside of the human body.

The unsporulated oocyst is easily recognized by its relatively large oval shape. Initially the oocyst will contain a single cell mass within the cyst wall called a sporoblast or sporont. The completion of sporogony involves a cell division of the sporoblast to form two sporoblasts. The sporoblasts will

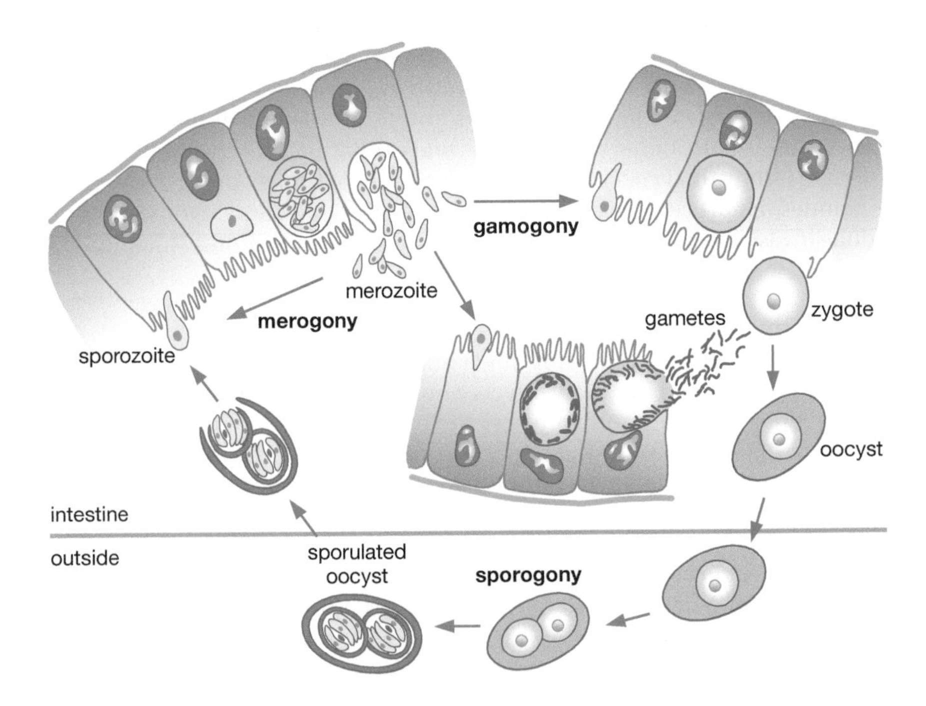

Figure 13.1 Life cycle of *Isospora belli*. The life cycle consists of merogony and gamogony (or gametogony) within intestinal epithelial cells and sporogony in the soil.

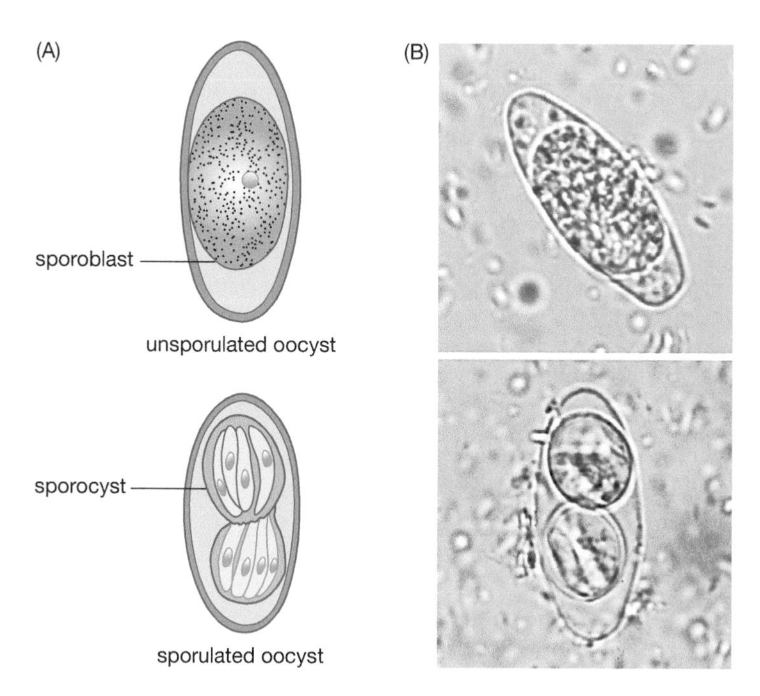

Figure 13.2 Oocyst structure. (A) Schematic figure illustrating an immature oocyst containing a single sporoblast and mature oocyst with two sporocysts. (B) Bright field microscopy of feces showing immature (top) and mature (bottom) oocysts. (Photos kindly provided by Marc Lontie, Medisch Centrum voor Huisartsen, Leuven, Belgium.)

undergo two more rounds of cell division to form four sporozoites each. Thus the mature sporulated oocyst consists of two **sporocysts** each with four sporozoites (Figure 13.2). A small percentage (2–5%) of the sporulated oocysts will contain a single sporocyst containing eight sporozoites and are referred to as *Caryospora*-like oocysts.

Symptoms and pathogenesis

Symptoms associated with *I. belli* infection include diarrhea, steatorrhea (i.e., excess fat in the feces), headache, fever, abdominal pain, nausea, dehydration, and weight loss. Blood is rarely present in the feces. In general, the symptoms are similar to those of cryptosporidiosis (Chapter 12). The disease is often self-limiting. However, it can become chronic with oocysts being detected in the feces for months to years and with recrudescences of the symptoms. The disease tends to be more severe in infants and young children than in adults. Pathology associated with *I. belli* infections are primarily villous atrophy, or blunting, and crypt hyperplasia as commonly seen in other intestinal infections (Chapter 2).

AIDS patients exhibit a higher prevalence of isosporiasis than immunocompetent persons and it has been noted as an opportunistic infection of the gastrointestinal tract, especially in patients with low $CD4^+$ lymphocyte counts. The diarrhea in AIDS patients is often very watery and can lead to dehydration requiring hospitalization. Fever and weight loss are also a common finding. Another common finding among AIDS patients is a chronic intermittent diarrhea lasting for months to years. The resulting excessive weight loss and electrolyte imbalance can lead to wasting and even death. There have also been a few reports of disseminated extraintestinal isosporiasis in AIDS patients in which parasites have been found in lymph nodes, the walls of the small and large intestines, liver, spleen, and bile duct.

Cyclospora cayetanensis

The first human cases of *Cyclospora cayetanensis* were reported in 1977. It was originally referred to as cyanobacteria-like bodies or coccidian-like bodies (CLB). The organism was confirmed to be a coccidian parasite with an oocyst structure similar to the genus *Cyclospora* and then named in 1994 after the Universidad Peruana Cayetano Heredia, where most of the early

Box 13.1 *Eimeria* and poultry production

Coccidiosis is a common disease found in many different animals and is especially important in poultry and cattle. *Eimeria* species are the primary cause of coccidiosis in animals. In general each *Eimeria* species is restricted to a single host species. Some of the more common pathogenic species are *E. necatrix* and *E. tenella* (in chickens), *E. meleagrimitis* (in turkeys), *E. stiedae* (in rabbits), and *E. bovis*, *E. ellipsoidalis*, and *E. zuernii* (in cattle). The life cycle of *Eimeria* is similar to other monoxenous coccidia. The oocyst structure consists of four sporocysts each containing two sporozoites. Several species of *Eimeria* infect chickens and each species infects different portions of the gastrointestinal tract and causes different lesions. Infection with multiple *Eimeria* species is quite common. In addition, infection with *Eimeria* predisposes the chickens to bacterial infections. The *Eimeria* of poultry have been extensively studied due to losses and high economic costs. Losses come from weight loss, failure to thrive, and death. Additional costs include drugs, veterinary services, and preventive efforts. Historically, the primary method for controlling coccidiosis has been to include anticoccidials in the feed. With their prolonged use, however, the parasite has become resistant to anticoccidials. In addition, consumer demand for poultry raised without drugs has increased. This has prompted more producers to use vaccination to control coccidiosis.

studies had been carried out. Molecular studies indicate a close relationship to *Eimeria*, an important veterinary parasite of poultry and other livestock (Box 13.1). *C. cayetanensis* has a worldwide distribution, but appears to be especially prevalent in Latin America, the Indian subcontinent, and southeast Asia. In developed countries the infections are usually associated with either food-borne outbreaks or traveler's diarrhea. Similar to *Isospora belli*, humans appear to be the only natural host and reservoir for *Cyclospora cayetanensis*.

Life cycle and transmission

The life cycle of *Cyclospora* is similar to *Isospora* (see above and Figure 13.1). The infection is acquired through the ingestion of oocysts. Sporozoites are released and infect epithelial cells of the upper small intestine (i.e., jejunum). The parasite undergoes merogony and the merozoites reinfect enterocytes and several more rounds of merogony can occur. Some of the merozoites undergo a sexual development resulting in the production of micro- and macrogametes. Fertilization of the macrogamete by the microgamete initiates sporogony and the formation of the oocyst. Like *Isospora*, sporulation is completed in the environment and immature noninfectious oocysts are excreted in the feces. The maturation of the oocysts to sporulated infectious oocysts generally takes 7–15 days. This extended maturation period in the environment suggests that a transmission vehicle is needed. In addition, the structure of the *Cyclospora* oocyst is slightly different from that of *Isospora*. The oocyst contains two sporocysts which each contain two sporozoites (Figure 13.3).

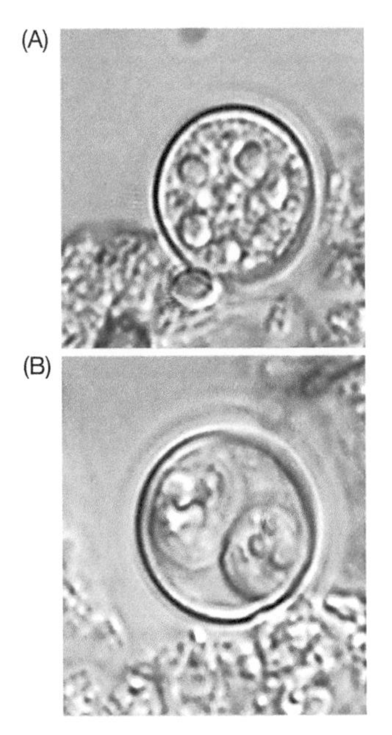

Figure 13.3 *Cyclospora* oocysts. Differential interference contrast micrographs of (A) unsporulated and (B) sporulated oocysts. (Photos kindly provided by Ynés Ortega, University of Georgia.)

Contaminated fresh produce, food, and water are major factors in the transmission of *Cyclospora*. Globalization of the food supply and international travel are the primary risk factors associated with transmission in developed countries. For example, several outbreaks in the United States and Canada have been associated with fresh produce imported from South and Central America (Table 13.1). In particular, berries and leafy vegetables have been identified as the probable contaminated item. These are food items that are typically eaten raw and are only rinsed. No outbreaks have been associated with frozen, processed, or peeled fruits or vegetables. A seasonality in the outbreaks has also been observed with most of the cases occurring in the spring and early summer. A similar seasonality has also been observed in endemic countries. In contrast to the United States and Canada, where food-borne transmission dominates, the majority of the cases in Europe and Australia have been associated with travel to endemic countries.

Table 13.1 Selected cyclosporiasis outbreaks in United States and Canada

Date	Location	Vehicle	Source
May 1995	Florida	Raspberries?	Guatemala?
May–June 1996	U.S., Canada	Raspberries	Guatemala
March–April 1997	Florida	Mesclun	Peru?
April–May 1997	U.S., Canada	Raspberries	Guatemala
June–July 1997	Washington, D.C.	Basil	Unknown
September 1997	Virginia	Fruit plate	Unknown
December 1997	Florida	Mesclun	Peru
May 1998	Ontario	Raspberries	Guatemala
	Georgia	Fruit salad?	Unknown
May 1999	Ontario	Berry dessert	Unknown
	Florida	Fruit (berry)	Unknown
July 1999	Missouri	Basil	Mexico or U.S.
June 2000	Pennsylvania	Raspberries	Guatemala
May 2001	British Columbia	Basil	Thailand
May–June 2004	Pennsylvania	Snow peas	Guatemala
March–May 2005	Florida	Basil	Peru

Modified from Herwaldt (2000) and Ortega and Sanchez (2010). Question marks (?) indicate probable vehicle or source, but not confirmed.

As a result of the high number of outbreaks associated with raspberries from Guatemala the United States temporarily restricted the import of raspberries during the spring of 1998 and required inspection of farms. This resulted in a subsequent drop in the number of outbreaks in the United States. Canada, which did not restrict importations, did not experience a drop in the number of outbreaks during the spring of 1998. Subsequent case controlled studies carried out in Guatemala revealed that infections were most common in children with the prevalence peaking in June. The major risk factor associated with the infection was drinking untreated water. In Peru, contact with soil was identified as another risk factor, especially among children less than 2 years old. Inadequate water treatment in these endemic countries may lead to contamination of the ground water and thus maintain the transmission cycle. Presumably food-borne transmission is due to irrigation or applying fertilizer with contaminated water, or washing and processing foods with poorly treated water. However, little is known about the routes of transmission and risk factors in endemic countries.

Symptoms

Cyclospora primarily infects epithelial cells in the upper portion of the small intestine. The incubation period is generally 1 to 2 weeks. Symptoms are similar to the gastroenteritis caused by *Isospora* and *Cryptosporidium* (Chapter 12) which typically includes cycles of watery diarrhea and periods of apparent remission. The diarrhea is characterized by frequent stools and can persist for up to 6 weeks, but is generally self-limiting in immunocompetent persons. Anorexia, malaise, nausea, and cramping are other frequent symptoms associated with cyclosporiasis. In some cases patients may experience vomiting, muscle aches, substantial weight loss, and explosive diarrhea. Previous exposure to *Cyclospora* appears to confer some resistance to infection with a lessening of the symptoms. Over time adults appear to develop immunity and asymptomatic carriers can be found in endemic areas.

As is also the case for *Cryptosporidium* and *Isospora*, the diarrhea caused by *Cyclospora* in AIDS patients is much more severe than in immunocompetent persons. The diarrhea can last for months and produce a syndrome that is debilitating and life threatening.

Diagnosis, Treatment, and Control of Intestinal Coccidia

Coccidiosis is diagnosed by demonstrating oocysts in the feces. Acid-fast staining is the preferred method for coccidia; the oocysts stain bright red against a blue background making them easy to detect. *Cryptosporidium*, *Cyclospora*, and *Isospora* are distinguished by size and oocyst structure (Table 13.2). *Cyclospora* and *Isospora* do not uniformly take up the stain resulting in a mixture of unstained, partially stained, and completely stained oocysts. *Cyclospora* and *Isospora* can also be detected via an autofluorescence associated with the cyst wall. Because of its relatively large size, *Isospora* is readily detected in unstained specimens. *Sarcocystis* is a rare human infection (Chapter 14) with oocysts similar to *Isospora* except that the sporocysts are generally released from the oocysts while still in the intestinal lumen.

As with many other apicomplexa, antifolates are effective in the treatment of intestinal coccidia with the exception of *Cryptosporidium*. The

Table 13.2 Distinguishing coccidian parasites found in human feces

Species	Excreted form	Size	Oocyst structure
Cryptosporidium	Sporulated oocysts	4–5 μm	4 sporozoites, no sporocysts
Cyclospora	Unsporulated oocysts	8–10 μm	2 sporocysts with 2 sporozoites each
Isospora	Unsporulated oocysts	30 × 12 μm	2 sporocysts with 4 sporozoites each
Sarcocystis	Sporulated sporocysts, occasional oocyst	13 × 10 μm	2 sporocysts with 4 sporozoites each

recommended treatment for *Cyclospora* and *Isospora* is the combination of trimethoprim-sulfamethoxazole (TMP-SMX). The treatment is generally effective, but relapses are observed in about half of the symptomatic cases. Continued treatment after the normal regimen is recommended for immunocompromised patients as a prophylaxis. Trimethoprim alone can be tried in patients allergic to sulfa drugs. There are no specific preventive measures because the nature of the transmission cycle is not completely understood. Precautions about food and water need to be taken and chemoprophylaxis is recommended for immunocompromised persons traveling to endemic areas.

Summary and Key Concepts

- *Isospora belli* and *Cyclospora cayetanensis* are monoxenous coccidia that infect humans and do not appear to infect other animals.
- Both species exhibit a typical coccidian life cycle with merogony and gametogony occurring primarily in the enterocytes of the small intestine.
- The completion of sporulation occurs outside of the body at ambient temperature and in the presence of oxygen, and thus these coccidia exhibit a soil transmission.
- *Cyclospora* has been noted for several food-borne outbreaks in the United States and Canada involving produce from Central and South America.
- The predominant symptom associated with both isosporiasis and cyclosporiasis in immunocompetent persons is a self-limiting watery diarrhea.
- AIDS patients exhibit a much more severe and longer lasting diarrhea.

Further Reading

Herwaldt, B.L. (2000) *Cyclospora cayetanensis*: A review focusing on the outbreaks of cyclosporiasis in the 1990s. *Clin. Inf. Dis.* 31: 1040–1057.

Jongwutiwes, S., Putaporntip, C., Charoenkorn, M., Iwasaki, T. and Endo, T. (2007) Morphologic and molecular characterization of *Isospora belli* oocysts from patients in Thailand. *Am. J. Trop. Med. Hyg.* 77: 107–112.

Lindsay, D.S., Dubey, J.P. and Blagburn, B.L. (1997) Biology of *Isospora* spp. from humans, nonhuman primates, and domestic animals. *Clin. Microbiol. Rev.* 10: 19–34.

Mansfield, L.S. and Gajadhar, A.A. (2004) *Cyclospora cayetanensis*, a food- and waterborne coccidian parasite. *Vet. Parasitol.* 126: 73–90.

Ortega, Y.R. and Sanchez, R. (2010) Update on *Cyclospora cayetanensis*, a food-borne and waterborne parasite. *Clin. Microbiol. Rev.* 23: 218–234.

Toxoplasma gondii and Tissue Cyst Forming Coccidia

Disease(s)	Toxoplasmosis
Etiological agent(s)	*Toxoplasma gondii*
Major organ(s) affected	Brain, eyes
Transmission mode or vector	Ingestion of food or water contaminated with oocysts from cat feces, ingestion of raw or undercooked meat containing tissue cysts, congenitally
Geographical distribution	Worldwide
Morbidity and mortality	Causes a benign disease in immunocompetent persons and only results in serious disease if obtained congenitally or if immunocompromised
Diagnosis	Serology
Treatment	Antifolates
Control and prevention	Avoid ingestion of contaminated food or water or raw meat

Toxoplasma gondii is a coccidian parasite which infects humans as well as a wide variety of mammals and birds. It was first described in 1908 in an African hamster-like rodent called the gundi and the species name reflects this original host. Subsequently *Toxoplasma* was identified in numerous other animals and the first human case was reported in 1939. Despite the wide host range (essentially all warm-blooded vertebrates) only one species is recognized within the genus *Toxoplasma*. *Toxoplasma* exhibits a predator–prey life cycle and felids are the only definitive host. Toxoplasmosis is found throughout the world (except extremely cold or dry climates) and tends to be more prevalent in tropical climates. It is estimated that approximately one-third of the world's population may be chronically infected with *Toxoplasma* based on serological evidence. The prevalence varies with location with the United States and the United Kingdom exhibiting prevalences ranging from 16 to 40%, whereas prevalences in continental Europe and Central and South America range from 50 to 80%. Despite the high incidence of infection, toxoplasmosis is often unrecognized because it most often causes a benign disease with few or no symptoms. Noted exceptions are in the cases of congenital infection or immunocompromised individuals.

Life Cycle and Transmission

Toxoplasma has a complex life cycle consisting of intestinal and tissue phases (Figure 14.1). Although the organism was first discovered in 1908 as a tissue parasite of the gundi, its complete life cycle was not determined until 1970. The life cycle is described as being **predator–prey** transmission in that the predator acquires the infection by eating a prey infected with the tissue stage of the parasite. Within the prey host the parasite undergoes an intestinal phase resulting in the excretion of the form infective for the predator. The intestinal phase of *Toxoplasma* only occurs in felids and is nearly identical to that of *Isospora* (Chapter 13). Cats acquire the infection by eating animals infected with the tissue stage of the parasite. This stage is often referred to as the tissue cyst and contains hundreds to thousands of bradyzoites. The bradyzoites invade intestinal epithelial cells and undergo an asexual replication called **merogony** that ends in the production of merozoites. The resulting merozoites are released from the infected cell and invade other intestinal epithelial cells and can then either undergo additional rounds of merogony or undergo **gametogony**. Similarly to other apicomplexa (Chapter 11), gametogony results in the production of macro- and microgametes. Thus, the cat is considered the **definitive host** since this is the host in which the sexual cycle occurs.

The biflagellated microgametes are released into the lumen of the intestine and fertilize the macrogametes within the host epithelial cells. Secretion of the oocyst wall begins shortly after fertilization. This sexual cycle culminates in the production of oocysts which are excreted in the feces. As with *Isospora*, these immature oocysts undergo sporogony at ambient temperature (i.e., **soil transmission**) and the mature oocysts contain two sporocysts, each with four sporozoites. Sporulation generally takes 1–4 days and the oocysts remain infective for months in shaded moist soil. Because this intestinal phase of the life cycle and the morphology of the oocyst are essentially identical to that of *Isospora*, it was initially felt that these parasites were a species of *Isospora* and the connection to the tissue cysts in other animals and complete life cycle of *Toxoplasma* were not determined until 1970.

Intermediate hosts, such as rodents and birds, become infected through the ingestion of sporulated oocysts. Following oocyst ingestion, the sporozoites are released and initially invade enterocytes followed by merogony (i.e., asexual replication). Hosts which cannot support the sexual phase of the life cycle are defined as intermediate hosts. The resulting merozoites are then capable of invading other cells and in particular macrophages which will disseminate the parasite throughout the intermediate host. The

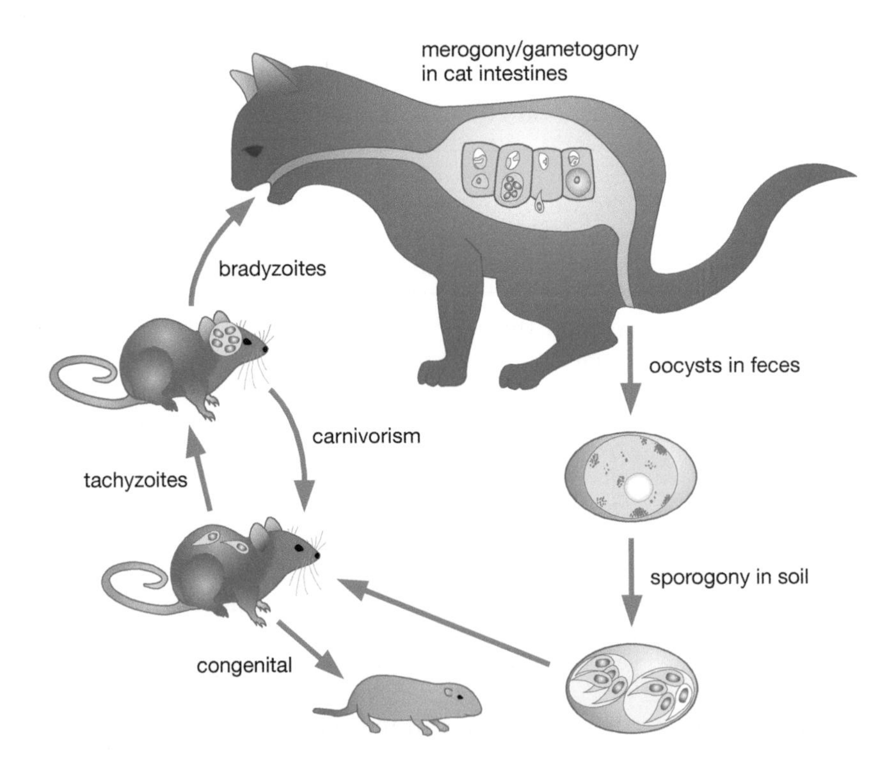

Figure 14.1 Life cycle of *Toxoplasma gondii*. *Toxoplasma* exhibits a predator–prey life cycle involving cats as the definitive host and a variety of intermediate hosts. The cat acquires the infection by ingesting an infected animal. Following an intestinal infection, oocysts are excreted with the feces which mature in the environment. Ingestion of oocysts by an intermediate host results in an initial acute infection (tachyzoites) followed by the formation of tissue cysts (bradyzoites). The infection can also be spread between intermediate hosts through carnivorism or congenitally.

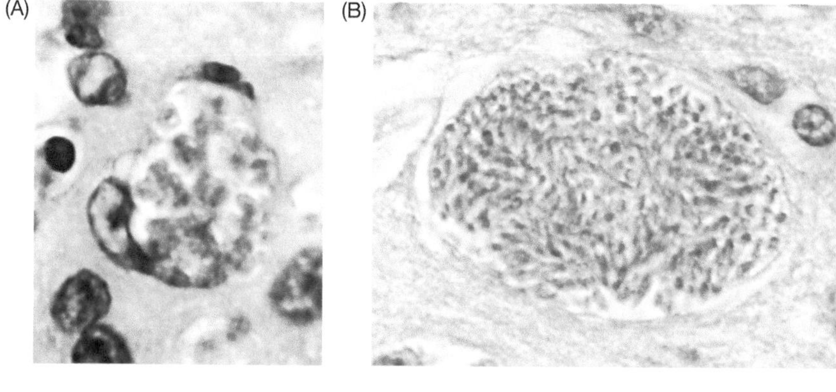

Figure 14.2 Tissue stages of *Toxoplasma gondii*. (A) Tachyzoites within an infected liver cell. (B) Tissue cyst with bradyzoites from brain section. (C) Schematic diagram showing conversion from tachyzoites to bradyzoites.

invasion process is typical of other apicomplexan parasites and involves the apical organelles (Chapter 11). The parasite literally crawls into a vacuole being formed as the parasite enters the host cell and upon completion of invasion the parasite lies within a parasitophorous vacuole. Thus, the parasite is not in direct contact with the host cell cytoplasm, but is surrounded by a parasitophorous vacuolar membrane. The parasitophorous vacuole is distinct from a phagosome and does not fuse with lysosomes. Because of this distinction between the parasitophorous vacuole and the phagolysosome, the parasite is able to survive within macrophages which would normally destroy intracellular pathogens. Within the parasitophorous vacuole the parasite undergoes binary fission by a unique process called endodyogeny. Endodyogeny is a specialized type of division in which the two daughter cells form within the mother cell (Chapter 11). These trophic forms are called **tachyzoites** (tachy- means rapid) in reference to their high level of replication (Figure 14.2). The host cell will rupture and release the tachyzoites which will invade new host cells and repeat the replicative cycle. The tachyzoites are capable of infecting any type of cell, but are generally found within the reticuloendothelial system (e.g., liver and spleen) during the early acute stage of the infection.

Host immunity influences the parasite stage within the intermediate host

Following the acute phase of the infection, characterized by rapid parasite replication and dissemination, the tachyzoites develop into slowly replicating forms called **bradyzoites** (brady- means slow). The bradyzoite-infected cell is encapsulated and forms a tissue cyst (Figure 14.2). Bradyzoites are considered to be metabolically quiescent—but remain viable—and thus represent a dormant or resting stage. Other changes that occur when tachyzoites convert to bradyzoites include secretion of chitin and other components of the tissue cyst wall and the accumulation of amylopectin granules (i.e., starch), reflecting changes in glucose metabolism and storage. The tissue cysts of *Toxoplasma* exhibit a range of sizes, but often obtain a size of 50–70 μm in diameter containing 1000–2000 bradyzoites.

The bradyzoites begin to appear approximately 10 days after the primary infection and are primarily found in brain and muscle tissue. In addition, the host immune system plays a role in the conversion of tachyzoites to bradyzoites in that bradyzoites begin to appear after an immune response by the host and immune suppression can cause the bradyzoites to become reactivated into tachyzoites. *Toxoplasma* initiates both innate and adaptive immune responses which lead to immunity from subsequent infection and disease resistance. The bradyzoite stage represents a chronic infection and probably persists for the life of the host. The mechanism for this persistence is unknown but the tissue cyst wall probably protects the parasite from the host immune system which effectively eliminates the tachyzoite stage of the parasite. Some investigators believe the tissue cysts periodically break down and release the bradyzoites which will invade new host cells and lead to the formation of more tissue cysts and promote parasite persistence.

The tissue phase of the infection can also be transmitted congenitally to offspring and to other intermediate hosts through carnivorism. Possibly all warm-blooded vertebrates can become infected with *Toxoplasma* and serve as intermediate hosts. Ingestion of an infected animal will release the bradyzoites from the tissue cysts which then infect cells in the new host. Because the new host is not immune, the parasites will go through an acute phase characterized by rapid replication and dissemination of tachyzoites followed by a chronic phase characterized by dormant tissue cysts containing bradyzoites. Both tachyzoites and bradyzoites can transmit the disease via carnivorism, but bradyzoites are much more infectious. This spread through carnivorism represents a horizontal transfer among the intermediate hosts.

Only felids support the sexual cycle of *Toxoplasma*

Ingestion of an infected intermediate host or oocysts by the cat will initiate the intestinal stage of the life cycle involving merogony and gametogony in the intestinal epithelial cells. The frequency of infection and the prepatent period is determined by the stage of the parasite ingested by the cat. Nearly all of the cats ingesting bradyzoites will become infected and oocysts can be detected in the feces 3–10 days after infection. In contrast, less than half of the cats ingesting tachyzoites or oocysts will become infected and the prepatent period will be approximately 3 weeks. Cats also exhibit the tissue stages (i.e., extraintestinal) of the parasite and spread the infection congenitally. Congenitally infected kittens can excrete oocysts.

Other coccidians exhibiting tissue cysts and a predator–prey type of life cycle include *Sarcocystis* (see below) and *Neospora caninum*. *N. caninum* is closely related to *Toxoplasma* and is also transmitted via a predator–prey type of life cycle with dogs as the definitive host (Box 14.1). It does not infect humans, but is an important pathogen in cows.

Human transmission

Obviously humans are not a natural part of this predator–prey life cycle and represent an incidental host that does not participate in the continuation of the transmission cycle (Figure 14.3). One source of infection is the ingestion of food or water contaminated with sporulated oocysts (Table 14.1). This implies some association with cats. However, since the oocysts need to mature in the environment before becoming infectious, transmission will include features of soil transmission similar to *Isospora* and *Cyclospora* (Chapter 13). For example, children of crawling and dirt-eating ages are believed to be at higher risk for infection. Oocysts can also be acquired through gardening activities or ingestion of unwashed fruits or vegetables. In addition, a few waterborne outbreaks have been documented. The high prevalence of toxoplasmosis in South and Central America is believed to be due to high levels of contamination of the environment with oocysts. Ironically, contact with dogs is often found to be more of a risk factor for becoming infected with *Toxoplasma* than contact with cats. This is likely due to dogs seeking out cat feces and becoming contaminated with the

Box 14.1 *Neospora caninum*

Neospora caninum is closely related to *Toxoplasma gondii* and exhibits a nearly identical morphology. As with *Toxoplasma*, *Neospora* infects many domestic animals and is a major cause of abortions and stillbirth in cattle worldwide. The definitive hosts are dogs which exhibit neuromuscular disease. Humans are not potential hosts. Dogs become infected after ingestion of infected tissues from intermediate hosts. Unsporulated oocysts are shed in the feces and they sporulate in the environment. Intermediate hosts acquire the infection by ingestion of sporulated oocysts. Unlike *Toxoplasma*, though, intermediate hosts cannot acquire the infection by ingestion of tissue forms from other intermediate hosts. However, the infection can be transmitted congenitally and the parasite is readily maintained in both cattle and dogs by vertical transmission. A sylvatic cycle involving white-tailed deer and coyotes has also been identified [1]. This implies there may be overlaps between the domestic cycle and the sylvatic cycle and will impact potential control measures.

1. Rosypal, A.C. and Lindsay, D.S. (2005) The sylvatic cycle of *Neospora caninum*: where do we go from here? *Tr. Parasitol.* 21: 439–440.

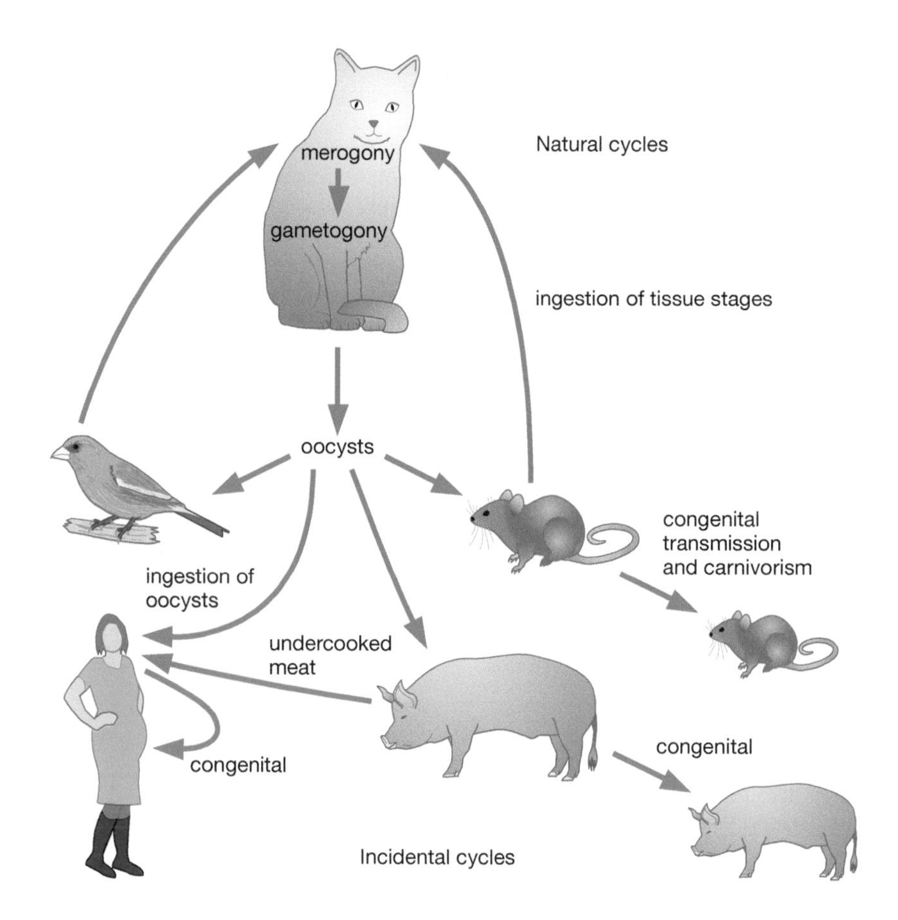

Figure 14.3 Transmission cycles of *Toxoplasma gondii*. The natural transmission cycle of *Toxoplasma gondii* is a predator–prey life cycle involving cats and rodents and birds. Other mammals, including humans, can be infected as incidental hosts with the oocysts excreted by cats. The infection can also be spread among intermediate hosts through carnivorism and congenitally. (Adapted from *Parasitology Today*, Volume 13, Number 10 with kind permission from Elsevier Ltd.)

feces by rolling in them and then transferring the sporulated oocysts to the clothes and hands of their owners.

Toxoplasmosis can also be acquired through the ingestion of undercooked meat containing tissue cysts and undercooked meat is generally considered to be the major source of infection. In fact, most infections in the United States and Europe among adults are probably acquired from undercooked meat. The especially high seropositive rate in France (approximately 85%) is likely due to a cultural predilection for lightly cooked or raw meat. Presumably livestock acquire the infection through grazing in areas contaminated with cat feces or via contaminated feed. Mutton and pork are more common sources than beef. In fact, there is no conclusive evidence demonstrating the presence of infectious parasites in beef. The decline in *Toxoplasma* seroprevalence in developed countries over the past decades is believed to be a result of modern farming practices and an increased use of frozen meat by consumers. However, an emerging risk is the increased popularity of game meat such as deer, boar, and kangaroo. This ability of the parasite to transfer between intermediate hosts may be a relatively recent evolutionary adaptation of the parasite that coincides with the domestication of the cat and the expansion of agriculture (Box 14.2).

Toxoplasma can also be transmitted from mother to fetus, often with dire consequences (see below). **Congenital transmission** can only occur during an acute infection (i.e., tachyzoite stage) acquired during pregnancy or very shortly before. Mothers with a chronic infection acquired before the pregnancy are not at a risk for transmitting *Toxoplasma*. Thus congenital transmission can only occur once with any given mother. Approximately one-third of mothers who seroconvert during pregnancy will pass on the infection to the fetus and congenital transmission is more likely later during the pregnancy. Chronically infected mothers can transmit the disease congenitally if the infection becomes reactivated through immunosuppression.

Table 14.1 Sources of human infection
Oocyst-contaminated food or water
Undercooked meat
Raw goat's milk
Congenital
Organ transplant
Blood transfusion

Box 14.2 Recent expansion and molecular epidemiology of *Toxoplasma*

Molecular analyses of *Toxoplasma* isolates from humans and domestic animals (primarily from North America and Europe) reveal a limited genetic diversity. This is in contrast to the sylvatic cycles (i.e., wild animals) in which a much higher degree of genetic diversity is observed. The majority (>94%) of the domestic isolates cluster into three distinct clonal lineages designated as type I, type II, and type III. These three clonal lineages are closely related (1–2% sequence difference at the nucleotide level) and are composed of various mixtures of just two alleles at the loci tested. The three types may have arisen from a limited number of genetic recombinations occurring within the last 10 000 years [1]. This time period is concurrent with the expansion of human agriculture and the adaptation of the domestic cat. Thus, changes in human behavior may have led to a selection and rapid propagation of *Toxoplasma*. Furthermore, these clonal types all exhibit the ability to be transmitted via a direct oral route between intermediate hosts, which may have not been a biological feature of the ancestral *Toxoplasma* or other closely related species like *Neospora* (Box 14.1). This acquisition of direct oral infectivity combined with domestication of animals (i.e., cats and agricultural animals) could have promoted a rapid, and primarily asexual, expansion of *Toxoplasma*.

The three *Toxoplasma* genotypes exhibit differences in geographical distribution and virulence [2]. For example, type I parasites are highly virulent in mice, whereas types II and III are not. Similarly, type I parasites appear disproportionately associated with severe atypical ocular toxoplasmosis in immunocompetent individuals. Severe disease associated with atypical strains (i.e., not types I, II, or III) has also been occasionally observed in immunocompetent persons. No clear relationships between parasite genotype and severe congenital toxoplasmosis or severe disease in immunocompromised persons have been observed. Type II infections tend to dominate, especially in the United States and France. However, there is some evidence suggesting that *Toxoplasma* exhibits more diversity in South America and that there may be an African clonal type. Improvements in our knowledge about *Toxoplasma* population biology may help resolve these issues and lead to better control and treatment.

1. Su, C. et al. (2003) Recent expansion of *Toxoplasma* through enhanced oral transmission. *Science* 299: 414–416.

2. Dardé, M.L. (2008) *Toxoplasma gondii*, "new" genotypes and virulence. *Parasite* 15: 366–371.

Transmission of *Toxoplasma* as a result of organ transplants is also possible. Human infections with *Toxoplasma* were noted in the early days of heart transplantations even before the life cycle of *Toxoplasma* was known. Tissue cysts from a chronically infected organ donor can reactivate when transplanted into a previously uninfected organ recipient. Such transmission from a seropositive donor to a seronegative recipient has been noted in heart, heart-lung, kidney, liver, and liver–pancreas transplant patients. In addition, the immunosuppressive therapy can reactivate a latent infection in a seropositive recipient. This is more likely to occur in bone marrow, hematopoietic stem cell, and liver transplant patients. Transmission via transplantation is quite rare now, because donors and recipients are screened for *Toxoplasma*. Acquisition of tachyzoites from an acutely infected person via blood transfusion is also possible, but extremely rare. In addition, there have also been a few isolated reports of *Toxoplasma* being transmitted via ingestion of tachyzoites in unpasteurized goat's milk.

Clinical Features

Toxoplasmosis in adults and children past the neonatal stage is usually benign and asymptomatic. Acquisition of the infection via either oocysts or tissue cysts results in an acute infection in which tachyzoites are disseminated throughout the body via the lymphatics and the circulatory system. The tachyzoites are especially prominent in the reticuloendothelial system (i.e., spleen, liver, lymphatic system). This acute stage will persist for several weeks as immunity develops. Antibody production requires 1–2 weeks and cellular immunity occurs 2–4 weeks post-infection. Both humoral and cellular immunity are important, but the cellular response appears critical for the conversion from acute (i.e., tachyzoites) to chronic (i.e., bradyzoites) infection. In particular, the parasites initially induce a strong innate response involving natural killer cells leading to a strong T-helper-1 cell response characterized by the production of pro-inflammatory cytokines including interleukin-12, interferon-γ, and tumor necrosis factor-α. These immune mediators control parasite growth and are associated with the transition from acute to chronic toxoplasmosis.

When acute symptoms do occur, they are generally mild and typically described as mononucleosis-like with chills, fever, headache, myalgia, fatigue, and swollen lymph nodes (particularly in the neck and back of the head). These symptoms are self-limiting and resolve within weeks to months. A chronic lymphadenopathy without fever persisting or recurring for up to a year has also been occasionally noted as a symptom of toxoplasmosis. Rarely do immunocompetent individuals exhibit severe symptoms and the acute infection almost always progresses to an asymptomatic chronic stage. This latent infection probably persists for the life of the patient without producing any progressive pathology. Exceptions to this include toxoplasmosis in the immunocompromised patient, ocular toxoplasmosis, and congenital toxoplasmosis (Figure 14.4). However, there is some evidence indicating that toxoplasmosis might be associated with subtle changes in behavior or mental diseases in some individuals (Box 14.3).

Reactivation of latent infection

Toxoplasmosis has been long noted as an opportunistic infection in regard to reactivation of latent infections due to immunosuppression associated with organ transplants and certain cancer treatments. During the 1980s toxoplasmic encephalitis emerged as a common complication associated with AIDS. An estimated 25–50% of AIDS patients with chronic toxoplasmosis develop toxoplasmic encephalitis. Reactivation of the infection typically occurs when CD4 cells drop below 100 cells per microliter. Early symptoms of toxoplasmic encephalitis can include headache, fever, lethargy, reduced alertness, and mood or personality changes with progression to focal neurological deficits and convulsions. The most common focal neurological signs are weakness and speech disturbances. Patients can also present with

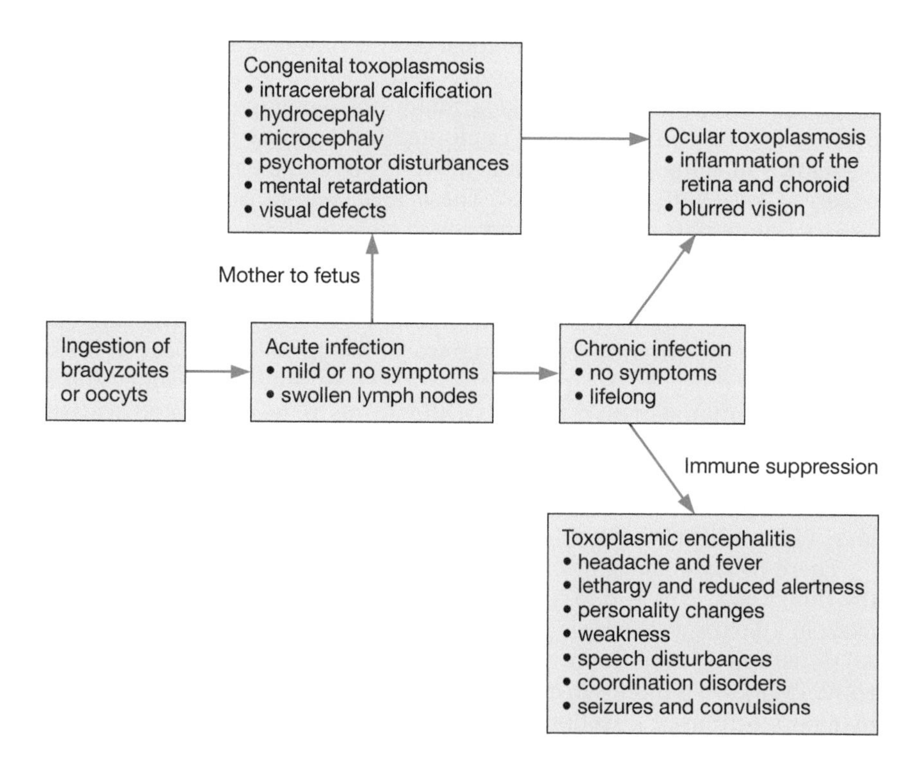

Figure 14.4 Major outcomes of toxoplasmosis. Schematic diagram showing major outcomes of human toxoplasmosis and major symptoms associated with these outcomes.

Box 14.3 Behavioral changes and schizophrenia

Studies in rodents have suggested that chronic toxoplasmosis may be associated with behavior changes that increase the potential for predation by cats [1]. For example, infected rodents exhibit diminished learning capacity and memory, increased activity levels, less fear of novelty (neophobia), and a decreased aversion to cat odor than uninfected controls. Increased activity levels and curiosity clearly increase predation rate since cats are attracted to moving and exposed objects. Possibly the most interesting behavior change is a lack of aversion to cat odor. Uninfected rats exhibit a strong aversion to cat odor, but do not avoid other animal odors such as rabbit. A proportion of the chronically infected rats not only exhibited a diminished odor aversion, but appeared to be attracted to the cat odor. These behavioral changes appear to be specific for *Toxoplasma* since other parasitic infections do not have the same effects. The mechanism of action in regard to how *Toxoplasma* alters rodent behavior is unknown. Presumably the cysts found in the brain have some destructive or neuromodulatory effect or the local immune response to the parasite affects brain function.

Associations between psychiatric disorders in humans and *Toxoplasma* infection have also been suggested [2]. In particular, several studies have indicated that the prevalence of seropositivity is higher among schizophrenic patients compared with control volunteers. The risk factor is relatively modest and behaviors associated with the disorder (e.g., lower personal hygiene) could increase the incidence of infection. Nonetheless, it is possible that *Toxoplasma* infection could be associated with some cases of schizophrenia. Similarly, other neuropsychological diseases, such as depression and epilepsy, have been associated with *Toxoplasma* infection. As is the case in the animal studies, the mechanism(s) by which the parasite leads to these psychiatric disorders is not known. The parasite could possibly directly affect neuronal function or the damage may be immune mediated.

1. Webster, J.P. (2007) The effect of *Toxoplasma gondii* on animal behavior: playing cat and mouse. *Schizophrenia Bull.* 33: 752–756.

2. Henriquez, S.A., Brett, R., Alexander, J., Pratt, J. and Roberts, C.W. (2009) Neuropsychiatric disease and *Toxoplasma gondii* infection. *Neuroimmunomodulation* 16: 122–133.

seizures, visual and other sensory disturbances, movement and coordination disorders, and other neuropsychiatric abnormalities. Typically the symptoms of toxoplasmic encephalitis are initially insidious and gradually worsen over time. Toxoplasmosis rarely presents as a rapidly fatal form of diffuse encephalitis spread throughout the brain. The disease is almost always due to a reactivation of a latent infection and tends to remain confined to the central nervous system. In other words, the tissue cysts rupture and the released bradyzoites transform into tachyzoites and invade neighboring cells resulting in the focal lesion. However, these tachyzoites normally do not spread to other organs and cause a disseminated disease.

The pathology is characterized by many foci of enlarging necrosis in the brain which can be detected by computed tomography (CT) or magnetic resonance imaging (MRI). A characteristic ring-enhancing lesion is often observed with MRI (Figure 14.5). These focal lesions are caused by the destruction of host cells in the immediate vicinity of the reactivated tissue cyst and an outward expansion of the tachyzoites. Sometimes these lesions are referred to as abscesses. However, the lesions have firm necrotic centers and do not contain pus or degenerating neutrophils, and thus are not abscesses in the strictest sense of the word. The presence of several brain lesions is the most characteristic feature of severe immunodeficiency and is especially common in AIDS patients. Other forms of the reactivated disease, especially retinochoroiditis (i.e., ocular toxoplasmosis), pneumonitis, myocarditis, and myositis, may occasionally occur in conjunction with immunosuppression. The predominance of lesions in the brain may be due to the loss of immunity in the brain before other tissues and organs.

There has been a significant decline in toxoplasmic encephalitis since the mid-1990s. This is due in part to the use of trimethoprim-sulfamethoxazole (Bactrim®) for prophylaxis against *Pneumocystis* infections in AIDS patients. In addition, the combination antiretroviral therapy (i.e., HAART) allows patients to maintain their CD4 cell levels and prevents many opportunistic infections including toxoplasmosis. However, HAART

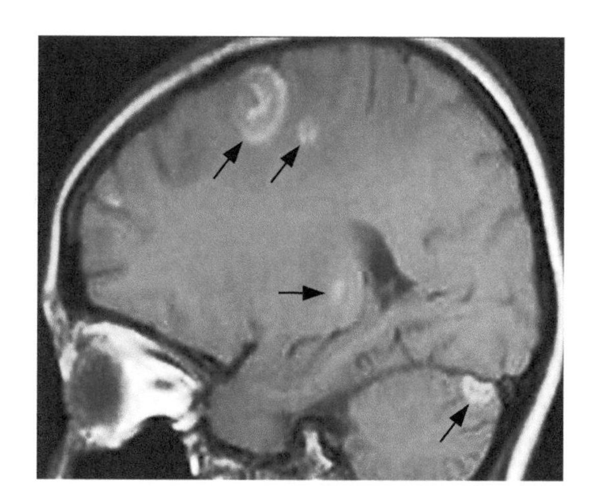

Figure 14.5 Toxoplasmic encephalitis. MRI image of ring-enhancing lesions associated with reactivation of a latent infection. Arrows denote lesions. (Photo kindly provided by Glenn A. Tung, Department of Diagnostic Imaging, Brown University School of Medicine.)

is quite expensive and in resource-poor settings toxoplasmosis and other opportunistic infections remain a major problem. In addition, persistent neurological problems are often present in patients successfully treated with HAART.

Ocular toxoplasmosis

A retinochoroiditis, inflammation of the retina and choroid (thick vascular area at back of eye), is another clinical manifestation of *Toxoplasma* infection. In fact, the majority of infectious uveitis (i.e., intraocular inflammation) may be due to *Toxoplasma*. Toxoplasmic retinochoroiditis can result from congenital infections or from infections acquired postnatally. Originally the ocular manifestations were more often associated with congenital infections or a late manifestation due to the reactivation of a congenital infection. However, ocular toxoplasmosis is being reported with increasing frequency in association with postnatally acquired infections in immunocompetent persons. It has been suggested that different genotypes of the parasite exhibit different levels of virulence, especially in regard to the expression of ocular disease (Box 14.2). In the case of congenital infection, the retinochoroiditis can develop weeks to years after birth. Approximately 20% of persons with congenital infections will exhibit retinochoroiditis at birth and by adolescence up to 85% will exhibit symptoms.

The lesions are focal in nature and generally self-limiting (Figure 14.6). They are believed to be the result of tissue cyst rupture in the retina in reactivated cases or tachyzoites in acute cases. The cells of the retina closely resemble those of the central nervous system. Granulomatous lesions may also be present in the choroid. The lesions are usually bilateral (i.e., both eyes) in congenital infections and unilateral (i.e., only one eye) if acquired postnatally. Animal studies provide evidence that the retinal necrosis associated with the lesion is attributable to proliferation of parasites, while hypersensitivity responses to toxoplasmic antigens are responsible for the accompanying inflammation. Symptoms can include blurred vision or other visual defects. Vision may improve with the resolution of the inflammation. Recurrences of the disease have been noted, but the frequency and factors influencing the recrudescence are not clear. The disease is rarely progressive in immunocompetent individuals, but can scar the retina. However, the disease can be quite severe in AIDS patients and the lesion continues to enlarge.

Congenital toxoplasmosis

Toxoplasma can also be transmitted congenitally (i.e., transplacentally) if the mother acquires the infection during pregnancy. In the United States between one per thousand and one per ten thousand live births are infected

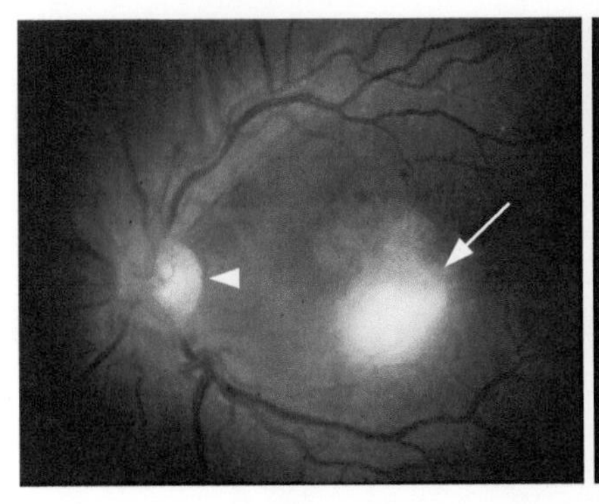
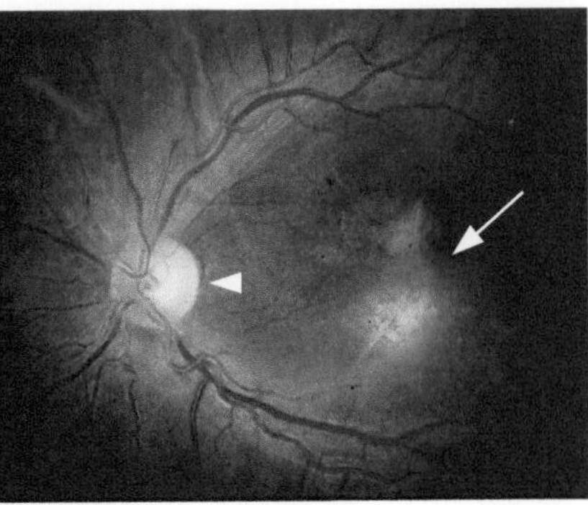

Figure 14.6 Ocular toxoplasmosis. Funduscopy of eye of patient with active lesion (left) and the same patient 3 weeks later (right). The optic nerve is denoted with arrowheads and the lesions with arrows. (Photos kindly provided by Mylan Van Newkirk, University of Otago, New Zealand.)

with *Toxoplasma* resulting in 500–5000 congenital toxoplasmosis cases per year. Infections acquired congenitally are more likely to be symptomatic than those acquired postnatally and can be particularly severe. The symptoms are generally more severe if the infection is acquired early in pregnancy. However, transmission is more frequent later in pregnancy (Table 14.2). Infection can result in spontaneous abortion, premature birth, or full-term birth with or without symptoms (Table 14.3). Typical disease manifestations include: retinochoroiditis, intracerebral calcification, hydrocephaly, microcephaly, psychomotor disturbances, mental retardation, or blindness and other visual defects. Many neonates exhibit subclinical infections (i.e., asymptomatic) at birth and will develop symptoms later in childhood or adolescence. Ocular disease (i.e., retinochoroiditis) is the most common sequela of congenital infection.

Diagnosis, Treatment, and Prevention

In contrast to most other protozoan infections, diagnosis is rarely made through the detection or recovery of organisms, but relies heavily on serologic procedures. Parasites can be detected in biopsied specimens, buffy coat cells, or cerebrospinal fluid. However, detecting tachyzoites from these materials may be difficult. These specimens can also be used to inoculate mice or tissue culture cells or be analyzed by PCR. Inoculation of animals is reliable, however time consuming. The results from inoculation of animals or PCR can be misleading though, since many individuals have been exposed to *Toxoplasma* and harbor tissue cysts (bradyzoites). Therefore, serologic tests are a recommended component of diagnosis. Imaging techniques may also be useful in the diagnosis of toxoplasmic encephalitis. MRI (Figure 14.5) has been shown to be superior to CT for the identification of the lesions.

Serology is useful in the diagnosis of *Toxoplasma*

Several serological tests are available for the detection of antibodies against *Toxoplasma*. In addition serology can be used to distinguish between the

Table 14.2 Risk of congenital toxoplasmosis

Gestational age (weeks)	Infection risk (95% CI)	Symptom risk (95% CI)
13	6% (3–9)	61% (34–85)
26	40% (33–47)	25% (18–33)
36	72% (60–81)	9% (4–17)

Modified from Montoya and Remington (2008). The table shows the percent risk (95% confidence interval) of becoming infected or exhibiting clinical symptoms according to the gestational age of the fetus when the mother seroconverted.

Table 14.3 Frequency of congenital toxoplasmosis outcomes

Frequency	Outcome
5–10%	Death (spontaneous abortion and stillbirth)
8–10%	Severe brain or eye damage
10–13%	Moderate to severe visual handicaps
58–72%	Asymptomatic at birth, but developing retinochoroiditis or neurological symptoms later

acute infection and the chronic infection. This is especially important in the case of pregnancy. High antibody titers by themselves are not definitive evidence of an acute infection and comparisons between IgM and IgG are usually necessary to distinguish between acute and chronic infections. The IgM levels rise during the first 8 months following infection and can persist for up to 18 months. IgG first appears 1–2 weeks after infection, peaks at 6–8 weeks, and persists for the life of the individual. Thus, indications for an acute infection are high IgM titers and rising total antibody titers in conjunction with symptoms. Chronic infections are characterized by low or no IgM and stable IgG levels. Tests of IgG avidity (functional affinity) are also useful in distinguishing chronic and acute infections and are becoming a standard in the diagnosis of toxoplasmosis. The presence of high affinity antibodies rules out a recent infection.

Because of the potentially dire consequences of a congenital infection, serological testing of pregnant women is recommended and required by law in some European countries. The tests can be carried out in nonreference clinical laboratories initially, but may need to be followed up with tests performed in a reference laboratory. Approximately 60% of the IgM positive samples identified in nonreference laboratories are found to be negative when retested in a reference laboratory. Both IgM and IgG should be tested early during the pregnancy and the results will determine the subsequent actions which may need to be taken (Table 14.4). Negative results for both IgM and IgG indicate that the patient is not infected and the woman is advised to take extra precautions to prevent becoming infected (see below) and should be retested throughout the pregnancy. Women positive for IgG and negative for IgM likely acquired the infection prior to the pregnancy and

Table 14.4 Interpretation of serological tests for toxoplasmosis

IgM result	IgG result	Clinical relevance and recommendations
Negative	Negative	Patient is not infected. Pregnant women should be retested throughout the pregnancy and avoid becoming infected
Negative	Positive	Probably represents an infection acquired before the pregnancy if the test is from the first or second trimester. No action needed
Positive or equivocal	Negative	If results are confirmed by a reference laboratory, the patient should be retested in 1–3 weeks to look for IgG seroconversion. If the results are still IgG negative the patient is probably not infected and the recommendations from above should be followed
Positive or equivocal	Positive	Confirmation by a reference laboratory indicates the patient is acutely infected and treatment should be initiated

present no danger of congenital infection. Women positive for IgM, with or without IgG, may have an acute infection and present a risk of infecting the fetus. These samples should be retested at a reference laboratory to confirm the results. Confirmation of the results is indicative of seroconversion and an acute infection.

Fetuses of pregnant women who are diagnosed with an acute infection also need to be tested. Congenital infections are difficult to diagnose serologically because maternal IgG crosses the placenta and persists for several months. Therefore it is necessary to test for IgM or IgA. In addition, PCR testing of the amniotic fluid has proven to be a reliable and safe means to determine if the fetus is also infected. Testing of the amniotic fluid allows for an earlier confirmation and is less invasive than testing fetal blood.

Treatment

Immunocompetent adults and children above 5 years of age are usually not treated unless the symptoms are severe or persistent. Such symptomatic patients, including those with active retinochoroiditis, should be treated until manifestations have subsided and there is evidence of acquired immunity (Table 14.5). The most typical drug combination is of pyrimethamine (Daraprim®), sulfadiazine, and folinic acid (Leucovorin®). Retinochoroiditis patients are usually supplemented with a corticosteroid such as prednisone as an anti-inflammatory agent. Corticosteroids should not be used without the antibiotics since the immune suppression will allow progression of the disease. Some physicians begin the corticosteroids 1–2 days after starting the antibiotics. The prognosis for acute toxoplasmosis in immunocompetent adults is excellent.

Pregnant women diagnosed with acute toxoplasmosis can be treated with spiramycin (Rovamycin®) to prevent infection of the fetus. It is not clear how effective spiramycin is at preventing congenital infections. However, pyrimethamine is not recommended during pregnancy since it is a folate antagonist and can cause bone marrow suppression. If the infection in the fetus is confirmed the treatment should be switched to standard treatment of pyrimethamine and sulfadiazine supplemented with folinic acid to minimize bone marrow suppression. Neonates with confirmed or suspected toxoplasmosis should also be treated to prevent the subsequent development of retinochoroiditis or other symptoms. Acute infections in the fetus or young children may be followed by repeated attacks of retinochoroiditis. Treatment does appear to reduce the frequency of these attacks.

Immunocompromised patients must be treated. If begun early enough, treatment with pyrimethamine, sulfadiazine, and folinic acid usually results in improvements, but recrudescences are common. Clindamycin (Cleocin®) can be used instead of sulfadiazine in patients intolerant to sulfonamides.

Table 14.5 Treatment of toxoplasmosis

Condition	Duration	Comments
Symptomatic disease	Until symptoms subside and evidence of immunity	
Active retinochoroiditis	Until symptoms subside and evidence of immunity	+ Corticosteroid as anti-inflammatory
Pregnant women (acute only)	Until term or fetal infection confirmed	Treat with spiramycin to prevent infection
Confirmed fetal infection	Starting 12–18 weeks after gestation until term	+ Folinic acid
Neonates and infants	12 months	Prevents retinochoroiditis
Immunocompromised	4–6 weeks after symptoms subside + continued prophylaxis	+ Folinic acid in AIDS

Table 14.6 Prevention of toxoplasmosis

Raw meat	Cook thoroughly (>66°C)
	Freeze meat before using (<–12°C for 24 hours)
	Wear gloves when handling
	Wash hands after handling
	Wash cutting boards, counters, utensils, etc.
Cat feces/soil	Clean litter box promptly (<24 hours)
	Wear gloves when gardening or cleaning litter box
	Wash hands
	Wash and peel fruits and vegetables
	Cover sand box
	Always keep cat in house
	Do not obtain new cat during pregnancy

Trimethoprim/sulfamethoxazole (Bactrim®) appears equally effective as pyrimethamine/sulfadiazine in AIDS patients. Therapy should continue for 4–6 weeks after cessation of symptoms, followed by prophylaxis for as long as the immunosuppression lasts.

Prevention

Prevention of toxoplasmosis depends on avoidance of the parasite found in cat feces, soil, water and food. This is especially important in pregnant women who have not previously been infected and immunocompromised persons. Control measures for toxoplasmosis focus on avoiding the two major sources of infection: raw or undercooked meat and food or water contaminated by cat feces (Table 14.6). Wearing gloves and washing hands frequently when contacting raw meat or items possibly contaminated with cat feces are simple preventive measures. In addition, since maturation of the oocysts takes at least 24 hours, daily cleaning of the cat litter box will minimize exposure. Cats which never have the opportunity to eat rodents or birds are also less likely to become infected.

Sarcocystis

Sarcocystis is a tissue cyst forming coccidian parasite with a predator–prey life cycle similar to *Toxoplasma*. These species exhibit a heteroxenous life cycle in which merogony only takes place in tissues of the intermediate host (prey) and gametogony takes place in the intestinal epithelium of the definitive host (predator). This predator–prey type of life cycle was not worked out until the 1970s. Prior to that, the intestinal infections in the predator were usually designated as an *Isospora* species and the tissue infections in the prey were usually designated as a *Sarcocystis* species. In many cases the definitive hosts have not been positively identified and the taxonomy of many *Sarcocystis* species is uncertain. The complete life cycle has been worked out in relatively few *Sarcocystis* species and often involves a relatively specific definitive host and a relatively specific intermediate host.

Generalized life cycle

The intermediate hosts, often herbivores, acquire the infection by ingesting food or water contaminated with sporulated **sporocysts** (Figure 14.7).

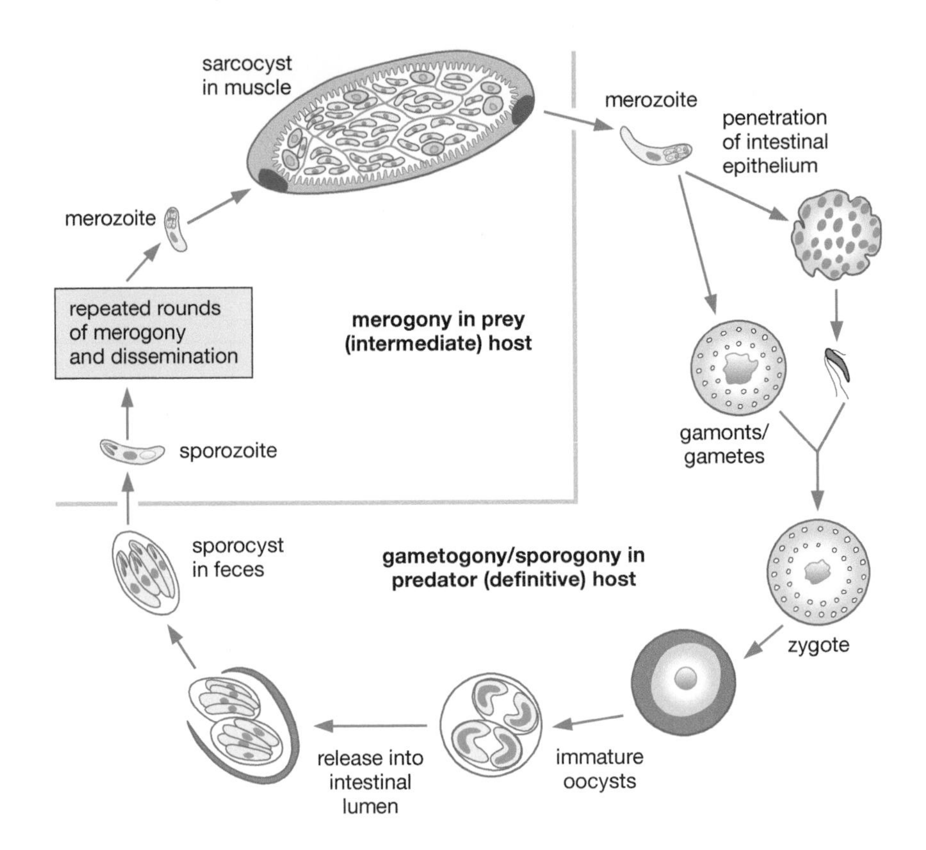

Figure 14.7 Life cycle of *Sarcocystis*. *Sarcocystis* species exhibit a predator–prey type of life cycle in which the prey host ingests sporocysts which will lead to repeated rounds of merogony and eventually the formation of tissue cyst in the muscle and nervous tissue. Ingestion of an infected prey by the predator will result in gametogony and sporogony within the lamina propria and excretion of sporulated sporocysts.

Sporozoites are released and penetrate the intestinal epithelium and eventually invade and replicate within endothelial cells. The resulting merozoites are then disseminated via the circulatory system to the muscles and invade myocytes. As in the case of *Toxoplasma*, initially the parasite exhibits a rapid rate of replication and the rapidly replicating parasites are referred to as **metrocytes**. The replication will slow resulting in the development of a stage analogous to the bradyzoites of *Toxoplasma*. These bradyzoites are often encapsulated and referred to as "tissue cysts" or "**sarcocysts**." Unlike *Toxoplasma*, the infection cannot be transferred to other intermediate hosts through ingestion of the sarcocyst stage.

Ingestion of the infected animal by the appropriate carnivore (i.e., definitive host) will release the merozoites which will invade intestinal epithelial cells. The merozoites develop into gamonts, which will either produce macro- or microgametes, within the lamina propria of the small intestine of the predator. The microgametes fertilize the macrogametes within the infected host cells and produce oocysts. The oocysts undergo sporogony forming two sporocysts each with four sporozoites. Sporulated sporocysts, and occasionally sporulated oocysts, are released from the intestinal epithelial cells and excreted in the feces. Thus, unlike *Isospora* (Chapter 13) and *Toxoplasma*, the parasite is immediately infectious upon excretion and is not soil transmitted. In addition, the intestinal stage of *Sarcocystis* is distinct from the intestinal stages of *Isospora* and *Toxoplasma* in that no asexual replication (i.e., merogony) takes place within the intestinal epithelium of the predator. Therefore, there is no amplification of the parasite during this stage and only sexual stages are formed from the merozoites in the tissue cysts.

Human infections

Sarcocystis infections in humans have been documented, but are quite rare. Humans can serve as either the intermediate or definitive host depending on the source of infection. Ingestion of undercooked beef or pork from infected animals produces an enteric infection which can cause intestinal symptoms (abdominal discomfort, nausea, diarrhea, and vomiting). Infections from beef are often designated as *S. hominis* or *S. bovihominis* and infections from pork are designated as *S. suihominis*. Sporocysts from human feces are infective to cows, pigs, and deer. However, there is some host specificity in regard to the intermediate hosts since *S. bovihominis* cannot infect pigs and *S. suihominis* cannot infect cows. Presumably these human–domestic animal transmission cycles reflect some evolutionary holdover from the early stages of animal domestication and the development of agriculture with humans being the definitive host. Improvements in animal husbandry and hygiene have probably nearly eliminated this parasite.

Infected individuals can shed sporocysts in the feces for weeks to months following the infection. Intestinal sarcocystosis is diagnosed by detection of sporocysts or oocysts in fresh stool specimens (Figure 14.8) and a recent history of eating raw or undercooked meat. *Sarcocystis* is distinguished from other intestinal coccidia (see Chapter 13) by the size, shape, and structure of the oocysts. The sporulated oocysts are of similar size and structure as *Isospora* oocysts. However, in *Isospora* the predominant form found in feces is the immature oocyst and in *Sarcocystis* the sporocysts containing four sporozoites are the predominant form with occasional oocysts containing two sporocysts. Because the parasite probably does not undergo an asexual replication (i.e., merogony) in the intestinal stage, the infection is self-limiting and no treatment is prescribed.

Ingestion of the sporocysts by humans will result in the formation of sarcocysts in the muscles. Muscle cysts in humans have only been sporadically reported and probably represent accidental infections. The sarcocysts are generally several 100 μm in size and cause little tissue damage. Clinical symptoms can include muscle tenderness or episodic painful inflammatory swellings. Biopsy is the most specific diagnostic procedure for the detection of muscle cysts. The morphology is similar to that of *Toxoplasma* tissue cysts except for generally being larger and containing septa, or distinct sections within the sarcocyst (Figure 14.8). As in the intestinal infection, no treatment is prescribed. A case of systemic sarcocystosis in an AIDS patient with concomitant intestinal sarcocystosis has been described. It was speculated that the systemic infection may be due to an autoinfection from sporocysts in the feces.

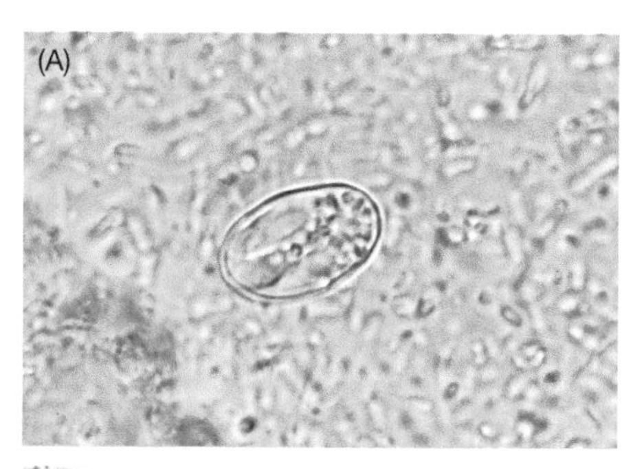

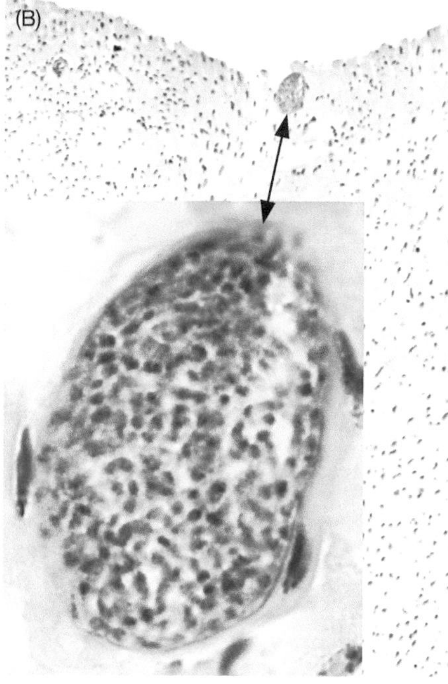

Figure 14.8 Sarcocystosis. (A) Bright field micrograph of unstained sporocyst in fecal sample. (Photo kindly provided by Marc Lontie, Medisch Centrum voor Huisartsen, Leuven, Belgium.) (B) A section of heart muscle at low magnification showing a sarcocyst (upper arrowhead) and a higher magnification of the sarcocyst (lower arrowhead) showing individual zooites.

Summary and Key Concepts

- *Toxoplasma gondii* is a cosmopolitan parasite that infects nearly all warm-blooded animals. It exhibits a predator–prey type of life cycle with felids as the definitive host.
- Cats acquire the infection by eating infected intermediate hosts and exhibit a typical coccidian life cycle with merogony and gametogony taking place in the intestinal epithelial cells. Immature oocysts are excreted in the feces and develop into infectious sporulated oocysts at ambient temperatures.
- Ingestion of sporulated oocysts by an intermediate host results in an acute infection by the tachyzoite stage primarily in the reticuloendothelial system followed by a chronic infection characterized by the bradyzoite stage within tissue cysts found primarily in neural and muscle tissue.
- The major sources of human infection are undercooked meat and ingestion of food or water contaminated with oocysts. Congenital infections are also a major source of human disease.
- Infection of immunocompetent adults and children generally results in an asymptomatic or a self-limiting mild disease characterized by lymphadenopathy.

- Immunosuppression, most notably in AIDS patients, can reactivate a latent infection generally resulting in toxoplasmic encephalitis.
- The infection can be acquired congenitally if the mother becomes infected during the pregnancy. Congenital infections are more likely to be symptomatic and symptoms can range from mild visual impairments to severe neurological manifestations, including death.
- Ocular toxoplasmosis, or retinochoroiditis, is another manifestation associated with *Toxoplasma* infection.
- Diagnosis of toxoplasmosis relies heavily upon serological methods.
- Treatment with antifolates is generally effective and prescribed in symptomatic infections. Prophylactic measures are also applied during pregnancy, and for neonates and immunocompromised persons.
- Preventive measures include avoidance of ingesting raw or undercooked meat, and food or water contaminated with sporulated oocysts; prevention is especially important for nonimmune pregnant women and immunocompromised persons.
- *Sarcocystis* species are a tissue cyst forming coccidian parasite similar to *Toxoplasma* which on rare occasions can infect humans.
- Humans can serve as either the intermediate host or the definitive host depending on the source of infection. Ingestion of raw or undercooked meat infected with sarcocysts will lead to an intestinal infection. Ingestion of food or water contaminated with sporocysts will lead to the formation of muscle cysts.
- Diagnosis of the infection is made by detecting sporocysts in feces or tissue cysts in a muscle biopsy and no treatment is prescribed.

Further Reading

Commodaro, A.G., Belfort, R.N., Rizzo, L.V., Muccioli, C., Silveira, C., Burnier, M.N. Jr and Belfort, R. Jr (2009) Ocular toxoplasmosis – an update and review of the literature. *Mem. Inst. Oswaldo Cruz* 104: 345–350.

Fayer, R. (2004) *Sarcocystis* spp. in human infections. *Clin. Microbiol. Rev.* 17: 894–902.

Grigg, M.E. and Sundar, N. (2009) Sexual recombination punctuated by outbreaks and clonal expansions predicts *Toxoplasma gondii* population genetics. *Int. J. Parasitol.* 39: 925–933.

Hill, D.E. Chirukandotha, S. and Dubey, J.P. (2005) Biology and epidemiology of *Toxoplasma gondii* in man and animals. *Anim. Health Res. Rev.* 6: 41–61.

Kijlstra, A. and Jongert, E. (2008) Control and risk of human toxoplasmosis transmitted by meat. *Int. J. Parasitol.* 38: 1359–1370.

Montoya, J.G. and Liesenfeld, O. (2004) Toxoplasmosis. *Lancet* 363: 1965–1976.

Montoya, J.G. and Remington, J.S. (2008) Clinical practice: Management of *Toxoplasma gondii* infection during pregnancy. *Clin. Infect. Dis.* 47: 554–566.

Sibley, L.D., Khan, A., Ajioka, J.W. and Rosenthal, B.M. (2009) Genetic diversity of *Toxoplasma gondii* in animals and humans. *Phil. Trans. R. Soc. Lond. B* 364: 2749–2761.

Tenter, A.M., Heckeroth, A.R. and Weiss, L.M. (2000) *Toxoplasma gondii*: from animals to humans. *Int. J. Parasitol.* 30: 1217–1258.

Velasquez, J.N., Di Risio, C., Etchart, C.B., Chertcoff, A.V., Mendez, N., Cabrera, M.G., Labbe, J.H. and Carnevale, S. (2008) Systemic sarcocystosis in a patient with acquired immune deficiency syndrome. *Hum. Pathol.* 39: 1263–1267.

Weiss, L.M. and Kim, K. (Eds) (2007) *Toxoplasma Gondii. The Model Apicomplexan: Perspectives and Methods.* New York: Elsevier Ltd.

Malaria

15

Disease(s)	Malaria
Etiological agent(s)	*Plasmodium falciparum, P. vivax, P. ovale, P. malariae*
Major organ(s) affected	Blood
Transmission mode or vector	Anopheline mosquitoes
Geographical distribution	Tropical and subtropical regions throughout the world
Morbidity and mortality	Causes an acute febrile disease that in the case of *P. falciparum* can develop into a complicated disease involving several organs and lead to death if not treated
Diagnosis	Detection of parasites in blood
Treatment	Chloroquine, mefloquine, quinine, Fansidar®, artemisinin derivatives combined with other drugs, plus others
Control and prevention	Mosquito avoidance such as use of bed nets, repellents, protective clothing; chemical or biological control of mosquitoes and destruction of breeding areas; case detection and treatment

Among the diseases caused by protozoa, malaria probably has the greatest impact upon humans. It has been, and still is, the cause of much human morbidity and mortality and an estimated 1.5–2.7 million people die of malaria every year. Although the disease has been eradicated in many temperate zones, it continues to be endemic throughout much of the tropics and subtropics. Forty percent of the world's population lives in endemic areas. Epidemics have devastated large populations and malaria poses a serious barrier to economic progress in many developing countries due to the severe clinical disease that often accompanies infection. There are an estimated 300–500 million cases of clinical disease per year. The impact of malaria on humans is further illustrated by some of the earliest known medical writings from China, Assyria, and India which accurately describe the malaria-like intermittent fevers. However, Hippocrates is generally credited with the first description of the clinical symptoms in 500 BC, more than 2000 years before the parasite and life cycle were described (Table 15.1).

Table 15.1 Some historical highlights in the description of malaria and the parasite

Year	Person	Feature described
500 BC	Hippocrates	Clinical symptoms
1880	Laveran	Blood-stage parasite
1898	Ross	Mosquito transmission
1948	Garnham	Liver stage of life cycle

Malaria is caused by apicomplexan parasites (Chapter 11) of the genus *Plasmodium*. The parasite exhibits a heteroxenous life cycle involving a vertebrate host and an arthropod vector. Vertebrate hosts include: reptiles, birds, rodents, monkeys, and humans and the most common arthropod vector is a mosquito. *Plasmodium* species are generally host and vector specific in that each species only infects a limited range of hosts and vectors. Four distinct species infect humans: *P. falciparum*, *P. vivax*, *P. ovale*, and *P. malariae*. These four species differ in regard to their morphology, details of their life cycles, and their clinical manifestations. *P. falciparum* and *P. vivax* are the most prevalent and each species accounts for about 40% of the total malaria cases. In addition, there have been some reports of humans naturally infected with the simian parasites on rare occasions. However, recently it has been recognized that *P. knowlesi*, a natural parasite of macaque monkeys, infects humans more frequently than previously believed. Foci of human infections have been observed in parts of Malaysia and other countries in Southeast Asia. Furthermore, these foci tend to be in areas bordering the jungle suggesting a zoonosis.

Life Cycle

The *Plasmodium* life cycle exhibits the general features of other apicomplexan parasites characterized by asexual and sexual replication and the formation of invasive stages with typical apical organelles (Chapter 11). Human and other mammalian *Plasmodium* species are transmitted by anopheline mosquitoes. The infection is acquired during mosquito feeding when the parasite is injected with the saliva. Following invasion of a liver cell the parasite first undergoes a round of asexual replication called **merogony**, or **schizogony**, followed by multiple rounds of merogony in the erythrocytes (Figure 15.1). **Gametogony**, or the formation of sexual forms, begins within the erythrocytes of the vertebrate host and is completed within the mosquito where another round of asexual replication called **sporogony** takes place. The nomenclature used to describe *Plasmodium* life cycles is usually different from that of the generalized apicomplexan life cycle. For example, merogony in *Plasmodium* species is almost always referred to as schizogony. Similarly, gametogony in *Plasmodium* is usually separated into two phases called gametocytogenesis and gametogenesis.

Liver stage

Sporozoites are injected with the saliva during mosquito feeding and are carried to the liver via the circulatory system. Within 30–60 minutes the sporozoites find the liver and invade a liver cell. The sporozoites gain access to the hepatocytes by first invading and traversing a macrophage within the liver called a Kupffer cell. Host cell entry and exit, as in all apicomplexa, is facilitated by the apical organelles (Chapter 11). After exiting the Kupffer cell, the sporozoite can traverse several hepatocytes before developing into an **exoerythrocytic** (or pre-erythrocytic) **schizont**. Schizogony refers to an asexual replicative process in which the parasite undergoes multiple rounds of nuclear division without cytoplasmic division followed by a budding, or segmentation, to form progeny called **merozoites**. The merozoites are released into the circulatory system following rupture of the host hepatocytes. Recent observations suggest that the merozoites are released as a membrane-bound aggregate, called a merosome, from the dying hepatocytes and the merosomes are delivered to the bloodstream before dispersal of the individual merozoites. This presumably protects the merozoites from phagocytosis by the Kupffer cells of the liver.

In *P. vivax* and *P. ovale* some of the sporozoites do not immediately undergo asexual replication, but enter a dormant phase known as the hypnozoite stage. The hypnozoite can reactivate and undergo schizogony at a later time resulting in a relapse. Relapse has a specific meaning in regard

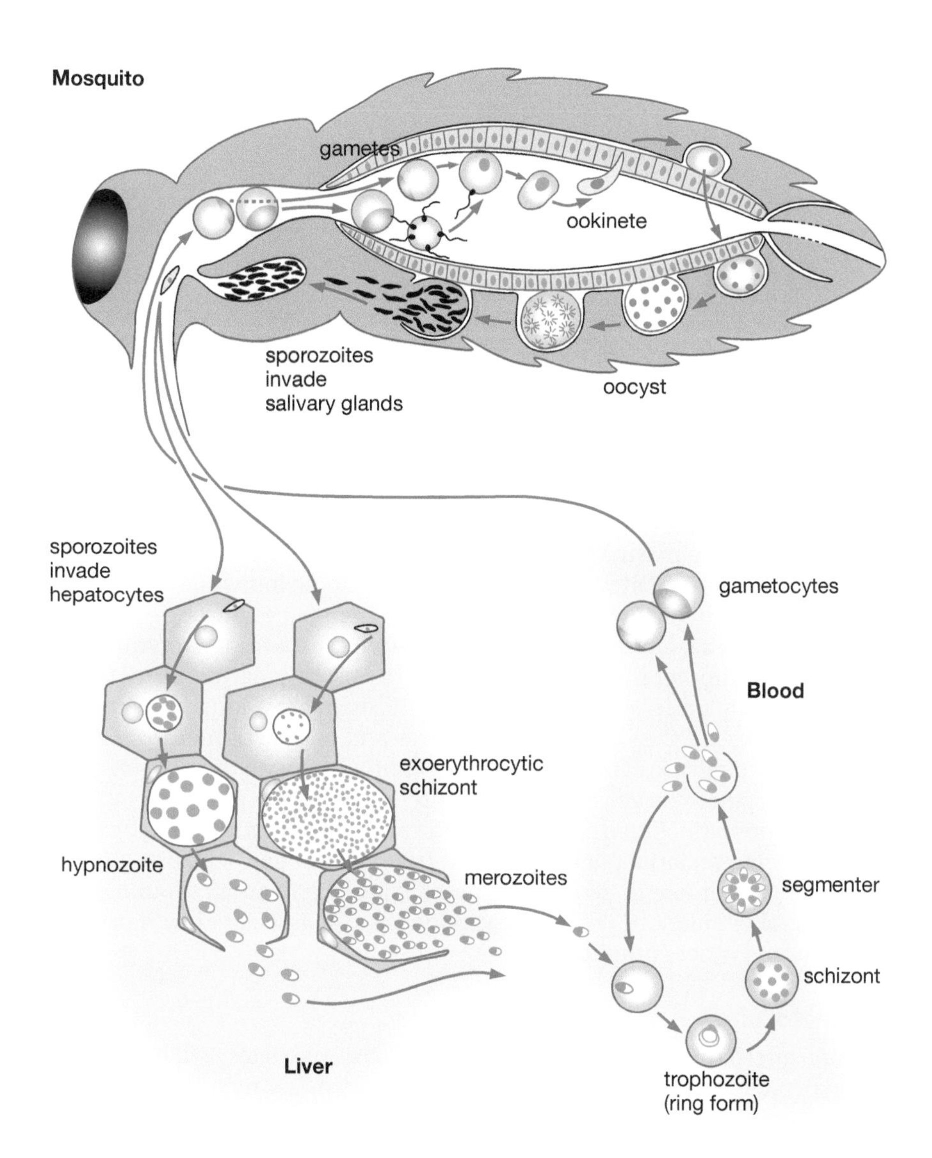

Figure 15.1 Life cycle of the malaria parasite. The infection in humans is acquired when sporozoites are injected with the saliva during mosquito feeding. The sporozoites invade liver cells and under an exoerythrocytic schizogony (i.e., merogony) resulting in the production of merozoites. In *P. vivax* and *P. ovale* some of the exoerythrocytic schizonts undergo a dormant period known as the hypnozoite stage. Merozoites are released from the infected liver cells and invade erythrocytes and undergo an erythrocytic stage schizogony producing more merozoites which can reinvade new erythrocytes. Alternatively, some of the merozoites will undergo sexual development and produce gametocytes which are infective for the mosquito. Gametes are formed when the blood is ingested by the mosquito and the resulting zygote develops into an ookinete. The ookinete penetrates the gut epithelial cells and develops into an oocyst and undergoes sporogony. The resulting sporozoites invade the salivary glands and are infective for the human host.

to malaria and refers to the reactivation of the infection via hypnozoites. Recrudescence is used to describe the situation in which parasitemia in the blood falls below detectable levels and then later increases to a patent parasitemia. Interestingly, strains isolated from temperate regions tend to exhibit a longer latent period between the primary infection and the first relapse than parasite strains from tropical regions with continuous transmission. This suggests that the hypnozoite stage provides a means for the parasite to survive during periods in which mosquitoes are not available for transmission.

Blood stage

Merozoites released from the infected liver cells invade erythrocytes. The merozoites recognize specific proteins on the surface of the erythrocyte and actively invade the cell in a manner similar to other apicomplexan parasites. After entering the erythrocyte the parasite undergoes a trophic period followed by an asexual replication. The young trophozoite is often called a **ring form** due to its morphology in Giemsa-stained blood smears (Figure 15.2). As the parasite increases in size this ring morphology disappears and it is called a **trophozoite**. During the trophic period the parasite ingests the host cell cytoplasm and breaks down the hemoglobin into amino acids. A by-product of the hemoglobin digestion is the malaria pigment, or **hemozoin**. (See Box 15.1) These golden-brown to black granules have been long recognized as a distinctive feature of blood-stage malaria parasites.

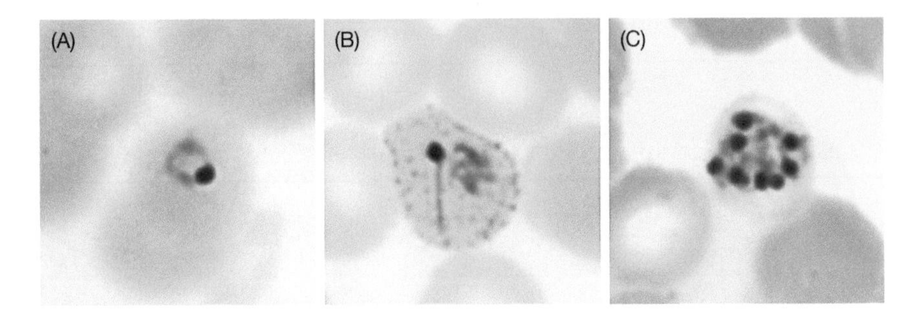

Figure 15.2 Blood stages of the malaria parasite. (A) Ring stage of *P. falciparum.* (B) Trophozoite of *P. vivax.* Note the stippling known as Schüffner's dots within the infected erythrocyte. (C) Multinucleated schizont stage of *P. malariae.*

Nuclear division marks the end of the trophozoite stage and the beginning of the **schizont** stage. Erythrocytic schizogony consists of three to five rounds (depending on species) of nuclear replication followed by a budding process. Budding involves the migration of the nuclei and other organelles to the plasma membrane where they are incorporated into the forming merozoite. Late stage schizonts in which the individual merozoites become discernable are called **segmenters**. The host erythrocyte ruptures and releases the merozoites. These merozoites invade new erythrocytes and initiate another round of schizogony. The blood-stage parasites within a host usually undergo a synchronous schizogony resulting in a simultaneous rupture of the infected erythrocytes and release of merozoites. The concomitant release of antigens and waste products with the rupture of the infected erythrocyte accounts for the intermittent fever paroxysms associated with malaria. Blood-stage schizogony in *P. falciparum* differs from the other human malarial parasites in that trophozoite- and schizont-infected erythrocytes adhere to capillary endothelial cells and are not found in the peripheral circulation. This sequestration may be associated with cerebral malaria and other complications.

Sexual stage

As an alternative to schizogony some of the parasites will undergo a sexual cycle and differentiate into either micro- or macrogametocytes. **Gametocytes** circulate in the peripheral bloodstream and are capable of forming gametes if ingested by a mosquito. They do not cause pathology in the human host and will disappear from the circulation if not taken up by a mosquito. The factors involved in the induction of **gametocytogenesis** are not known. Host immunity and drug treatment may result in a higher proportion of gametocytes as compared with asexual forms. However, it is not clear if this is due to an induction of gametocytogenesis or a loss of asexual forms. Interestingly, commitment to the sexual stage occurs during the asexual erythrocytic cycle that immediately precedes gametocyte formation because all of the daughter merozoites from a single schizont will develop into either asexual forms or sexual forms.

The next step in sexual development is **gametogenesis**, or the formation of micro- and macrogametes. After ingestion by the mosquito, the gametocytes escape from the erythrocyte and undergo gametogenesis within the mosquito gut. Following release from the erythrocyte, the microgametocyte undergoes three rounds of nuclear replication. Then these eight nuclei become associated with flagella that emerge from the body of the microgametocyte to form the **microgametes**. This process is readily observable by light microscopy due to the thrashing flagella and is called **exflagellation**. Similarly, the macrogametocytes mature into **macrogametes**. However, at the morphological level this is much less dramatic than the exflagellation exhibited by the microgametocytes in that the changes are primarily biochemical and not morphological.

Exflagellation occurs spontaneously when infected blood is exposed to air. Critical factors involved in the induction of this gametogenesis are

Box 15.1 The food vacuole and the digestion of hemoglobin

The malaria parasite takes up the host erythrocyte cytoplasm and breaks down the hemoglobin into amino acids. These amino acids are then used for the synthesis of parasite proteins or possibly as an energy source. The breakdown of hemoglobin may also function to make space for the growing parasite or to assist in controlling osmolarity. During the early ring stage the parasite takes up the host cell stroma by pinocytosis. As the parasite matures, it develops a special organelle, called the cytostome, for the uptake of host cytoplasm. The hemoglobin-containing vesicles fuse to form a food vacuole. The food vacuole is an acidic compartment (pH 5.0–5.4) that contains protease activities, and in this regard resembles a lysosome.

Several distinct protease activities, representing three of the four major classes of proteases, have been identified in the food vacuole [1]. The digestion of hemoglobin probably occurs by a semi-ordered process involving the sequential action of different proteases (Figure 15A). Aspartic acid proteases designated as plasmepsins make the initial cleavages of the native hemoglobin causing it to dissociate into large globin fragments and free heme. The globin fragments are then further digested by plasmepsins and falcipains (cysteine proteases) into peptides. A metalloprotease designated as falcilysin then digests these peptides into smaller peptides of six to eight residues. A dipeptidyl aminopeptidase will then generate dipeptides which are then converted into amino acids by aminopeptidases. Some of the smaller peptides may also be transported into the cytoplasm and converted to amino acids through the action of amino peptidases in the cytoplasm. Pfmdr-1, a member of the ATP-binding cassette transporter family, is localized to the food vacuole membrane and may function as a peptide transporter as evidenced by its ability to transport peptides in yeast.

Digestion of hemoglobin also releases heme. Free heme is toxic due to its ability to destabilize and lyse membranes, as well as inhibit the activity of several enzymes. Some of this free heme can be degraded in oxidative processes. However, the majority of the heme is sequestered into hemozoin or the malarial pigment [2].

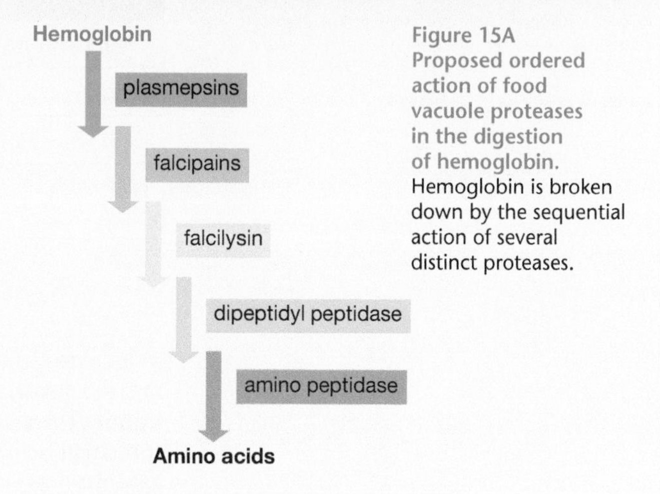

Figure 15A Proposed ordered action of food vacuole proteases in the digestion of hemoglobin. Hemoglobin is broken down by the sequential action of several distinct proteases.

The deposition of hemozoin has long been known as a characteristic feature of malaria, even before the parasite was discovered. X-ray crystallography and spectroscopic analysis indicate that hemozoin has the same structure as β-hematin which is a heme dimer. These dimers interact through hydrogen bonds to form crystals of hemozoin. Therefore, pigment formation is best described as a biocrystallization or biomineralization process. Recently a protein that may catalyze the formation of hemozoin has been described [3]. Lipids may also participate in the process.

Chloroquine and other 4-aminoquinolines inhibit pigment formation, as well as the heme degradative processes, and thereby prevent the detoxification of heme. The free heme destabilizes the food vacuolar membrane and other membranes and leads to the death of the parasite. The fact that the biocrystallization of heme is a unique process to the parasite and not found in the host accounts for the high therapeutic index of such drugs in the absence of drug resistance. Many other antimalarials target the food vacuole indicating the importance of this organelle and its various functions (Figure 15B) to the survival of the parasite.

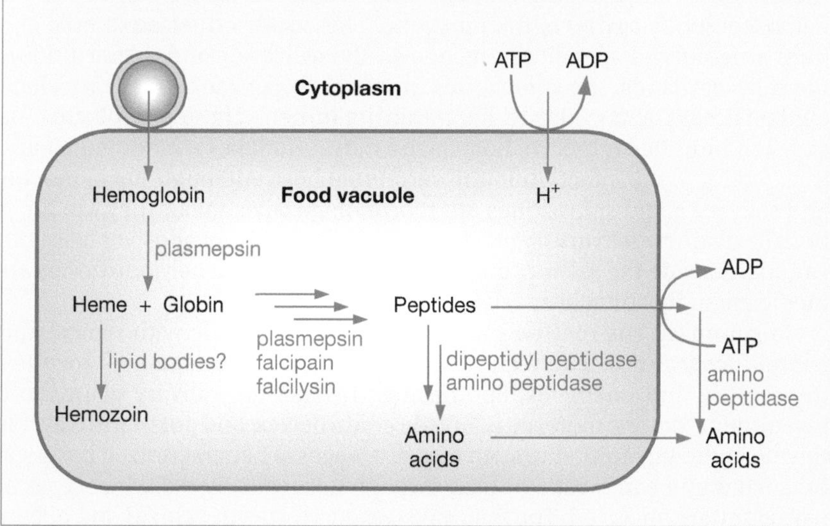

1. Goldberg, D.E. (2005) Hemoglobin degradation. *Curr. Top. Microbiol. Immunol.* 295: 275–291.

2. Egan, T.J. (2008) Haemozoin formation. *Mol. Biochem. Parasitol.* 157: 127–136.

3. Jani, D., Nagarkatti, R., Beatty, W., Angel, R., Slebodnick, C., Andersen, J., Kumar, S. and Rathore, D. (2008) HDP—A novel heme detoxification protein from the malaria parasite. *PLoS Pathogens* 4(4): e1000053.

Figure 15B Summary of activities in the food vacuole of the malaria parasite. Hemoglobin is taken up by the parasite via pinocytosis and delivered to the food vacuole where proteases break down the globin into peptides and amino acids that are transported to the host erythrocyte. The heme is crystallized into hemozoin resulting in its detoxification.

Table 15.2 Factors associated with exflagellation (i.e., gametogenesis)	
↓ temperature (2–3°C)	Occurs in mosquito gut or spontaneously after exposure to air
↓ pCO_2	
↑ pH (8–8.3)	
MEF	A mosquito-derived exflagellation factor (xanthurenic acid) decreases permissive pH

a decrease in temperature and a decrease in the dissolved carbon dioxide and the subsequent increase in pH to above 8.0 (Table 15.2). This somewhat mimics the environmental changes experienced by the gametocytes in that there will be a change to ambient temperature and the gut of the mosquito exhibits a pH of approximately 7.8 as compared to a pH of 7.4 for blood. In addition, a mosquito-derived exflagellation factor also promotes the formation of gametes at a lower pH. This factor has been identified as xanthurenic acid, a metabolite from insects. Xanthurenic acid lowers the permissive pH for exflagellation and is probably a biological cue for the parasite to undergo gametogenesis.

The highly mobile microgametes seek out and fuse with macrogametes. Within 12–24 hours after gamete fusion (i.e., fertilization), the resulting zygote develops into an **ookinete**. The ookinete is a motile and invasive stage which can traverse both the peritrophic membrane and the midgut epithelium of the mosquito. Traversing the peritrophic membrane probably involves secretion of chitinases. Penetration of the midgut epithelium involves invading and exiting several epithelial cells before emerging on the basal side of the epithelium. The invasion process is similar to other apicomplexa except that the ookinete does not have rhoptries and does not form a parasitophorous vacuole after invading the host cell.

Sporogony

After reaching the extracellular space between the epithelial cells and the basal lamina, the ookinete develops into an **oocyst**. Oocysts undergo an asexual replication, called sporogony, which culminates in the production of several thousand sporozoites. This generally takes 10–28 days depending on species and ambient temperature. Upon maturation, the oocyst ruptures and releases the sporozoites which cross the basal lamina into the hemocoel (body cavity) of the mosquito. These sporozoites are motile and have an ability to specifically recognize the salivary glands. After finding the salivary glands, the sporozoites invade and traverse the salivary gland epithelial cells and come to lie within its lumen. These sporozoites are expelled into the vertebrate host as the mosquito takes a blood meal, and thus reinitiate the infection in the vertebrate host. Although the hemocoel and salivary gland sporozoites are morphologically similar, they are functionally distinct. Salivary gland sporozoites efficiently invade liver cells, but cannot reinvade the salivary glands, whereas the hemocoel sporozoites are inefficient at invading liver cells.

In summary, the malaria parasite exhibits a life cycle with typical apicomplexan features. There are three distinct invasive stages: sporozoites, merozoites, and ookinetes. Sporozoites traverse the salivary glands and invade hepatocytes; merozoites invade erythrocytes; and ookinetes traverse the mosquito gut epithelium. All invasive stages are characterized by apical organelles and can invade or pass through host cells. Two distinct types of merogony are observed. The first, called exoerythrocytic schizogony, occurs in the liver and is initiated by the sporozoite. The resulting merozoites then invade erythrocytes and undergo repeated rounds of merogony called erythrocytic schizogony. Some of the merozoites produced from the erythrocytic schizogony will undergo gametogony. *Plasmodium* gametogony is described in two phases: gametocytogenesis occurring in the bloodstream of the vertebrate host, and gametogenesis taking place in the mosquito gut. The gametes fuse to become a zygote which first develops into an ookinete and then becomes an oocyst where sporogony takes place.

Clinical Manifestations

The pathology and clinical manifestations associated with malaria are almost exclusively due to the asexual erythrocytic stage parasites. Tissue schizonts and gametocytes cause little, if any, pathology. *Plasmodium* infection causes an acute febrile illness which is most notable for its periodic fever paroxysms occurring at either 48- or 72-hour intervals. The severity of the attack depends on the *Plasmodium* species as well as other circumstances such as the state of immunity and the general health and nutritional status of the infected individual. Malaria also develops into a chronic disease which has a tendency to relapse or recrudesce over months or even years.

Symptoms of malaria usually start to appear 10–20 days after the bite of an infected mosquito. The prepatent and incubation periods vary according to species (Table 15.3). The prepatent period is defined as the time between sporozoite inoculation and the appearance of parasites in the blood and represents the duration of the liver stage and the number of merozoites produced. The incubation period tends to be a little longer and is defined as the time between sporozoite inoculation and the onset of symptoms. Symptoms will appear when the blood stage parasitemia reaches a sufficient level and is determined in part by the number of sporozoites successfully infecting liver cells, the number of merozoites produced per exoerythrocytic schizont, and the time it takes for the merozoites to form. Sometimes the incubation periods can be prolonged for several months in *P. vivax*, *P. ovale*, and *P. malariae*. Mechanical transmission of infected blood via blood transfusions or sharing syringes will result in a shorter incubation period because there is no liver stage. There is also an increased risk of fatality with mechanically transmitted *P. falciparum*. The lack of the liver stage infection as a result of mechanical transmission also precludes relapses in *P. vivax* or *P. ovale* infections.

Table 15.3 Exoerythrocytic schizogony and prepatent and incubation periods

Feature	*P. falciparum*	*P. vivax*	*P. ovale*	*P. malariae*
Prepatent period (days)	6–9	8–12	10–14	15–18
Incubation period (days)	7–14	12–17	16–18	18–40
Merozoite maturation (days)	5–7	6–8	9	12–16
Merozoites produced	40 000	10 000	15 000	2000

Malaria is characterized by febrile attacks

All four species can exhibit nonspecific **prodromal** symptoms a few days before the first febrile attack. These prodromal symptoms are generally described as flu-like and include: headache, slight fever, muscle pain, anorexia, nausea, and lassitude. The symptoms tend to correlate with increasing numbers of parasites in the blood. These prodromal symptoms will be followed by febrile attacks also known as malarial **paroxysms**. These paroxysms exhibit periodicities of 48 hours for *P. vivax*, *P. ovale*, and *P. falciparum*, and a 72-hour periodicity for *P. malariae*. Initially the periodicity of these paroxysms may be irregular as the broods of merozoites from different exoerythrocytic schizonts synchronize. This is especially true in *P. falciparum*, which may not exhibit distinct paroxysms, but instead may exhibit a continuous fever, daily attacks, or irregular attacks (e.g., 36–48 hour periodicity). Patients may also exhibit splenomegaly, hepatomegaly, slight jaundice, and hemolytic anemia during the acute stage of the disease.

The malarial paroxysm (Table 15.4) usually lasts 4–8 hours and begins with a sudden onset of chills in which the patient experiences an intense feeling of cold despite having an elevated temperature. This is often referred to as the cold stage, or rigor, and is characterized by a vigorous shivering. Immediately following this cold stage is the hot stage. The patient feels an intense heat accompanied by severe headache. Fatigue, dizziness, anorexia, myalgia, and nausea will often be associated with the hot stage. Next a period of profuse sweating will ensue and the fever will start to decline. The patient is exhausted and weak and will usually fall asleep. Upon awakening the patient generally feels well, other than being tired, and does not exhibit symptoms until the onset of the next paroxysm.

The periodicity of these paroxysms is due to the synchronous development of the malarial parasite within the human host. In other words, all of the parasites within a host are at approximately the same stage (i.e., ring, trophozoite, schizont) as they proceed through schizogony. The malarial paroxysm corresponds to the rupture of the infected erythrocytes and the release of merozoites (Figure 15.3). The 72-hour periodicity in *P. malariae* is due to its slower growth and maturation during blood-stage schizogony. Studies in *P. vivax* have demonstrated a correlation between fever and serum tumor necrosis factor-alpha (TNF-α) levels (Figure 15.4). Presumably antigens or toxins are released into the bloodstream and tissues when the infected erythrocyte ruptures. These released parasite products then stimulate macrophages and other immune effector cells to produce TNF-α and other cytokines leading to the febrile attacks. It has been suggested that a major component of this presumptive toxin might be glycosylphosphatidylinositol derived from various parasite proteins.

Typically the acute infection characterized by the febrile illness and paroxysms is controlled by the host immune response and a chronic infection is established. The malarial paroxysms become less severe and irregular in periodicity as immunity develops and the infection enters a chronic phase characterized by lower parasitemias. This immunity, however, is not a sterilizing immunity in that the infection persists and individuals can exhibit relapses or recrudescences or become reinfected. In addition, intermittent episodes of fever can still occur during the chronic phase and are usually associated with transient increases in parasitemia (i.e., recrudescence).

Table 15.4 Malarial paroxysm

Cold stage	Feeling of intense cold
	Vigorous shivering
	Lasts 15–60 minutes
Hot stage	Intense heat
	Dry burning skin
	Throbbing headache
	Lasts 2–6 hours
Sweating stage	Profuse sweating
	Declining temperature
	Exhausted and weak → sleep
	Lasts 2–4 hours

The species exhibit different ranges of symptoms

The severity of the paroxysms and duration of the symptoms varies according to species (Table 15.5). In general, the severity of the disease correlates with the average and maximum parasitemia exhibited by the various species. *P. falciparum* is capable of producing a severe and lethal infection, whereas the other species are rarely mortal. Patients infected with *P. vivax*, especially for the first time, can be quite ill. However, *P. vivax* rarely causes

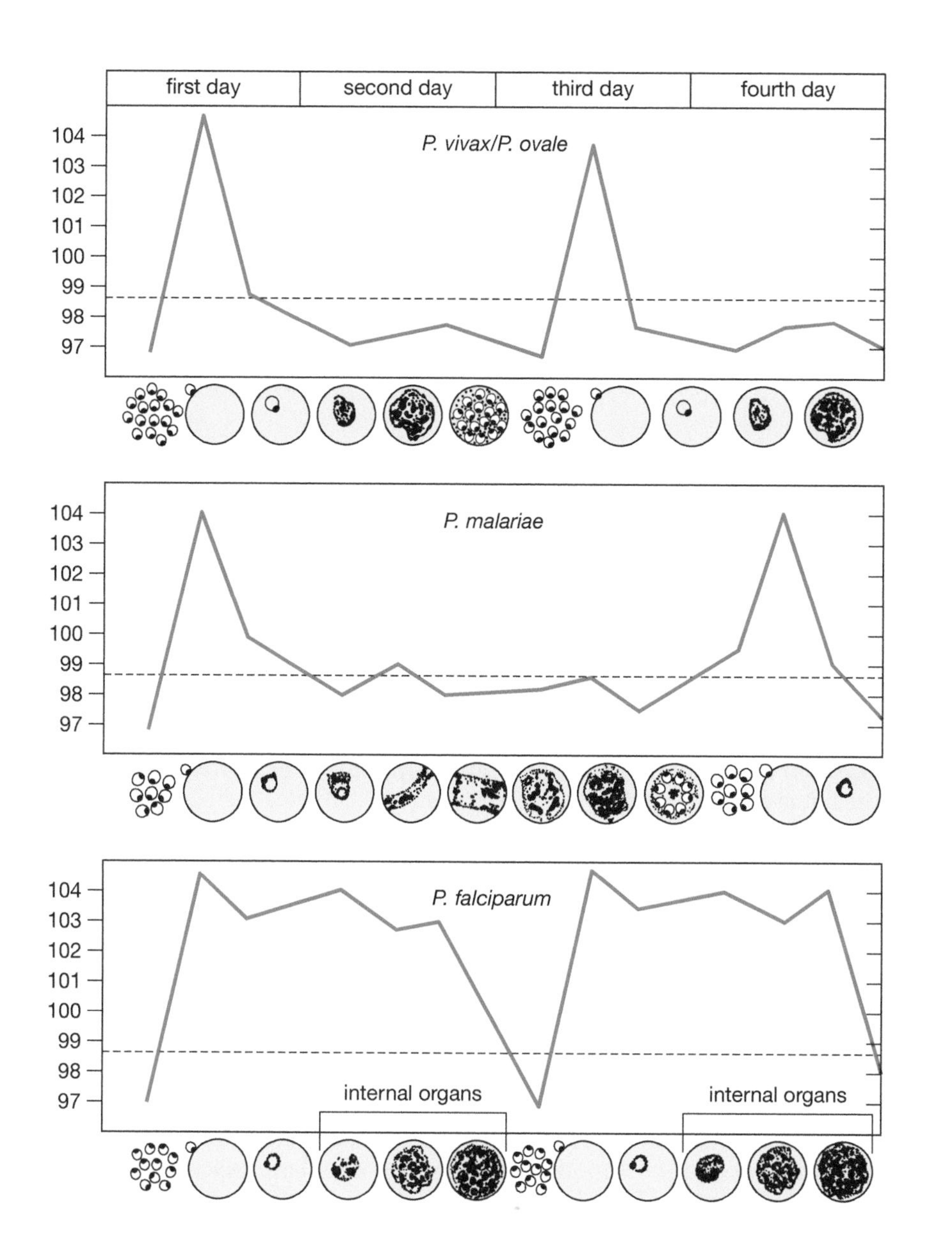

Figure 15.3 A typical pattern of temperature (fever) in relation to blood-stage schizogony for the human malarial parasites. The fever paroxysm corresponds to the period of infected erythrocyte rupture and merozoite invasion.

severe complications or results in death. Relapses due to the activation of *P. vivax* hypnozoites can occur for several years. *P. ovale* is the most benign in that the paroxysms tend to be mild and of short duration and relapses seldom occur more than 1 year after the initial infection. *P. malariae* generally

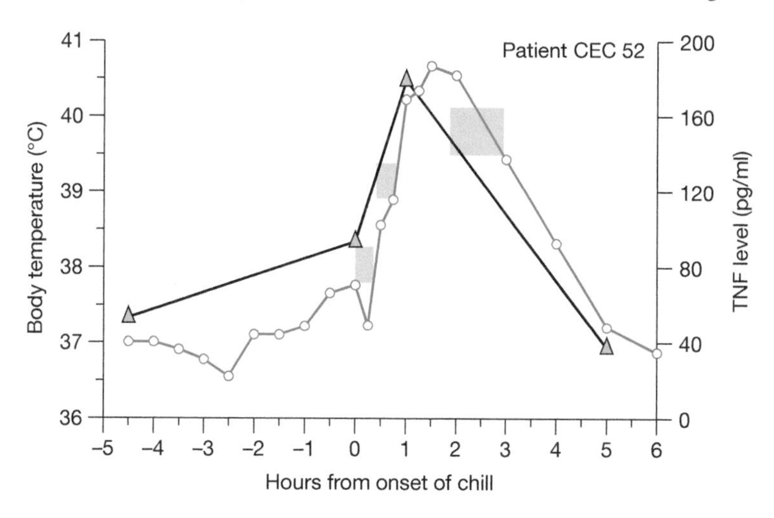

Figure 15.4 Correlation between fever and TNF-α levels during *P. vivax* infection. Body temperature (circles) and serum TNF-α levels (triangles) were measured during the malarial paroxysm. The boxes denote the cold stage (left), the hot stage (middle), and sweating stage (right) of the malarial paroxysm. (Figure kindly provided by Kamini Mendis and reprinted, with permission, from Karunaweera, N.D., Grau, G.E., Gamage, P., Carter, R. and Mendis, K.N., 1992, Dynamics of fever and serum levels of tumor necrosis factor are closely associated during the clinical paroxysms of *Plasmodium vivax* malaria, *Proc. Natl Acad. Sci.* 89: 3200–3203.)

Table 15.5 Disease severity and duration

	P. vivax	*P. ovale*	*P. malariae*	*P. falciparum*
Severity of initial paroxysms	Moderate to severe	Mild	Mild to moderate	Severe
Average parasitemia (per mm^3)	20 000	9 000	6 000	50 000–500 000
Maximum parasitemia (per mm^3)	50 000	30 000	20 000	2 500 000
Symptom duration (untreated)	3–8+ weeks	2–3 weeks	3–24 weeks	2–3 weeks
Maximum infection duration (untreated)	5–8 years*	12–20 months*	20–50+ years	6–17 months
Anemia	++	+	++	++++
Other complications			Renal	Cerebral

* Includes relapses from the hypnozoite stage.

produces a mild disease, but the initial paroxysms can be moderate to severe. *P. malariae* is the most chronic, though, and recrudescences have been documented several decades after the initial infection. This chronicity is sometimes associated with renal complications, which are probably due to the deposition of antigen–antibody complexes in the glomeruli of the kidney.

In contrast to the other three species, *P. falciparum* can produce serious disease with mortal consequences. This increased morbidity and mortality is due in large part to the high parasitemias associated with *P. falciparum* infections. These potentially high parasitemias are partly the result of the large number of merozoites produced and the ability of *P. falciparum* to invade all erythrocytes. In contrast, *P. vivax* and *P. ovale* prefer reticulocytes (i.e., immature erythrocytes), whereas *P. malariae* prefers senescent erythrocytes. Higher parasitemias are also associated with the cytoadherence and sequestration (see below) of *P. falciparum*. This sequestration in the tissues minimizes removal of infected erythrocytes by the spleen and can be viewed as an avoidance of the immune system. In addition, the cytoadherence to the endothelial cells of the capillaries allows for a more efficient erythrocyte invasion because the blood flow is much slower. The low oxygen tension found in the deep tissues is also metabolically favorable for parasite growth.

Severe falciparum malaria

Approximately 10% of falciparum malaria cases may develop into complicated or severe disease with a mortality of 10–50%. The high parasitemia and sequestration can result in various complications and severe malaria encompasses a complex syndrome affecting many organs and resulting in biochemical and hematological abnormalities (Table 15.6). Patients often exhibit several of these manifestations either simultaneously or sequentially and these complications can develop rapidly and progress to death within hours or days. The three most common syndromes associated with severe malaria and most often correlated with death are: cerebral malaria, severe anemia, and respiratory distress.

Cerebral malaria is characterized by an impaired consciousness and other neurological symptoms (Table 15.7). Patients typically present with fever for several days followed by a loss of consciousness. The onset of cerebral malaria can be gradual or rapid. The presenting symptoms are usually severe headache followed by drowsiness, confusion, and ultimately an unrousable coma. Convulsions, vomiting, and respiratory distress are also frequently associated with cerebral malaria, especially in children. The

Table 15.6 Indicators of severe malaria and poor prognosis

Manifestation	Features
Cerebral malaria	Unrousable coma not attributable to any other cause
Severe anemia	Hematocrit <15% or hemoglobin <50 g l^{-1} in the presence of parasite count >10 000/μl
Respiratory distress	Defined by labored breathing and pulmonary edema that can progress to an acute respiratory distress syndrome
Renal failure	Low urine output and high serum creatinine despite adequate volume repletion
Circulatory collapse (shock)	Systolic blood pressure <70 mmHg in patients with cold clammy skin
Acidemia/acidosis	Arterial pH <7.25 or plasma bicarbonate <15 mmol l^{-1}
Hypoglycemia	Whole blood glucose concentration <2.2 mmol l^{-1} (<40 mg dl^{-1})
Impaired consciousness	Impaired consciousness less marked than unrousable coma, can localize a painful stimulus
Repeated generalized convulsions	≥3 convulsions observed within 24 hours
Prostration or weakness	Patient unable to sit or walk, with no other obvious neurological explanation
Abnormal bleeding and/or coagulation	Spontaneous bleeding from gums, nose, gastrointestinal tract, or laboratory evidence of disseminated intravascular coagulation
Malarial hemoglobinuria	Need to exclude hemoglobinuria due to antimalarial medications and to G6PD deficiency
Jaundice	Yellow skin color, bilirubin >43 μmol l^{-1} (>2.5 mg dl^{-1})
Hyperparasitemia	>5% parasitized erythrocytes or >250 000 parasites/μl (interpreted in light of immune status and prior exposure)
Hyperpyrexia	Core body temperature >40°C

neurological manifestations are believed to be due to the sequestration of the infected erythrocytes in the cerebral microvasculature.

The severe anemia is due in part to the destruction of erythrocytes during parasite maturation and merozoite release. In addition, noninfected erythrocytes are destroyed at higher rates in infected persons as a result of complement-mediated lysis and phagocytosis. This increase in complement activation and phagocytosis may be due to the high levels of antibodies observed during infection and the deposition of antibody complexes on the uninfected erythrocytes. Accordingly, severe anemia is more common in children from highly endemic areas because of repeated or chronic infections. Furthermore, there is a decreased production of erythrocytes during infection. This decrease in erythropoiesis may be due to the higher levels of cytokines and other mediators of inflammation which are known to affect the maturation of erythroid cells. In addition, other infections (e.g., hookworm, bacteremia, and HIV), vitamin deficiencies, and glucose-6-phosphate dehydrogenase deficiency can contribute to the anemia associated with malaria.

More recently it has been recognized that metabolic acidosis (i.e., low blood pH) as manifested by respiratory distress has emerged as a central feature of severe malaria and is a better predictor of death than cerebral malaria or severe anemia. The first signs of lung injury are rapid and difficult breathing and are indicative of pulmonary edema (i.e., fluid in the lungs). This pulmonary edema can progress to an acute respiratory distress syndrome and even respiratory failure.

Table 15.7 Cerebral malaria
Complication of severe falciparum malaria
A diffuse encephalopathy with loss of consciousness
Consciousness ranges from stupor to coma
Onset can be gradual or rapid
Unresponsive to pain, visual, and verbal stimuli
Associated with sequestration in cerebral microvasculature

Pathogenesis

Pathology associated with all malarial species is related to the rupture of infected erythrocytes and the release of parasite material and metabolites, hemozoin (i.e., malaria pigment), and cellular debris. In addition, there is an increased activity of the reticuloendothelial system as evidenced by macrophages containing ingested infected and normal erythrocytes and hemozoin. Coincident with this increased activity in the reticuloendothelial system, the liver and spleen are often enlarged during malaria. Except for *P. falciparum*, the pathology associated with malaria tends to be benign with little mortality. As discussed above, proinflammatory cytokines, and especially TNF-α, are believed to play a role in the disease manifestations. Higher levels of TNF-α and other proinflammatory cytokines are also associated with severe anemia, cerebral malaria, and respiratory distress. However, it is not clear the precise role pro- and anti-inflammatory immune responses play in the resolution of the disease and its pathogenesis.

Cytoadherence and sequestration are also believed to contribute to the pathology of severe falciparum malaria. Sequestration refers to the cytoadherence of trophozoite- and schizont-infected erythrocytes to endothelial cells of deep vascular beds in vital organs, especially brain, lung, gut, heart, and placenta. As discussed above, this sequestration provides several advantages for the parasite and contributes to the high parasitemias. In addition, this cytoadherence and sequestration can have pathological effects on the specific organs, most notably the brain.

Receptor–ligand interactions between protein ligands expressed on the surface of trophozoite and schizont-infected erythrocytes and various receptors found on endothelial cells mediate cytoadherence (see next section). The parasite ligands are focused in electron-dense protuberances, or "knobs," on the surface of the infected erythrocyte. Among human *Plasmodium* species, knobs are restricted to *P. falciparum* and this suggests that the knobs play a role in cytoadherence. In addition, there is also a good correlation between animal *Plasmodium* species which express knobs and exhibit sequestration. Electron microscopy also shows that the knobs are contact points between the infected erythrocyte and the endothelial cell.

Pathophysiology of cerebral malaria

Early observations of the pathology of cerebral malaria suggested a relationship between large numbers of infected erythrocytes in the microvasculature and the development of the syndrome (Figure 15.5). Initially it was assumed that the cytoadherence would lead to a mechanical blockage (i.e., cerebral ischemia) and subsequently hypoxia. In addition, the parasite

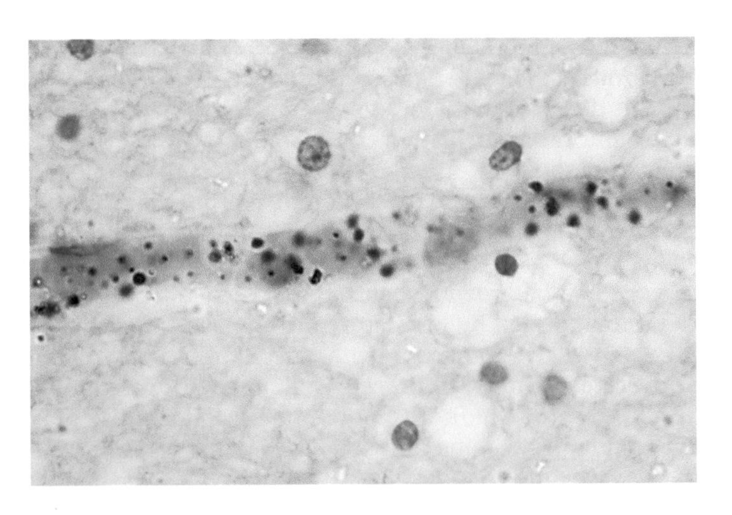

Figure 15.5 Brain section from autopsy of cerebral malaria patient. A capillary is filled with infected erythrocytes as evidenced by the numerous hemozoin granules. Relatively little inflammation is associated with the adherence of the infected erythrocytes.

exhibits a high level of glycolysis which could cause localized metabolic effects such as hypoglycemia and lactic acidosis. The hypoxia and metabolic effects would then disrupt normal brain function and lead to coma and subsequent death.

However, long-term neurological sequelae among survivors of cerebral malaria are lower than would be expected for ischemia and the coma associated with cerebral malaria is rapidly reversible upon treatment. In addition, the phenomenon of sequestration occurs in all *P. falciparum* infections and is not limited to cerebral malaria cases. Thus, other factors likely contribute to the development of cerebral malaria. For example, cytokines, such as TNF-α, or short-lived intermediates, such as nitric oxide, are speculated to be responsible for the pathology. In this cytokine theory, malarial antigens stimulate the production of TNF-α and other cytokines which could then induce nitric oxide or have other pathological effects. Nitric oxide is known to affect neuronal function and it could also lead to intracranial hypertension through its vasodilator activity. It is unlikely, though, that the systemic release of cytokines would cause coma and there is minimal lymphocyte infiltration or inflammation associated with the blocked capillaries (Figure 15.5). Thus questions have been raised in regard to whether these mediators would reach sufficiently high local concentrations in the brain. At this time, the exact mechanisms leading to the development of cerebral malaria are not known.

The sequestration hypothesis and cytokine theory are not mutually exclusive, and both phenomena are likely to be involved in the pathophysiology of cerebral malaria. For example, parasite exo-antigens, which are released during erythrocyte rupture, are known to stimulate macrophages to secrete TNF-α. TNF-α up-regulates the expression of adhesion molecules, such as ICAM1, on the surface of brain endothelial cells which would lead to increased binding of infected erythrocytes and amplify the effects whether they are due to vascular blockage, soluble mediators, metabolic effects, or a combination of these (Figure 15.6). TNF-α and other cytokines also stimulate the production of nitric oxide which affects neuronal function as well as increasing intracranial hypertension through its vasodilator activity. The pathophysiology of cerebral malaria is not completely understood, but likely involves multiple factors and complex interactions between the host and parasite.

Rosetting

Rosetting is another adhesive phenomenon exhibited by *P. falciparum*-infected erythrocytes which may play a role in pathogenesis. Some infected erythrocytes bind to multiple uninfected erythrocytes resulting in a large clump of erythrocytes, or a rosette. Some studies have shown an association between the rosetting phenotype and severe malaria. These clumps could restrict microvascular flow in a similar fashion as cytoadherence to endothelial cells and thus contribute to the pathology.

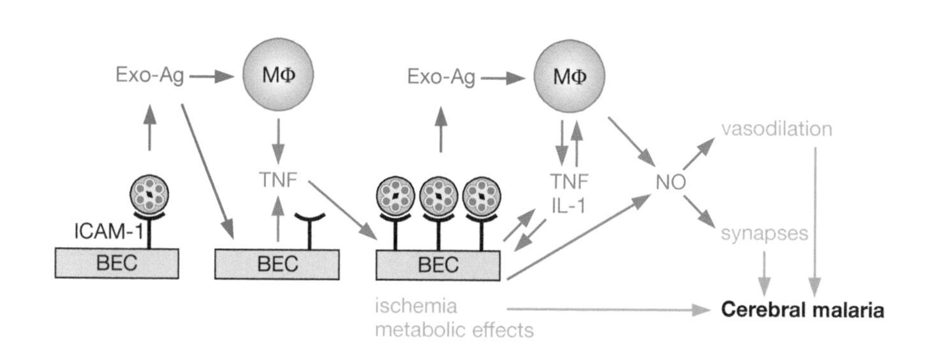

Figure 15.6 Model depicting some possible mediators of cerebral malaria. The cytoadherence of infected erythrocytes to brain endothelial cells (BEC) and the release of exo-antigens (Exo-Ag) could stimulate the BEC and immune effector cells such as macrophages (MΦ) to secrete cytokines. These cytokines, such as tumor necrosis factor-α (TNF), would lead to an increased expression of possible endothelial cell receptors (e.g., ICAM-1) and promote increased cytoadherence of infected erythrocytes. Large numbers of bound infected erythrocytes could lead to vascular blockage, hypoxia, and have localized metabolic effects (e.g., hypoglycemia, lactic acidosis). TNF-α is also known to stimulate nitric oxide (NO) which can affect neuronal function by interfering with neurotransmission and causing vasodilation.

Modification of the Host Erythrocyte by the Malarial Parasite

Another aspect of the pathophysiology of malaria is the alteration of the host erythrocyte by the parasite. During the trophic period the parasite induces many changes in the erythrocyte which affect its structure and function. For example, the malarial parasite is a rapidly growing organism that exhibits a high metabolic rate and has a large demand for small molecular metabolites that will serve as precursors for the synthesis of nucleic acids, proteins, and lipids. Thus, the host erythrocyte with its comparatively low level of metabolism and limited transport capabilities poses some potential problems for the actively growing parasite. Accordingly, the infected erythrocyte exhibits a substantial increase in its permeability to low molecular weight solutes as compared with the uninfected erythrocyte. Much of this increase in permeability can be attributed to a new permeation pathway induced by the parasite with characteristics quite distinct from the transporters of the host erythrocyte.

Ultrastructural alterations of the host erythrocyte have also been noted. For example, caveola–vesicle complexes are found on the surface of *P. vivax*-infected erythrocytes and electron-dense protrusions, or knobs, are found of the surface of *P. falciparum*-infected erythrocytes (Figure 15.7). The function of the caveola–vesicle complexes is not known, but they are responsible for the Schüffner's dots seen in stained thin blood smears. The stain accumulates in the induced cavities on the surface of the infected erythrocyte and gives the appearance of stippling over the erythrocyte cytoplasm (Figure 15.2B). In addition, membrane-bound compartments such as Maurer's clefts are found within the cytoplasm of the infected erythrocyte. The knobs of *P. falciparum* play an important role in the cytoadherence and sequestration of the infected erythrocyte which, as discussed above, plays an important role in the survival of the parasite and the pathogenesis associated with the disease.

Parasite proteins are associated with knobs

The knobs are induced by the parasite and several parasite proteins are associated with the knobs. Two proteins which probably participate in knob formation are the knob-associated histidine rich protein (KAHRP) and erythrocyte membrane protein-2 (PfEMP2). Neither KAHRP nor PfEMP2 are exposed on the outer surface of the erythrocyte, but are localized to the cytoplasmic face of the host membrane (Figure 15.8). Their exact roles in knob formation are not known, but may involve reorganizing the

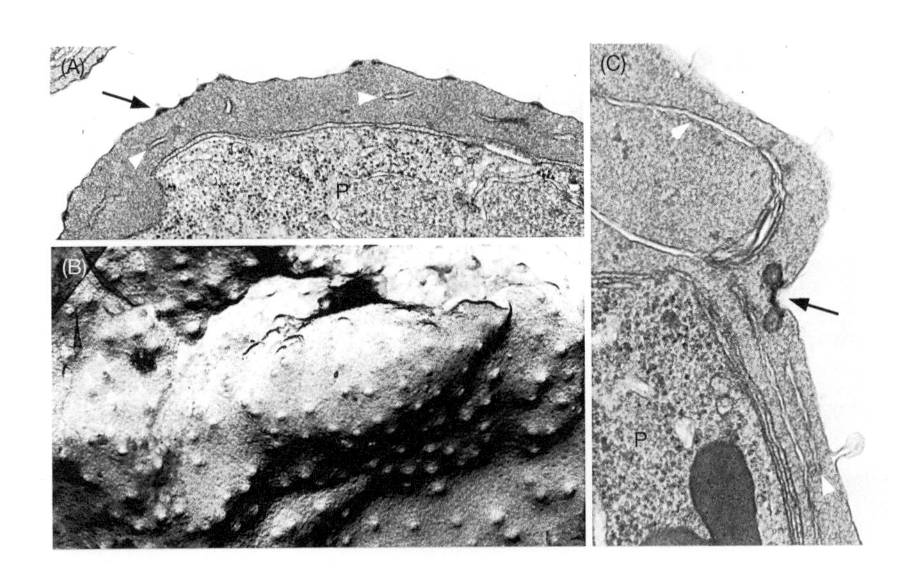

Figure 15.7 Ultrastructural alterations of the host erythrocyte by the malaria parasite. (A) A *P. falciparum*-infected erythrocyte showing knobs (one of several denoted with black arrow) and Maurer's clefts (white arrowheads) within the erythrocyte cytoplasm. (B) A carbon replica of a *P. falciparum*-infected erythrocyte showing the distribution of knobs protruding from the erythrocyte surface. (C) A *P. vivax*-infected erythrocyte showing caveola–vesicle complexes (black arrow) and membrane-bound clefts in erythrocyte cytoplasm (white arrowheads). (Electron micrographs kindly provided by H. Norbert Lanners.)

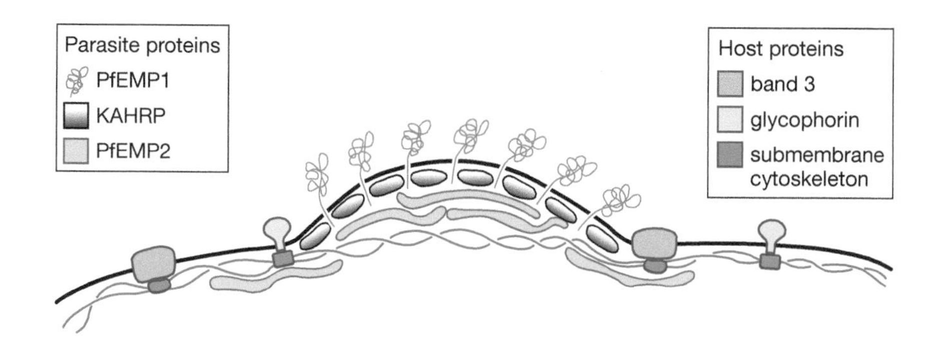

Figure 15.8 Schematic representation of knob structure. Parasite proteins are associated with the knobs found on the surface of infected erythrocytes. These proteins probably cause a reorganization of the submembrane cytoskeleton of the erythrocyte and are important in the cytoadherence of the infected erythrocyte to host endothelial cells.

submembrane cytoskeleton. A polymorphic protein, called PfEMP1, has also been localized to the knobs and is exposed on the host erythrocyte surface and probably functions as a ligand which binds to receptors on host endothelial cells.

PfEMP1 corresponds to members of the *var* gene family. The 40–50 *var* gene paralogs (i.e., alleles) present in the *P. falciparum* genome exhibit a high degree of variability, but have a similar overall structure (Figure 15.9). PfEMP1 has a large extracellular N-terminal domain, a transmembrane region and a C-terminal intracellular domain. The C-terminal region is conserved between members of the *var* family and is believed to anchor PfEMP1 to the erythrocyte submembrane cytoskeleton. In particular, this acidic C-terminal domain may interact with the basic KAHRP of the knob as well as spectrin and actin of the submembrane cytoskeleton. The extracellular domain is characterized by one to five copies of Duffy-binding-like (DBL) domains. These DBL domains are similar to the receptor-binding region of the ligands involved in merozoite invasion (Chapter 11). The DBL domains exhibit a conserved spacing of cysteine and hydrophobic residues, but otherwise show little homology. Phylogenetic analysis indicates that there are five distinct classes (designated as α, β, γ, δ, and ε) of DBL domains. The first DBL is usually a DBLα type and this is followed by a cysteine-rich interdomain region (CIDR). A variable number and types of DBL domains in various orders make up the rest of the extracellular domain of PfEMP-1.

The various alleles of the *var* gene family bind to different host receptors

Several possible endothelial receptors (Table 15.8) have been identified by testing the ability of infected erythrocytes to bind in static adherence assays. These receptors include a wide range of proteins, proteoglycans and complex carbohydrates. Interestingly, many of these receptors also normally function in the adhesion of cells to the extracellular matrix or to other cells. Detailed analysis of the binding activity of PfEMP1 to the receptors indicates that different domains of PfEMP1 are responsible for binding to different receptors. For example, CD36, an 88 kDa integral membrane protein found on monocytes, platelets, and endothelial cells binds to the CIDR domain. Similarly, the different classes of DBL domains bind to different endothelial cell receptors. For example, DBLα binds to the receptors

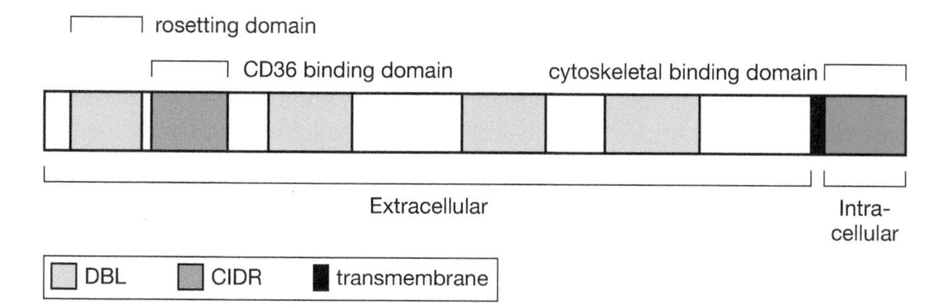

Figure 15.9 Schematic representation of a typical PfEMP1 (= *var* gene). PfEMP1 is a transmembrane protein composed of one to five different Duffy-binding-like (DBL) domains and one to two cysteine-rich interdomain regions (CIDR). These variable domains are exposed on the surface of the infected erythrocyte. The intracellular domain is conserved and binds to the submembrane cytoskeleton of the erythrocyte.

Table 15.8 Possible cytoadherence receptors identified by *in vitro* binding assays

Host receptor	PfEMP1 domain	Comments
CD36	CIDR	
Intracellular adhesion molecule-1 (ICAM1)	DBLβ	Implicated in cerebral malaria
Chondroitin sulfate A	DBLγ	Receptor in the placenta
Heparan sulfate		
Hyaluronic acid		
E-selectin		
Thrombospondin		
Complement receptor-1	DBLα	Rosetting receptor
Blood group A antigen	DBLα	Rosetting receptor
Glycosaminoglycan	DBLα	Rosetting receptor

associated with rosetting; DBLβ binds to intracellular adhesion molecule-1 (ICAM1); and DBLγ binds to chondroitin sulfate. Thus, the different alleles of PfEMP1 exhibit different phenotypes in regard to receptor binding activity depending upon their domain structure and organization.

Different parasite isolates exhibit different phenotypes in regard to which receptors are recognized as a result of the particular *var* gene being expressed. For example, most parasite isolates bind to CD36 which is likely due to the CIDR domain being relatively conserved and present on all PfEMP1 alleles. However, CD36 has not been detected on endothelial cells of the cerebral blood vessels and parasites from clinical isolates tend to adhere to both CD36 and ICAM1. ICAM1 is a member of the immunoglobulin superfamily and functions in cell–cell adhesion. As discussed above, binding of infected erythrocytes to ICAM1 is speculated to play a role in the development of cerebral malaria. Similarly, chondroitin sulfate A has been implicated in the cytoadherence within the placenta and may contribute to the adverse effects of *P. falciparum* during pregnancy.

Antigenic variation

Although sequestration offers many advantages to the parasite, the expression of antigens on the surface of the infected erythrocyte provides a target for the host immune system. The parasite counters the host immune response by expressing antigenically distinct PfEMP1 molecules on the erythrocyte surface, thus allowing the parasite to evade the immune system. Normally, though, major changes in the sequence of a protein will result in structural changes that affect the function of the protein—especially in the case of receptor–ligand interactions which are generally highly specific. Several studies have indicated that the expression of different PfEMP1 genes correlates with different receptor-binding phenotypes. Thus, the cytoadherence function is preserved through its ability to recognize multiple receptors (Figure 15.10). This allows the parasite to avoid clearance by the host immune system but maintain the cytoadherent phenotype.

This antigenic switching may occur as frequently as 2% per generation in the absence of immune pressure. The exact molecular mechanism of antigenic switching is not known. Experimental evidence indicates that the mechanism is not associated with gene rearrangements into specific

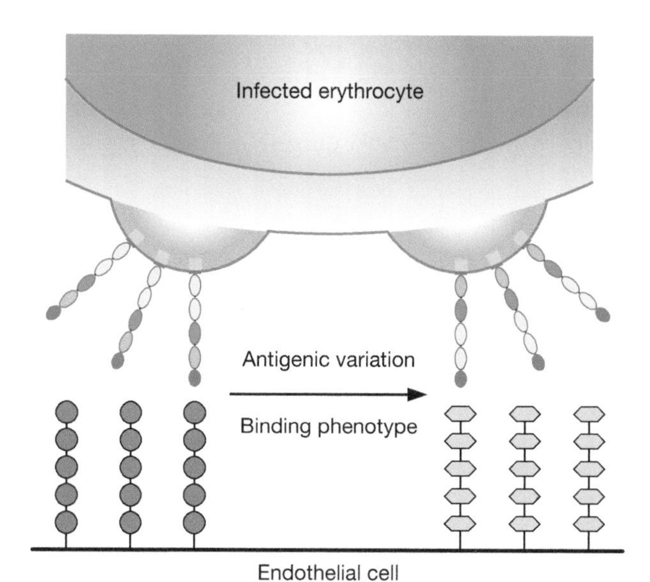

Figure 15.10 Antigenic variation. Changes in the expression of PfEMP1 alleles will result in an infected erythrocyte with a different binding phenotype as well as a different antigenic type.

expression-linked sites as found in African trypanosomes (Chapter 8). However, as in the case of African trypanosomes, only a single *var* gene is expressed at a time (i.e., allelic exclusion). Expression appears to occur at a distinct expression site within the nucleus that can only accommodate a single *var* gene promoter. Silent copies (i.e., nonexpressed) of the *var* gene are localized to regions of the nucleus containing primarily heterochromatin. Heterochromatin is a region of the chromosome that is highly condensed and transcriptionally inactive. Activation of gene expression is associated with chromatin modification which results in a repositioning of the gene to be expressed in a region of euchromatin, which is more open and available for transcription. In addition, the nonexpressed *var* genes are silenced by proteins which bind to the promoter regions. Thus the *var* promoter is sufficient for both the silencing and the mono-allelic transcription of a PfEMP1 allele. Once the transcriptional status of a gene has been established (i.e., either active or silent), this state is maintained through several replication cycles. The switching is thought to be controlled epigenetically in that activation or silencing is not accompanied by changes in the DNA sequence or the presence or absence of specific transcription factors. The exact nature of these epigenetic factors is not known but may involve post-translational modification of the histones, which are proteins involved in the formation of chromosomes.

The expression of a particular PfEMP1 will result in a parasite with a distinct cytoadherent phenotype and this may also affect pathogenesis and disease outcome. For example, binding to ICAM1 is usually implicated in cerebral pathology. Therefore, parasites expressing a PfEMP1 which binds to ICAM1 may be more likely to cause cerebral malaria. In fact, higher levels of transcription of particular *var* genes are found in cases of severe malaria as compared with uncomplicated malaria. Similarly, a higher proportion of isolates which bind to chondroitin sulfate are obtained from the placenta as compared with the peripheral circulation of either pregnant women or children. Furthermore, placental malaria is frequently associated with higher levels of transcription of a particular *var* gene which encodes a PfEMP that binds to chondroitin sulfate. This phenomenon is not restricted to the placenta in that there is a dominant expression of particular *var* genes in the various tissues (Figure 15.11). This tissue-specific expression of particular *var* genes implies that different tissues are selecting out different parasite populations based on the particular PfEMP1 allele being expressed on the surface of the infected erythrocyte.

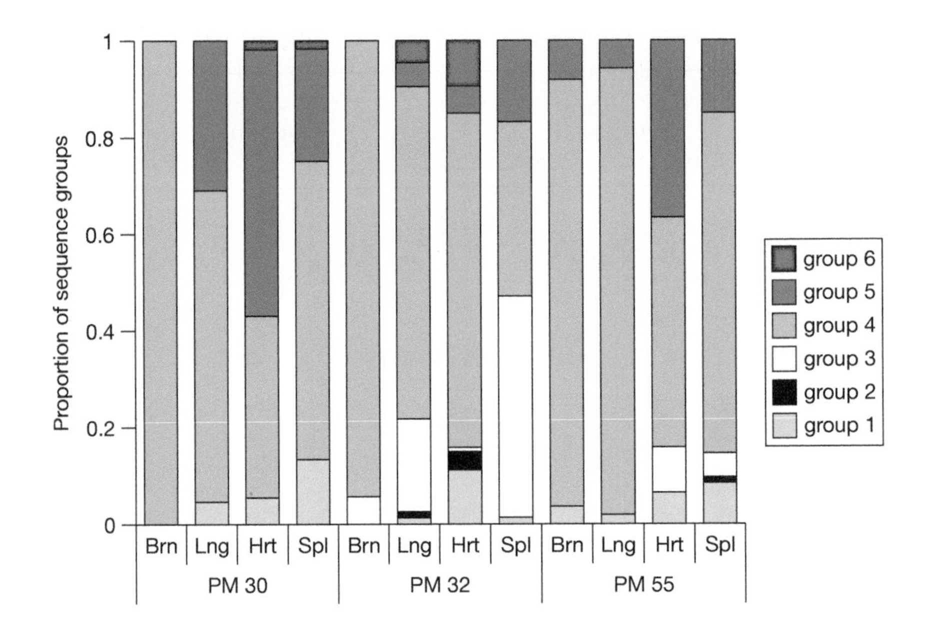

Figure 15.11 Frequency of PfEMP1 types in organs of malaria patients. The organs of patients dying from severe malaria were analyzed for transcripts of *var* genes. The various types of PfEMP1 were grouped according to their sequences at positions of limited variability. Different types of PfEMP1 predominated in the various organs. (Figure from Montgomery et al., 2007, Differential *var* gene expression in the organs of patients dying of falciparum malaria, *Molecular Microbiology* 65: 959–967. Reprinted with permission from Jacqui Montgomery of the College of Medicine, Blantyre, Malawi and Wiley-Blackwell Publishing.)

Immunity

Persons living in endemic areas do develop immunity against malaria. Almost always a person will exhibit symptoms during their initial exposure to malaria and the symptoms associated with subsequent exposures to malaria are usually less severe. However, the immunity against malaria is slow to develop and requires multiple exposures. In highly endemic areas only young children are at a high risk of developing severe falciparum malaria, whereas older children and adults are essentially protected from severe disease and death. However, this immunity is not a sterilizing immunity in that persons can still become infected. In addition, the immunity is short lived and in the absence of repeated exposures the level of immunity decreases. For example, previously semi-immune adults will often develop severe malaria upon returning to an endemic area after being in a non-endemic area for 1–2 years. This state of partial immunity in which parasitemia is lowered, but not eliminated, and infection is better tolerated is sometimes referred to as premunition. Premunition refers to an immunity that is contingent upon the pathogen being present.

Immunity against the blood-stage parasite

The immune response could be directed at either the pre-erythrocytic or erythrocytic stages of the parasite's life cycle. However, the erythrocytic stage of the life cycle is probably the most important in terms of clearing the parasite and lessening the disease. Because of the lack of MHC molecules on the surface of the parasite or the erythrocyte, it is usually assumed that antibody will play a key role in blood-stage immunity. Possible effector mechanisms for antibody include: blocking erythrocyte invasion by merozoites, antibody-dependent cellular killing mediated by cytophilic antibodies, or increased clearance of infected erythrocytes due to binding of antibodies to parasite antigens exposed on the erythrocyte surface. All of these potentially result in a lower parasitemia. The relative importance of these various mechanisms is not clear and immunity probably requires the generation of antibodies against numerous antigens. This, along with antigenic variation and polymorphisms in many *Plasmodium* antigens, could explain the slow development of immunity.

The observation that asymptomatic individuals can exhibit high levels of parasitemia has led to the concept of "antidisease" immunity. This would be in addition to the "antiparasite" immunity which lowers the parasitemia. Severe malaria and death are correlated with TNF-α and other proinflammatory cytokines. As discussed for the paroxysms and cerebral malaria, antigens or toxins released by the infected erythrocyte might stimulate the production of proinflammatory cytokines. Antibodies against these exo-antigens could neutralize their toxic effects and thus lead to an antidisease immunity. People from endemic areas develop antibodies specific for the parasite glycosylphosphatidylinositol, a proposed parasite toxin, and some studies show a correlation between these antibodies and improved clinical outcomes.

Vaccine development

Because of the difficulties in controlling malaria by other means there is much interest in developing a vaccine against malaria. Currently there is no available vaccine, but there is substantial research directed at identifying vaccine candidates and testing potential vaccines for safety and efficacy. Thus far, though, only a single vaccine candidate, based on a sporozoite surface protein, has progressed significantly in clinical trials and no vaccine is on the immediate horizon.

The complex life cycle and biology of the parasite provide several potential targets (Table 15.9). For example, vaccination against the sporozoite or exoerythrocytic stage could prevent infection. However, the induced immunity would need to be highly effective because the escape of a single sporozoite could lead to a blood-stage infection and disease. Vaccines targeted against merozoites or the infected erythrocyte would lower parasitemia by interfering with merozoite invasion or increasing the elimination of infected erythrocytes. Such a vaccine could potentially alleviate much of the pathogenesis associated with malaria even if it were not completely effective. In addition, infection may serve to boost the immune response. It may also be possible to vaccinate against the disease by immunizing against potentially toxic antigens. Antibodies neutralizing antigens or toxins that stimulate a proinflammatory immune response may lessen the pathogenesis associated with malaria even though such a vaccine may not lower the parasitemia. Sexual stages of the parasite such as gametocytes and gametes could also be targeted. Antibodies directed against gamete antigens can prevent infection of the mosquito and the subsequent sporogony. Such a vaccine would be altruistic in that it would not protect the individual against disease, but would protect others in the community by lowering the person-to-person transmission by decreasing the number of infected mosquitoes. These different vaccine strategies are not mutually exclusive and an ideal vaccine would combine several of these elements.

Table 15.9 Potential vaccine strategies

Target	Protection	Mechanism
Sporozoite	Anti-infection	Prevent or eliminate liver stage
Merozoite	Antiparasite	Decrease efficiency of merozoite invasion
Infected erythrocyte	Antiparasite	Increase clearance of infected erythrocytes
Exo-antigens	Antidisease	Lower production of inflammatory cytokines
Sexual stages	Antitransmission	Eliminate gametes or prevent infection of mosquitoes

Epidemiology and Transmission

Malaria is primarily a disease of the tropics and subtropics and is widespread in hot humid regions of Africa, Asia, and South and Central America. Currently malaria is endemic in more than 100 countries and 40% of the world's population lives in areas at risk for infection. The disease was also previously common in many temperate areas including the United States (see Box 15.2) and northern Europe, which are now virtually free of autochthonous transmission. However, many areas which previously had malaria under control are now experiencing resurgences.

The four human malarial species exhibit an overlapping geographical distribution. *P. vivax* and *P. falciparum* are the most commonly encountered species and both are found throughout tropical and subtropical areas. In addition to being found in tropical and subtropical areas, *P. vivax* is endemic in some temperate areas. *P. falciparum* is the predominant species in sub-Saharan Africa, whereas *P. vivax* is less common in Africa. *P. malariae* exhibits a similar distribution as *P. falciparum*, but is not as extensive and exhibits a more spotty distribution. *P. ovale* also exhibits a patchy distribution and is most prominent in western Africa. Mixed infections are common in many endemic areas.

Stable versus unstable malaria

The epidemiology of malaria can be viewed in terms of being stable (i.e., endemic) or unstable (i.e., epidemic). Stable malaria refers to a situation in which there is a measurable incidence of natural transmission over several years and this incidence remains somewhat constant. This would also include areas which experience seasonal transmission. Different areas

Box 15.2 Malaria in the United States

Malaria was previously more widespread in temperate areas including North America and Europe. It is believed that malaria was introduced to the Americas by the European colonists (*P. vivax* and *P. malariae*) and African slaves (*P. falciparum*) during the 16th and 17th centuries. Malaria became endemic in many parts of the United States excluding deserts and mountainous areas and the incidence probably peaked around 1875. A population shift from rural to urban areas, drainage of swamps to create farmland, improved housing and nutrition, better socioeconomic conditions and standards of living, greater access to medical services, and the availability of quinine for treatment all contributed to the decline in the prevalence of malaria even before the introduction of specific control measures (Table 15A). Some control activities, such as case detection and treatment, larviciding and house spraying, were introduced during the 1940s and led to the eradication of malaria in the United States. Since the 1950s nearly all cases of malaria in the United States have been imported. The major factors contributing to this eradication appear to be a population shift from rural to urban areas and an increase in the standard of living, which resulted in improved housing, better nutrition, and greater access to medical services.

The vast majority of malaria cases diagnosed in the United States are acquired by persons while traveling to countries where malaria is endemic. However, there have been several outbreaks of autochthonous malaria transmission in the United States [1, 2]. More than 80% of the cases were *P. vivax*. In some cases, the outbreaks were associated with large numbers of immigrants suggesting that infected persons can import the infection and then transmit it to local *Anopheles*. Many of these outbreaks were associated with unusually hot and humid weather, which may increase anopheline survival and decrease the duration of the sporogonic cycle, thus allowing for the development of infective sporozoites.

1. Zucker, J.R. (1996) Changing patterns of autochthonous malaria transmission in the United States: A review of recent outbreaks. *Emerg. Infect. Dis.* 2: 37–43.

2. CDC (2003) Local transmission of *Plasmodium vivax* malaria—Palm Beach County, Florida, 2003. *MMWR* 52: 908–911.

Table 15A Factors leading to a decline in malaria in the United States
Population shift from rural to urban areas
Improved socioeconomic conditions
Drainage of breeding grounds
Availability of quinine
Mosquito control activities

Table 15.10 Endemicity of malaria

Endemicity	Definition	Prevalence
Hypoendemic	Areas with low levels transmission of malaria	≤10%
Mesoendemic	Usually small rural communities with varying intensity of malaria	11–50%
Hyperendemic	Areas with intense, but seasonal, transmission of malaria	51–75%
Holoendemic	Areas with a high degree of year-long malaria transmission	≥75%

can experience different levels of incidence rates and this is often denoted by: hypoendemic, mesoendemic, hyperendemic, and holoendemic (Table 15.10). Persons living in highly endemic areas usually exhibit a high level of immunity and tolerate the infection well. Symptomatic disease is usually confined to children with relatively low levels of disease in the adult population despite high rates of infection. Paradoxically, the risk of severe disease—especially in children—tends to be lower among populations exposed to high transmission intensities, and the highest disease risk is found in populations exposed to low-to-moderate levels of transmission. This observation is consistent with the concept of antidisease immunity developing after repeated exposures to the parasite.

Unstable, or epidemic, malaria refers to an increase in malaria in areas of low endemicity or to outbreaks in areas previously without malaria or among nonimmune persons. Morbidity and mortality can be quite high in these nonimmune populations. These outbreaks are often attributed to changes in human behavior or effects on the environment. For example, human migration and resettlement can either introduce malaria into an area of hypoendemicity or expose a previously nonimmune population to endemic transmission. Changes in the ecology caused by natural disasters or public works projects, such as building roads or dams, can also impact malaria transmission and lead to epidemics.

Host and parasite factors influencing the transmission of malaria

The epidemiology of malaria is quite complex and highly dependent on the local conditions. The intricate interactions between parasite, host, and vector are the major factors in this epidemiological complexity. For example, as with most vector-transmitted diseases, the parasite must be able to establish a chronic infection within the host to maximize the opportunities for transmission. This is especially true in the case of seasonal transmission and in areas of low endemicity, and in general malaria infections are characterized by an initial acute phase followed by a longer relatively asymptomatic chronic phase. This is due in part to the ability of the parasite to avoid complete clearance by the immune system. For example, *P. falciparum* exhibits an antigenic variation that allows it to stay one step ahead of the immune system. Similarly, *P. vivax* and *P. ovale* exhibit the hypnozoite stage and are capable of relapses. This allows the parasite to maintain the infection within the human host even after the blood stage of the infection has been cleared. The relative long interval between relapses in some *P. vivax* isolates probably explains its ability to maintain transmission cycles in some temperate climates.

In regard to the host, humans are the only significant reservoir for the parasite and sustained transmission depends upon maintaining a pool of infected individuals and contact between humans and anopheline

Table 15.11 Human genetics and innate resistance

Polymorphism	Comment
Duffy negative	Erythrocytes resistant to *P. vivax* infection
Ovalcytosis	Erythrocytes resistant to merozoite invasion
Sickle-cell anemia	Heterozygous individuals exhibit lower parasitemias and less severe disease
Thalassemia	Erythrocytes unfavorable for *P. falciparum* development
G6PD deficiency	Erythrocytes not able to handle increased oxidative stress associated with parasite metabolism

mosquitoes. Several factors influence the susceptibility of humans to infection. Obviously the immune status of the individual and their prior experience with malaria will influence the course of the infection. Pregnant women, especially during the first pregnancy, are more susceptible to falciparum malaria as illustrated by a higher prevalence of infection and higher parasitemias. In addition, certain genetic diseases and polymorphisms have been associated with decreased infection or disease (Table 15.11). For example, individuals lacking the Duffy blood-group antigen are refractory to *P. vivax*. A large proportion of the populations in western Africa are Duffy negative, thus accounting for the low levels of *P. vivax* in western Africa. This innate resistance led to the identification of the Duffy antigen as the erythrocyte receptor for merozoite invasion (Chapter 11).

Several inherited erythrocyte disorders are found predominantly in malaria endemic areas and at frequencies much higher than expected. This has led to speculation that these disorders confer some protection against malaria. For example, southeast Asian ovalcytosis is caused by a mutation in an erythrocyte membrane protein called band 3. This mutation causes the erythrocyte membrane to become more rigid and more refractory to merozoite invasion. The mechanisms by which the other diseases might confer protection against malaria are not known. In most cases it is presumed or speculated that the combination of the defect and infection leads to premature lysis or clearance of the infected erythrocyte. For example, glucose-6-phosphate dehydrogenase deficient erythrocytes would have an impaired ability to handle oxidative stress. The additional oxidants produced as a result of parasite metabolism and the digestion of hemoglobin (Box 15.1) may overwhelm the infected erythrocyte and lead to its destruction before the parasite is able to complete schizogony. Sickle cell anemia and thalassemia are also speculated to make the infected erythrocyte more susceptible to oxidative stress.

Vectorial capacity

The potential of the mosquito to serve as a vector depends on the ability to support sporogony, mosquito abundance, and contact with humans, which are all influenced by climatic and ecological factors (Table 15.12). The ability to support sporogony is largely dependent upon species in that not all species of *Anopheles* are susceptible to *Plasmodium* infection. Ambient temperature and mosquito longevity are other key factors affecting the parasite's interaction with the vector. Development of *P. falciparum* requires a minimum temperature of 16–19°C, whereas the minimum temperature for *P. vivax* is 14–16°C. Temperature also affects the time of development in that the duration of sporogony is substantially shorter at higher temperatures. A shorter duration of sporogony increases the chances that the mosquito will transmit the infection within its lifespan.

Mosquito density and feeding habits also influence the transmission of malaria. Temperature, altitude, rainfall, and the availability of breeding places are factors that affect mosquito density, whereas human–mosquito contact will be influenced by the mosquito behavior. For example, the degree to which a particular mosquito species is anthropophilic influences the probability of the mosquito becoming infected and the subsequent transmission to another human. These anthropophilic tendencies are not necessarily absolute in that many zoophilic mosquitoes will switch to humans if densities reach high levels or the preferred animal source is diminished. The preferred feeding time and whether the mosquito feeds predominantly indoors or outdoors will influence the transmission dynamics. For example, outdoor feeding mosquitoes are more likely to find a human blood meal in the early evening than those feeding late at night when most people are inside. The behavior of the local mosquito vector also needs to be considered in control activities.

Figure 15.12 Factors influencing vectorial capacity

Sporogony	Temperature
	Mosquito longevity
	Mosquito species
Mosquito density	Temperature
	Altitude
	Rainfall
	Breeding places
Human contact	Anthropophilic
	Indoor vs. outdoor
	Feeding time

Prevention and Control

Strategies for preventing and controlling malaria involve three different general approaches: reduce human–mosquito contact, reduce vector density, and reduce the parasite reservoir (Table 15.13). Prevention of malaria in individuals will generally involve the reduction of human–mosquito contact through the use of bed nets, repellents, protective clothing, and mosquito avoidance. This can also include reducing the number of mosquitoes in houses through the use of screens and insecticides. Chemoprophylaxis is another strategy used primarily for persons visiting an endemic area for a relatively short period of time. However, chemoprophylaxis only suppresses parasitemia and does not prevent infection. Furthermore, chemoprophylaxis should not be used on a mass scale to control malaria because it promotes the development of drug resistance.

Control activities at the community level can utilize approaches which directly reduce human–mosquito contact as well as approaches which reduce the total number of mosquitoes in an area. Such approaches include the reduction in mosquito breeding grounds (e.g., environmental modification), targeting the larva stages with chemical or biological agents, and massive insecticide spraying for the adult mosquitoes. Biological control methods include the introduction of fish which eat the mosquito larvae or bacteria (e.g., *Bacillus thuringiensis*) which excrete larval toxins. Case detection and treatment is another potential control method. Identifying and treating infected persons, especially asymptomatic individuals, will reduce the size of the parasite reservoir within the human population and can lower transmission rates. However, this can be a relatively expensive approach.

Table 15.13 Prevention and control strategies

Reduce human–mosquito contact	Impregnated bed nets
	Repellents, protective clothing
	Screens, house spraying
Reduce vector density	Environmental modification
	Larvicides/insecticides
	Biological control
Reduce parasite reservoir	Case detection and treatment
	Chemoprophylaxis

Malaria control is complex and multifactorial

These various approaches are not mutually exclusive and can be combined. Many of the successful control programs include both measures to control mosquitoes and treatment of infected individuals. There is no standard method of malaria control that has proven universally effective. The epidemiologic, socioeconomic, cultural, and infrastructural factors of a particular region will determine the most appropriate malaria control. Some of the factors which need to be considered include: infrastructure of existing health care services and other resources; intensity and periodicity (e.g., seasonality) of transmission; mosquito species (ecological requirements, behavioral characteristics, insecticide sensitivity, etc.); parasite species and drug sensitivities; cultural and social characteristics of the population; and presence of social and ecological change.

The control of malaria in tropical Africa has been particularly problematic because of the high transmission rates and the overall low socioeconomic level. Several studies have shown that insecticide-treated bed nets reduce the morbidity and mortality associated with malaria. Introducing bed nets does not require large promotional programs and their use is readily accepted in most areas. This may be in part due to the reduction in mosquito nuisance biting. Some questions have been raised in regard to the economic sustainability of bed net programs. It is necessary to retreat the bed nets with insecticide periodically and the bed nets need to be repaired and replaced as they become torn and wear out. In addition, some have raised concerns about the long-term benefits of bed nets because they reduce exposure, but do not eliminate it. This reduction in exposure may delay the acquisition of immunity and simply postpone morbidity and mortality to older age groups.

Diagnosis and Species Identification

Malaria is suspected in persons with a history of being in an endemic area and presenting symptoms consistent with malaria, such as fever, chills, headache, and malaise. These symptoms, especially in the early stages of the infection, are nonspecific and often described as flu-like. As the disease progresses, the patient may exhibit an enlarged spleen and/or liver and anemia. Diagnosis is confirmed by microscopic demonstration of the parasite in blood smears. Thick blood smears are generally more sensitive for the detection of parasites, whereas thin smears are preferable for species identification. If parasites are not found on the first blood smear it is recommended to make additional smears every 6–12 hours for 2 days. The level of parasitemia can also be quantified by counting the percent infected erythrocytes in thin smears or converting the ratio of parasites to white blood cells detected in thick smears to parasites per microliter.

Rapid diagnostic tests based on immunochromatography, or "dipsticks," are also available. These tests are based on detecting parasite proteins within blood samples. Several commercially available tests are available. Most are based on the detection of parasite lactate dehydrogenase or the histidine rich protein-2 of *P. falciparum* combined with parasite aldolase. Many of the tests can distinguish *P. falciparum* from the other three species, but cannot distinguish between those three species. Molecular methods using DNA probes or PCR can also be used for diagnosis. However, these are usually used in epidemiological studies and not for routine diagnosis.

Distinguishing features of blood-stage parasites

The blood-stage parasites of human *Plasmodium* species exhibit differences in their morphology and modify the host erythrocyte differently (Figure 15.12). These differences can be used to distinguish the four species. The typical features of a *P. falciparum* infection are that only ring forms are observed in the circulation because the mature forms sequester. In addition, the gametocytes of *P. falciparum* are elongated and crescent shaped and readily distinguished from the gametocytes of the other species. The ring stages of *P. falciparum* tend to be slightly smaller than the other species and are generally more numerous.

The most distinctive features of *P. vivax* are the enlarged infected erythrocytes and the appearance of granules, called "Schüffner's dots," over the erythrocyte cytoplasm. These granules are manifestation of caveola–vesicle complexes that form on the erythrocyte membrane. *P. ovale* also exhibits Schüffner's dots and an enlarged erythrocyte, making it difficult to distinguish from *P. vivax*. The growing trophozoite of *P. vivax* often has an ameboid appearance and the schizonts can have more than 20 merozoites. Whereas, *P. ovale* trophozoites are more compact than *P. vivax* and the mature schizonts have fewer merozoites. *P. ovale* also has more of a tendency to form elongated host erythrocytes. *P. malariae* is characterized by a compact parasite and does not alter the host erythrocyte or cause enlargement. Mature schizonts (i.e., segmenters) will typically have eight to 10 merozoites that are sometimes arranged in a rosette pattern with a clump of pigment in the center.

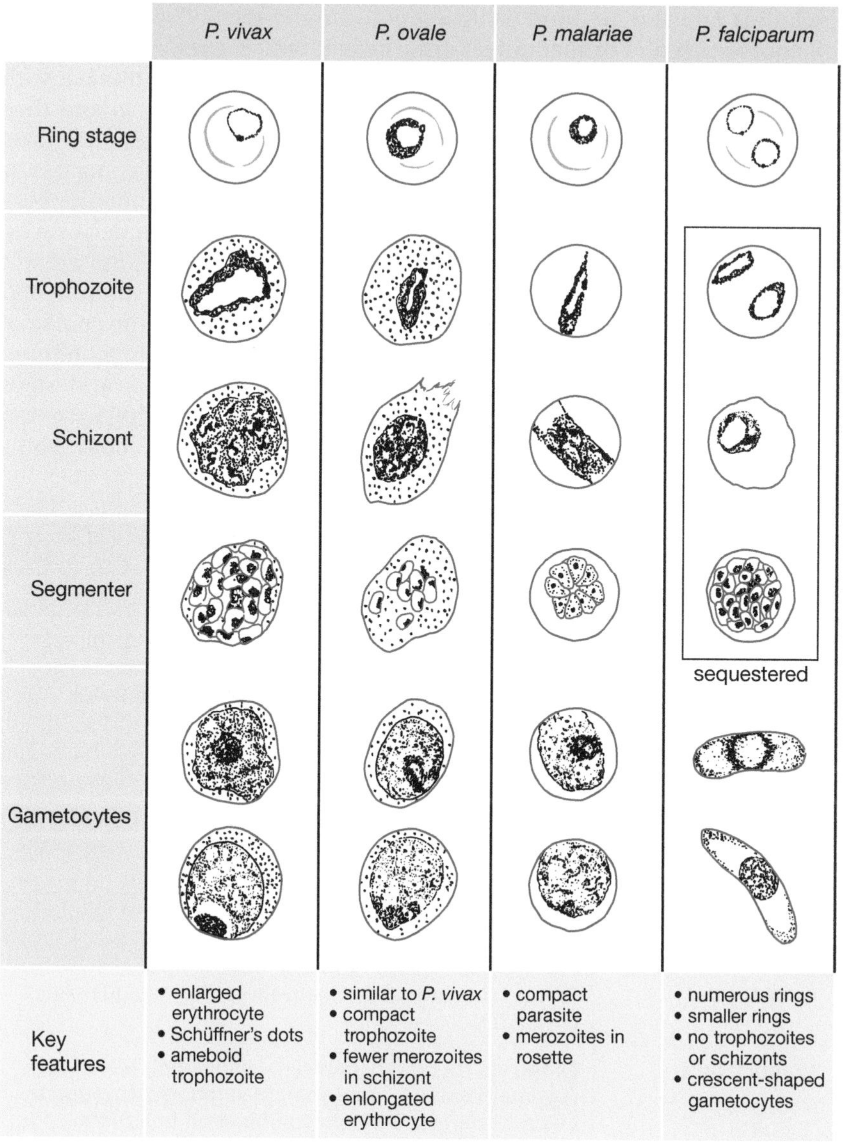

	P. vivax	P. ovale	P. malariae	P. falciparum
Key features	• enlarged erythrocyte • Schüffner's dots • ameboid trophozoite	• similar to *P. vivax* • compact trophozoite • fewer merozoites in schizont • enlongated erythrocyte	• compact parasite • merozoites in rosette	• numerous rings • smaller rings • no trophozoites or schizonts • crescent-shaped gametocytes

Figure 15.12 Key morphological features of human malaria parasites. The ring forms of all four species are very similar and difficult to distinguish. The presence of a large number of rings in the absence of trophozoite and schizont stages, as well as multiply infected erythrocytes, is highly suggestive of *P. falciparum*. Erythrocytes infected with *P. vivax* and *P. ovale* are enlarged and exhibit Schüffner's dots as the rings mature into trophozoites. The typical number of merozoites produced per schizont is: *P. vivax* 14–20 (up to 24), *P. ovale* 6–12 (up to 18), *P. malariae* 8–10 (up to 12), and *P. falciparum* 16–24 (up to 36). (Adapted from *Tropical Medicine and Parasitology* by R. Goldsmith and D. Hyneman, 1989 with kind permission from The McGraw Hill Companies, Inc.)

Chemotherapy and Drug Resistance

Several antimalarial drugs are available. Many factors are involved in deciding the best treatment for malaria. These factors include the parasite species, the severity of disease, the patient's age and immune status, the parasite's susceptibility to the drugs (i.e., drug resistance), and the cost and availability of drugs. Therefore, the exact recommendations will often vary according to geographical region. In addition, the various drugs act differentially on the different life cycle stages (Table 15.14).

Fast-acting blood schizontocides, which act upon the blood stage of the parasite, are used to treat acute infections and to quickly relieve the clinical symptoms. Chloroquine is generally the recommended treatment for patients with *P. vivax*, *P. ovale*, *P. malariae*, and uncomplicated chloroquine-sensitive *P. falciparum* infections. After its introduction near the end of World War II, chloroquine quickly became the drug of choice for the treatment and prevention of malaria. Not only is chloroquine an effective drug—probably due to its site of action in the food vacuole and its interference with hemozoin formation (Box 15.1)—but it is also relatively nontoxic and cheap. Side effects may include pruritus (i.e., itching), nausea, or agitation. Patients infected with either *P. vivax* or *P. ovale*, and who are not at a high risk for reinfection, should also be treated with primaquine (a tissue schizontocide). Primaquine is effective against the liver stage of the parasite, including hypnozoites, and will prevent future relapses. The combination of chloroquine and primaquine is often called "radical cure."

Severe, or complicated, falciparum malaria is a serious disease with a high mortality rate and must be regarded as life threatening, and thus requires urgent treatment. Treatment typically requires parenteral drug administration (i.e., injections), because the patients are often comatose or vomiting and thus cannot take the drugs orally. Parenteral formulations are available for chloroquine, quinine, quinidine, and artemisinin derivatives. The artemisinin derivatives are generally the preferred choice, but are not yet approved everywhere. For example, in the United States quinine and quinidine are the approved drugs for severe malaria. During the course of treatment for severe malaria patients need to be continuously monitored for hematocrit, parasitemia, hydration levels, hypoglycemia, and signs of drug toxicity and other complications. A switch to oral administration should be made as soon as the patient is able. Unfortunately, most deaths

Table 15.14 Selected antimalarial drugs

Drug class	Examples
Fast-acting blood schizontocide	Choloroquine (+ other 4-aminoquinolines), quinine, quinidine, mefloquine, halofantrine, antifolates (pyrimethamine, proguanil, sulfadoxine, dapsone), artemisinin derivatives
Slow-acting blood schizontocide	Doxycycline (+ other tetracycline antibiotics)
Blood + mild tissue schizontocide	Proguanil, pyrimethamine, tetracyclines
Tissue schizontocide (antirelapsing)	Primaquine, tafenoquine
Gametocidal	Primaquine, artemisinin derivatives, 4-aminoquinolines (limited?)
Combinations	Fansidar® (pyrimethamine (+ sulfadoxine), Maloprim® (pyrimethamine (+ dapsone), Malarone® (atovaquone (+ proguanil), artemisinin combination treatment

Box 15.3 Genes associated with chloroquine resistance

Chloroquine resistant *P. falciparum* were first detected in Colombia and at the Cambodia–Thailand border during the late 1950s. During the 1960s and 1970s, resistant parasites spread through South America, Southeast Asia, and India. Resistance was first reported in East Africa in 1978 and spread throughout the continent during the 1980s. Chloroquine resistant *P. vivax* was first reported in 1989 in Papua New Guinea and is now found in several foci in southeast Asia and perhaps South America.

It is generally accepted that the site of action of chloroquine is in the food vacuole and the mechanism of action is the inhibition of hemozoin formation (Box 15.1). The food vacuole concentrates the chloroquine up to several thousand-fold above the plasma concentration. Possible mechanisms for this selective accumulation of chloroquine in the food vacuole are: (1) protonation and ion trapping of the chloroquine due to the low pH of the food vacuole; (2) active uptake of chloroquine by a parasite transporter(s); and/or (3) binding of chloroquine to a specific receptor in the food vacuole. The exact contributions of these three postulated mechanisms are not clear.

Chloroquine resistance is associated with a decrease in the amount of chloroquine that accumulates in the food vacuole. The mechanism for this decreased accumulation is controversial. However, some studies have shown that the decrease in drug accumulation is due to an increase in drug efflux. The observation that verapamil and related drugs can reverse the chloroquine resistant phenotype has led to speculation that an ATP-dependent transporter plays a role in drug efflux and chloroquine resistance, similar to the multidrug resistance (MDR) in cancer. An MDR-like transporter, designated PfMDR1, has been identified on the food vacuole membrane. However, no definitive correlations between PfMDR1 and chloroquine resistance could be demonstrated. An ancillary role for PfMDR1 in chloroquine resistance cannot be ruled out though.

A genetic cross and mapping studies between a chloroquine resistant clone and a chloroquine sensitive clone resulted in the identification of a 36 kb region on chromosome 7 associated with chloroquine resistance. One of the 10 genes in this 36 kb region encodes a protein with 10 transmembrane domains and resembles a transporter protein similar to chloride channels. The gene has been designated as PfCRT (for *P. falciparum* chloroquine resistance transporter) and the protein is localized to the food vacuole membrane. Several mutations in the PfCRT gene show correlations with the chloroquine resistance phenotype and one mutation, a substitution of a threonine (T) for a lysine (K) at residue 76 (K76T) shows good correlation with chloroquine resistance. Presumably these mutations affect the accumulation of chloroquine in the food vacuole, but the exact mechanism of chloroquine resistance is not known. Furthermore, the observation that chloroquine resistance has arisen relatively few times and then subsequently spread has lead to speculation that multiple genes are involved in the development of resistance.

due to severe malaria occur at or close to home in situations where the patients cannot be taken to the hospital. Artemisinin suppositories which can be administered by village health workers have also been developed and have proven to be safe and effective.

The efficacy of chloroquine is greatly diminished by the widespread chloroquine resistance of *P. falciparum* and the emergence of chloroquine-resistant *P. vivax* (Box 15.3). If chloroquine therapy is not effective, or if in an area with chloroquine-resistant malaria, common alternative treatments include: mefloquine, quinine in combination with doxycycline, or Fansidar®. Derivatives of artemisinin (dihydroartemisinin, artesunate, and artemether) are increasingly used in Asia and Africa and are now recommended as the first line of treatment by the World Health Organization. These drugs were originally derived from the wormwood plant (*Artemesia annua*) and have been used for a long time in China as a herbal tea called quinhaosu to treat febrile illnesses. To prevent the high recrudescence rates associated with artemisinin derivatives and to slow the development of drug resistance it is recommended that treatment be combined with an unrelated antimalarial. Drugs used in combination with artemisinin include mefloquine, lumefantrine, Fansidar®, and amodiaquine.

Chemoprophylaxis

Chemoprophylaxis is especially important for persons from nonmalarious areas who visit areas endemic for malaria. Such nonimmune persons can quickly develop a serious and life-threatening disease. As in the case of treatment there is no standard recommendation and the choices for chemoprophylaxis are highly dependent upon the conditions associated with the travel and the individual person. Chemoprophylaxis requires the use of nontoxic drugs because these drugs will be taken over extended periods of time. Generally the patient will start to take the drug before traveling and then continue taking the drug during the stay in the endemic area and continue taking the drug after returning. This is to insure the drug is maintained at sufficient levels throughout the visit and to protect against any infection obtained during the visit. Unfortunately, many of the effective and nontoxic drugs (e.g., chloroquine, pyrimethamine, proguanil) are of limited use because of drug resistance. Another strategy is presumptive (or "standby") treatment to be used in conjunction with prophylaxis. In this case a person either forgoes prophylaxis or takes chloroquine or another relatively nontoxic drug for prophylaxis and carries a drug such as Fansidar®, mefloquine, or quinine, which they will take if they start to exhibit symptoms associated with malaria.

The use of mefloquine for malaria chemoprophylaxis is somewhat controversial. Mefloquine is efficacious at preventing malaria with a single dose per week, thus offering advantages to drugs that need to be administered daily. At this dosage mefloquine is tolerated by most individuals. However, some people experience neuropsychiatric adverse affects such as sleep disturbances and nightmares. This could be exacerbated by stress associated with international travel. Randomized, blinded, and controlled trials indicate that neuropsychiatric adverse affects are only slightly higher with mefloquine than with other antimalarials.

Killing the exoerythrocytic stage (i.e., liver stage) would prevent the blood infection and is known as causal prophylaxis. This is highly desirable in that it limits the amount of time the prophylactic drug needs to be taken before and after travel to an endemic area. The only currently available drug for causal prophylaxis is primaquine. However, malaria prophylaxis is not an approved use of primaquine and should only be prescribed for prophylaxis on a case-by-case basis. For example, for persons who frequently have trips of short duration to highly endemic areas and if the person does not exhibit glucose-6-phosphate dehydrogenase deficiency. Tafenoquine is currently undergoing field evaluation for its use in causal prophylaxis.

Drug resistance

Drug resistance, and in particular chloroquine resistance, is a major public health problem in the control of malaria. Drug resistance is defined by a treatment failure and can be graded into different levels depending on the timing of the recrudescence following treatment. Traditionally these levels of drug resistance have been defined as sensitive (no recrudescence), RI (delayed recrudescence), RII (early recrudescence), and RIII (minimal or no antiparasite effect). Drug resistance by this protocol is determined by monitoring patients for parasitemia for 28 days following standard drug treatment. Because of the difficulties in using this protocol in endemic areas, a modified protocol based on clinical outcome was introduced by the World Health Organization in 1996. In this protocol the level of resistance is expressed as adequate clinical response (ACR), late treatment failure (LTF), or early treatment failure (ETF). ACR is defined as the absence of parasitemia (irrespective of fever) or absence of clinical symptoms (irrespective of parasitemia) during 14 days following treatment. LTF is the reappearance of symptoms or the presence of parasitemia during days 4 to 14 of follow-up. ETF is the persistence of clinical symptoms or the presence of parasitemia during the first 3 days of follow-up.

Either protocol can be used to determine drug resistance, but the clinical outcome protocol is more practical in areas of intense transmission where it may be difficult to distinguish reinfection from recrudescence and where parasitemia in the absence of clinical symptoms is common. In addition,

Table 15.15 Proteins and mutations possibly involved in drug resistance

Protein[a]	Subcellular location	Primary function	Major drugs affected	Major polymorphisms[b]
CRT	Food vacuole	Putative transporter	Chloroquine	K76T
MDR1	Food vacuole	Transporter	Mefloquine, quinine (?)	Amplification, D86Y[c]
DHFR	Cytoplasm	Folate metabolism	Pyrimethamine, proguanil	S108N, N51I, C59R, I164L
DHPS	Cytoplasm	Folate metabolism	Sulfadoxine, dapsone	A437G, K540E, A581G
Cytochrome b	Mitochrondria	Electron transport	Atovaquone	Y268S/N/C
ATPase6	Endoplasmic reticulum	Calcium transport	Artemisinins	S769N

[a]CRT, chloroquine resistance transporter; MDR1, multidrug resistance (P-glycoprotein homolog); DHFR, dihydrofolate reductase; DHPS, dihydropteroate synthetase; ATPase6, sarco/endoplasmic reticulum calcium-dependent ATPase ortholog.
[b]Polymorphisms associated with drug resistance where the number refers to amino acid (i.e., codon) and the first letter is the wild-type residue and the second letter is the polymorphism(s) associated with resistance.
[c]Associated with increased sensitivity to mefloquine and dihydroartemisinin, but a decreased sensitivity to chloroquine.

the clinical outcome protocol requires a shorter period of hospitalization. There are also *in vitro* tests that can determine the efficacy of the drugs against *P. falciparum* grown in culture. The *in vivo* and *in vitro* tests do not always correspond because host immunity and other factors can affect the *in vivo* outcomes. For some drugs the mechanism of resistance is known and specific genetic polymorphisms are associated with drug resistance (Table 15.15). For these drugs it is possible to screen for particular polymorphisms using molecular markers. In some cases the protein involved in resistance is the target of the drug and reflects decreased affinity of the drug for the target. In other cases the protein involved in resistance is not the target of drug action. For example, chloroquine resistance correlates with specific mutations in a gene for a putative transporter protein which presumably exports the drug out of the food vacuole (Box 15.3).

Drug resistance develops when parasites with decreased sensitivities to antimalarial drugs are selected under drug pressure. Decreased drug sensitivity can be conferred by several mechanisms and reflects genetic mutations or polymorphisms in the parasite population. The drug-resistant parasites will have a selective advantage over the drug-sensitive parasites in the presence of drug and will be preferentially transmitted. Major factors in the development of drug resistance are the use of subtherapeutic doses of drugs or not completing the treatment regimen (Table 15.16). The lower drug levels will eliminate the most susceptible parasites, but those which

Table 15.16 Factors contributing to development and spread of drug resistance

Factor	Comments
Self-treatment	Individuals may only take the drug until symptoms clear or will take lower doses to save money
Poor compliance	Individuals may not complete the full course of treatment because of drug side effects
Mass administration	The widespread use of a drug in an area of intense transmission increases drug pressure by exposing a larger parasite population to the drug
Long drug half-life	Drugs that are slowly eliminated will lead to a longer exposure of the parasite to subtherapeutic drug concentrations
Transmission intensity	High levels of transmission may allow reinfection while drugs are at subtherapeutic levels

can tolerate the drug will recover and reproduce. Over time this will lead to a continued selection for parasites which can tolerate even higher doses of the drug. This is likely due to the accumulation of mutations in the target proteins that confer higher levels of drug resistance. It is crucial to maintain an adequate concentration of the drug for a sufficient time to completely eliminate the parasites from any given individual.

Summary and Key Concepts

- Malaria is a major human disease caused by apicomplexan parasites of the genus *Plasmodium* and is found throughout tropical and subtropical regions.
- Four distinct *Plasmodium* species with different biological and epidemiological characteristics as well as different disease etiologies infect humans: *P. falciparum*, *P. vivax*, *P. malariae*, and *P. ovale*.
- The parasite is transmitted to humans by anopheline mosquitoes and exhibits a complex life cycle in both the human host and the vector.
- During one stage of its life cycle the parasite infects erythrocytes and this blood stage is responsible for the clinical manifestations associated with the disease.
- Infection with the malaria parasite causes an acute febrile illness which is most notable for its periodic fever paroxysms coinciding with the rupture of the infected erythrocyte and release of merozoites.
- *P. falciparum* can cause a severe and fatal disease involving multiple organs. The most notable manifestations of severe malaria are: cerebral malaria, severe anemia, and acute respiratory distress syndrome.
- The higher morbidity associated with *P. falciparum* is due in part to the potentially high levels of parasitemia associated with *P. falciparum* infections and the cytoadherence of infected erythrocytes to endothelial cells within various organs.
- Cytoadherence is mediated by ligands expressed on parasite induced structures located on the surface of *P. falciparum*-infected erythrocytes which bind to various receptors found on endothelial cells.
- Immunity against malaria is slow to develop and requires repeated exposures to the parasite.
- The epidemiology and transmission of malaria varies according to the intricate interactions between the human host, mosquito vector, and parasite in any particular location.
- Control and prevention measures should consider the epidemiologic, socioeconomic, cultural and infrastructural factors of a particular region. Generally control and prevention will involve reducing mosquito–human contact, reducing vector density, treatment or chemoprophylaxis, or some combination of approaches.
- The choice of antimalarial drugs for treatment or chemoprophylaxis depends on the parasite species, the severity of disease, drug resistance, and the cost and availability of drugs. Currently artemisinin derivatives combined with other drugs are generally the recommended first line of treatment, but use is often restricted by cost or availability.
- Widespread and increasing drug resistance to antimalarial drugs is leading to increased morbidity associated with malaria as well as limiting the ability to control the disease and its spread.

Further Reading

Clark, I.A. and Cowden, W.B. (2003) The pathophysiology of falciparum malaria. *Pharmacol. Ther.* 99: 221–260.

Dzikowski, R. and Deitsch, K.W. (2009) Genetics of antigenic variation in *Plasmodium falciparum. Curr. Genet.* 55: 103–110.

Greenwood, B.M., Fidock, D.A., Kyle, D.E., Kappe, S.H.I., Alonso, P.L., Collins, F.H. and Duffy, P.E. (2008) Malaria: progress, perils, and prospects for eradication. *J. Clin. Invest.* 118: 1266–1276.

Hyde, J.E. (2007) Drug-resistant malaria—an insight. *FEBS J.* 274: 4688–4698.

Idro, R., Jenkins, N.E. and Newton, C.R.J.C. (2005) Pathogenesis, clinical features, and neurological outcome of cerebral malaria. *Lancet Neurol.* 4: 827–840.

Langhorne, J., Ndungu, F.M., Sponaas, A.M. and Marsh, K. (2008) Immunity to malaria: more questions than answers. *Nat. Rev. Immunol* 9: 725–732.

Pasvol, G. (2006) The treatment of complicated and severe malaria. *Br. Med. Bull.* 75/76: 29–47.

Phillips, R.S. (2001) Current status of malaria and potential for control. *Clin. Microbiol. Rev.* 14: 208.

Walther, B. and Walther, M. (2007) What does it take to control malaria? *Ann. Trop. Med. Parasitol.* 101: 657–672.

Wellems, T.E., Hayton, K. and Fairhurst, R.M. (2009) The impact of malaria parasitism: from corpuscles to communities. *J. Clin. Invest.* 119: 2496–2505.

White, N.J. (2008) Qinghaosu (Artemisinin): the price of success. *Science* 320: 330–334.

Williams, T.N. (2006) Human red blood cell polymorphisms and malaria. *Curr. Opin. Microbiol.* 9: 388–394.

Wongsrichanalai, C., Barcus, M.J., Muth, S., Sutamihardja, A. and Wernsdorfer, W.H. (2007) A review of malaria diagnostic tools: Microscopy and rapid diagnostic test (RDT). *Am. J. Trop. Med. Hyg.* 77: 119–127.

World Health Organization (2000) Severe falciparum malaria. *Trans. R. Soc. Trop. Med. Hyg.* 94: S1–90.

Babesiosis

Disease(s)	Babesiosis
Etiological agent(s)	*Babesia microti, B. duncani, B. divergens*
Major organ(s) affected	Blood
Transmission mode or vector	Ixodid ticks
Geographical distribution	Northeastern United States and sporadic cases in Midwestern and west coast of U.S. and Europe
Morbidity and mortality	Generally associated with splenectomy or other immunosuppression. Can be lethal
Diagnosis	Detection of parasites in blood
Treatment	Atovaquone plus azithromycin or clindamycin plus quinine
Control and prevention	Avoid tick bites, especially if immunocompromised

Babesia species are common blood parasites of wild and domestic animals and cause significant losses in livestock and companion animals throughout the world. Human babesiosis is a relatively rare zoonotic disease transmitted by ticks and may be an emerging disease in some parts of the world. This emergence is due in part to better awareness and the increasing numbers of immunocompromised patients. Of historical note, Smith and Kilborne reported that ticks transmit *B. bigemina*, the cause of Texas cattle fever, in 1893. This was the first demonstration of an arthropod-transmitted disease and likely inspired the subsequent discovery of other vector-transmitted diseases such as yellow fever and malaria. In addition, it has been speculated that the plague of the Egyptians' cattle described in the biblical book of Exodus may have been red water fever caused by *B. bovis*.

Life Cycle

Babesia species are members of a group called the **piroplasms**, in reference to intraerythrocytic forms that are teardrop shaped in some species. The other major genus of piroplasms is *Theileria* (Box 16.1), which have not been reported to infect humans, but are a major pathogen of cattle, particularly in Africa. Both *Babesia* and *Theileria* species are important veterinarian pathogens and cause tremendous losses of livestock in endemic areas. Two distinct groups of *Babesia* species have been informally recognized as being "large" *Babesia* and "small" *Babesia* based on the size of the intraerythrocytic trophozoites. Molecular phylogenetic data also support this grouping. However, the molecular phylogenetic data also indicate that *Theileria* species form a sister clade with the small *Babesia* and not the large *Babesia*. Thus the large and small *Babesia* species do not form a monophyletic clade and therefore the taxonomy within the piroplasms is not completely congruent with the phylogenetics of the group.

Box 16.1 *Theileria* and lymphoblastogenesis

Theileria species infect and cause disease in livestock, especially cattle, in many parts of the world. The most serious is East Coast fever of cattle, caused by *T. parva.* It has 90–100% mortality in Africa. *T. annulata* causes a milder disease of cattle along the Mediterranean and in the Middle East known as tropical theileriosis. *Theileria* are closely related to *Babesia* and exhibit a similar life cycle. The major difference is a pre-erythrocytic stage exhibited by *Theileria* species. Sporozoites from the tick invade lymphocytes instead of erythrocytes and induce proliferation of the host lymphocytes by an unknown mechanism. Similar to other Apicomplexa, and *Plasmodium* in particular, the parasite develops into a multinucleated schizont (i.e., meront). The schizont also undergoes a division coincident with the replication of the proliferating lymphocyte, and thus a schizont is transferred to each of the daughter lymphocytes. Therefore, not only does the parasite replicate within the lymphocytes, but the lymphocytes are also induced to replicate leading to an expansion of the parasite. Some of the schizonts will develop into merozoites which are released by lysis of the infected lymphocyte and the free merozoites invade erythrocytes. In contrast to *Babesia,* the intraerythrocytic trophozoites of *Theileria* do not replicate. Instead they ultimately develop into gamonts which are infectious for the tick and undergo a similar developmental process as *Babesia* within the tick. It is the lymphoproliferative process that leads to the severe disease manifestations associated with theilerioses. In other words, a cancer-like disease is induced by *Theileria* and *Theileria* have been used as a model system to study the transformation of lymphocytes. This lymphocyte transformation is reversible in that treatment of *Theileria* leads to parasite clearance and the lymphocyte proliferation subsequently ceases.

Babesia exhibits a typical apicomplexan life cycle characterized by merogony, gametogony, and sporogony (Figure 16.1). The infection is acquired by the reservoir and human hosts when **sporozoites** are transferred from the salivary glands during tick feeding. The sporozoites invade erythrocytes utilizing a mechanism of invasion that is similar to other apicomplexa (Chapter 11). In contrast to *Plasmodium* and many other apicomplexa, the parasitophorous vacuolar membrane disintegrates after invasion and the parasite is in direct contact with the host erythrocyte cytoplasm. *Babesia* **trophozoites** divide by binary fission and produce **merozoites.** In the large

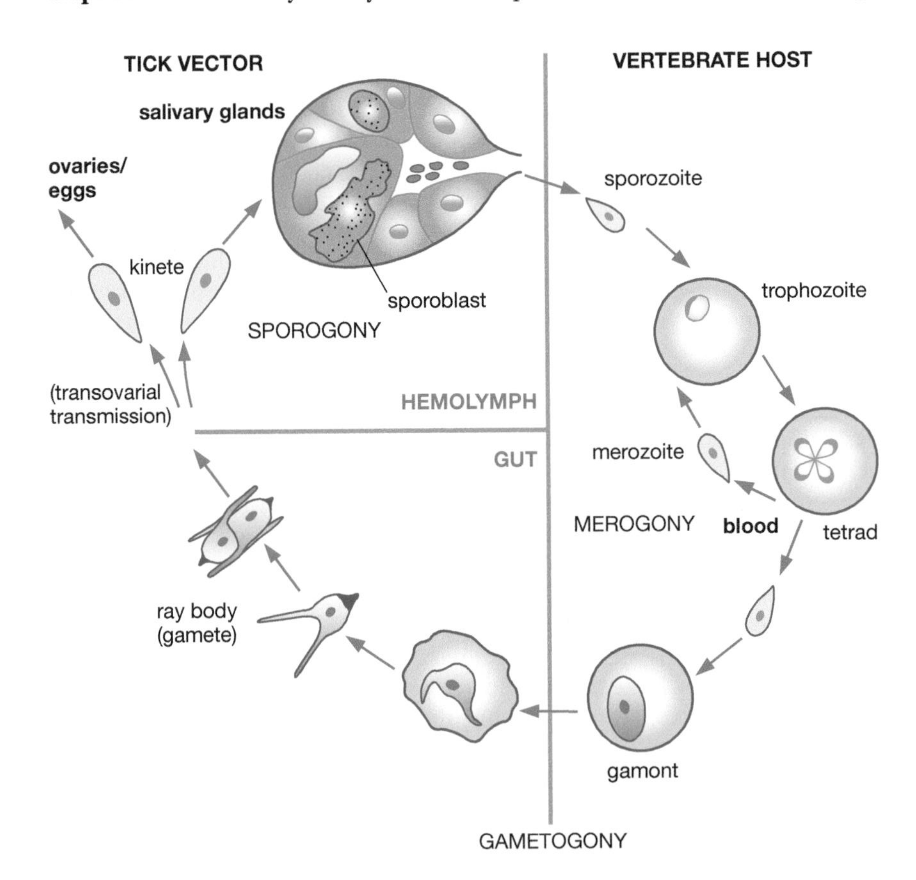

Figure 16.1 Life cycle of *Babesia. Babesia* species are intracellular parasites of erythrocytes in a wide variety of mammals. The infection is transmitted by ticks and the parasite replicates within erythrocytes. The parasite undergoes a more complex cycle in the tick involving sexual stages and motile forms (kinetes) that invade the salivary glands and undergo sporogony. Some *Babesia* species can also invade the ovaries resulting in a transovarial transmission.

Babesia species paired trophozoites connected at the pointed end are often observed. The small *Babesia* species sometimes form a mini-schizont with four merozoites. This tetrad is often referred to as the Maltese cross form. The merozoites escape from the infected erythrocyte and reinvade other erythrocytes to reinitiate the replicative cycle.

Some of the trophozoites will, instead of replicating, develop into gamonts (or gametocytes), which are responsible for initiating the infection in the ixodid tick vector. The gametocytes undergo morphological changes within the tick's gut and develop into **ray bodies** (aka Strahlenkörper). Two ray bodies (i.e., gametes) will fuse to form a zygote which then develops into a kinete. The kinete is a mobile form that penetrates the peritrophic membrane and intestinal epithelium of the tick to gain access to the hemolymph. The kinetes invade various organs and undergo a few cycles of asexual replication. Large *Babesia* species can invade the ovaries and eggs leading to a transovarial (i.e., vertical) transmission to the tick's offspring. The small *Babesia* species are not capable of this vertical transmission.

Sporogony is initiated when kinetes invade the salivary glands. The parasite expands and fills a hypertrophied host cell and develops into a multinucleated **sporoblast**, or sporont. Mature sporozoites, possessing apical organelles, will bud from this undifferentiated sporoblast when the tick feeds again on a new host. A single sporoblast can produce 5000–10 000 sporozoites. The sporozoites are then injected into the host with the saliva, thus completing the life cycle.

Human Babesiosis

At least seven distinct *Babesia* species have been documented to infect humans. However, traditionally only two predominant *Babesia* species were recognized as infecting humans: *B. microti* in the United States and *B. divergens* in Europe. More recently a new species, designated as *B. duncani*, has been identified from several patients on the west coast of the United States (Table 16.1). In addition, molecular analyses have revealed several isolated cases that are distinct from these species suggesting that the range of *Babesia* species capable of infecting humans may be more extensive and it is likely that human babesiosis has been overlooked in many parts of the world due to lack of awareness, the nonspecific symptoms, and the limited diagnostic methods.

B. divergens is a large *Babesia* that naturally infects cattle and wild ruminants. Accordingly human cases are predominantly found in European countries with an extensive cattle industry and farmers and people vacationing in rural areas have a higher risk of acquiring the infection. The infection with *B. divergens* is generally fulminant with a high level of mortality. Molecular analysis has revealed a few cases of a distinct divergens-like parasite in both Europe and the United States.

The most predominant etiological agent of human babesiosis is *B. microti*. *B. microti* is a small *Babesia* and is probably a species complex composed of several closely related species. The largest focus of human infections in the United States has been along the northeastern coastal region, giving rise to the name Nantucket fever. Microtine rodents, especially

Table 16.1 Human babesiosis

Species	Location	Reservoir host	Vector	Cases	Mortality
B. divergens	Europe	Cattle and other ruminants	*Ixodes ricinus*	>30	42%
B. microti	Eastern U. S.	Microtine rodents	*Ixodes scapularis*	>1000	5%
B. duncani	Western U. S.	?	?	9	11%

the white footed mouse (*Peromyscus leucopus*) and voles (*Microtus pennsylvannicus*), have been identified as the primary reservoir. *Ixodes scapularis* (formerly known as *I. dammini*) is the vector. Interestingly, the distribution *B. microti* overlaps with *Borrelia burgdorferi*, the causative agent of Lyme disease, and both are transmitted by the same vector. *B. microti* infections have also been identified in the northern Midwestern region of the United States. Along the west coast of the United States human babesiosis has been attributed to the newly designated *B. duncani*. *B. duncani* is morphologically identical to *B. microti* and the two species form sister clades. In addition, a few cases of microti-like babesiosis have been identified in Europe and Asia.

Most *Babesia* infections are acquired through tick bites. However, a substantial number of cases associated with blood transfusions from asymptomatic donors have been documented. Splenectomy is the most common risk factor shared among babesiosis patients. Additional risk factors for severe disease include age over 55, depressed cellular immunity, immunocompromising conditions such as AIDS, or immunosuppressive drugs administered after organ transplants or cancer chemotherapy. Although symptomatic immunocompetent persons infected with *Babesia* have been described, infected immunocompetent persons are generally asymptomatic. Serological surveys suggest that the infection may be under diagnosed and in this regard human babesiosis may be an opportunistic disease.

Clinical Features

The clinical features of babesiosis patients vary from asymptomatic to life threatening depending on the species and the immune status of the patient. Most of the *B. divergens* cases in Europe involved splenectomized, immunocompromised, or elderly persons and resulted in severe disease. These patients exhibited an acute febrile hemolytic disease within a few weeks of tick transmission. Headache, chills, intense sweating, myalgia, and abdominal pain were common symptoms as well a high level of intravascular hemolysis resulting in jaundice and hemoglobinuria. More than half the patients exhibited a rapid onset of kidney failure and pulmonary edema (excessive fluid in the lungs) and more than a third of the patients died within a week of symptoms.

Cases from the United States due to *B. microti* exhibit a wider range of clinical outcomes. Many infected people never experience symptoms as suggested by the disparity between seroprevalence and the number of reported cases. Asymptomatic infection may persist for months to years. In persons with intact spleens the infection is generally self-limited and characterized by a gradual onset of malaise, fever, headache, chills, sweating, myalgia, fatigue, and weakness. A mild to moderate hemolytic anemia can also accompany these symptoms. Many of the infections will resolve on their own without treatment, but the parasites can persist for months. The disease tends to be more fulminant and severe in splenectomized or immunosuppressed individuals and can be life threatening. In such cases parasitemias of greater than 25% and severe anemia can occur. In addition, immunocompromised patients are at higher risk for a persistent and relapsing disease. Accordingly, the mortality rate is significantly higher in immunocompromised patients and, as seen in *B divergens* infections, complications such as renal failure, acute respiratory failure, congestive heart failure, and splenic rupture are common.

Diagnosis and Treatment

Symptoms and case histories suggestive of babesiosis include living or traveling in an endemic area and a viral-like illness particularly in late spring,

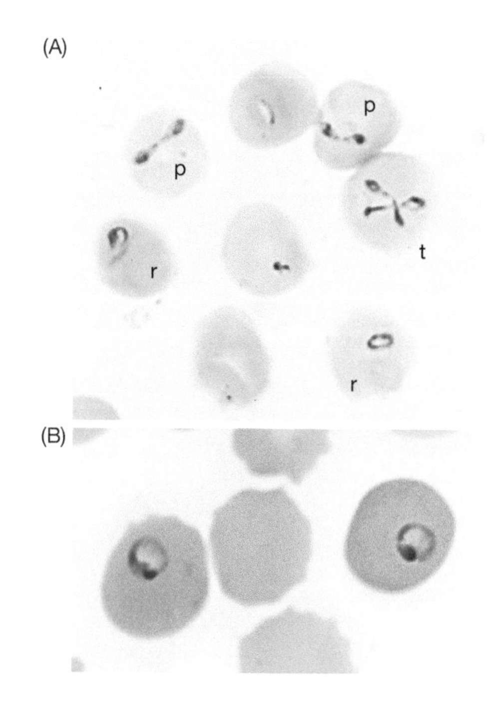

Figure 16.2 Blood smear of *Babesia*. (A) A *B. divergens*-like parasite showing ring forms (r), paired piriforms (p), and tetrads (t). (B) *B. microti* showing ring forms resembling those of *Plasmodium falciparum*. (Photos kindly provided by Marc Lontie, Medisch Centrum voor Huisartsen, Leuven, Belgium.)

summer, or fall. Diagnosis is confirmed by detecting the parasite in Giemsa-stained blood smears. Three basic blood-stage morphologies are observed: ring forms, tetrads, and paired piriforms (Figure 16.2). The tetrads are more common in small *Babesia* and paired piriforms are more common in large *Babesia*. *B. microti* expresses primarily ring forms and tetrads are quite rare, whereas tetrads are more common in *B. duncani*. Paired piriforms can be observed in addition to the rings and tetrads in *B. divergens*. In general, speciating *Babesia* infections based on blood stages can be difficult because of the similar morphologies exhibited by the species.

There is a possibility of *Babesia* being confused with the malaria parasite because of morphological similarities to *Plasmodium falciparum* ring-stage parasites (Table 16.2). The lack of response to conventional anti-malarials and no travel history to malaria endemic areas are factors to rule out malaria. Rodents such as mice, gerbils, or hamsters can be inoculated and examined for parasitemia. However, this generally takes 2–4 weeks and is not useful in the diagnosis of acute cases. Serological tests can also be used to confirm *Babesia* infections. However, the diagnosis of active infections based solely on serological results may be unreliable. PCR provides a highly sensitive and accurate test that can also be sequenced to positively identify the species. Such molecular diagnoses can only be carried out in selected laboratories, though.

Although many drugs have been developed for the treatment of babesiosis in animals, none have been adopted for human use and there are no clearly effective drugs against human babesiosis. The standard recommended drug treatment is clindamycin (Cleocin®) plus quinine (Quinamm®). However, atovaquone (Mepron®) plus azithromycin (Zithromax®) has been shown to be as effective as clindamycin plus quinine, but with fewer adverse affects. Therefore, atovaquone plus azithromycin is now the recommended treatment for mild cases. However, clindamycin plus quinine is still preferred in severe cases. Pentamidine (Pentam®) has also been shown to suppress, but not eliminate, parasitemia. Chloroquine, although it does not appear to affect parasitemia, does provide some symptomatic relief, which may be due to its anti-inflammatory properties. Exchange transfusion to reduce the parasitemia and remove toxic metabolic products of the parasite is indicated as a life-saving effort in patients with high parasitemia (>10%) and severe symptoms.

Summary and Key Concepts

- Babesiosis is a rare zoonotic infection transmitted by ticks.
- The most predominant species in humans are *Babesia microti* in northeastern United States and *B. divergens* in Europe.
- The infection can produce a wide range of symptoms from asymptomatic to an acute febrile illness to hemolytic anemia to death.
- Splenectomy and other immunocompromising conditions are risk factors for severe disease.
- Diagnosis is made by detecting the parasite in blood smears and can be confused with malaria due to morphological similarities.
- Treatment involves chemotherapy and in severe cases exchange transfusion.

Table 16.2 Factors distinguishing *Babesia* and falciparum malaria
Presence of tetrads and paired piriforms
Lack of travel to malaria endemic area
Lack of response to anti-malarials
Infection of rodents
Serology
Molecular analyses

Further Reading

Allsopp, M.T.E.P. and Allsopp, B.A. (2006) Molecular sequence evidence for the reclassification of some *Babesia* species. *Ann. N.Y. Acad. Sci.* 1081: 509–517.

Gray, J., Zinti, A., Hildebrandt, A., Hunfeld, K.P. and Weiss, L. (2010) Zoonotic babesiosis: overview of the disease and novel aspects of pathogen identity. *Ticks Tick-borne Dis.* 1: 3–10.

Homer, M.J., Aguilar-Delfin, I., Telford, S.R., Krause, P.J. and Persing, D.H. (2000) Babesiosis. *Clin. Microbiol. Rev.* 13: 451–469.

Hunfeld, K.-P., Hildebrandt, A. and Gray, J.S. (2008) Babesiosis: Recent insights into an ancient disease. *Int. J. Parasitol.* 38: 1219–1237.

Kjemtrup, A.M. and Conrad, P.A. (2000) Human babesiosis: an emerging tick-borne disease. *Int. J. Parasitol.* 30: 1323–1337.

Vannier, E. and Krause, P.J. (2009) Update on babesiosis. *Interdiscip. Perspect. Infect. Dis.*, article ID 984568.

Free-Living Protozoa Affecting Human Health

Most protozoa are free living and have little impact on human health. Free-living protozoa can be found throughout the environment and are particularly abundant in soil and water. Examples of free-living protozoa affecting human health are free-living amebae which can cause pathology if introduced into the human host. Another means by which some free-living protozoa can affect humans is indirectly through their effect on the environment. For example, some free-living dinoflagellates produce toxins that accumulate in seafood and adversely affect various organs when ingested.

Pathogenic Free-Living Amebae

Disease(s)	Primary amebic meningoencephalitis, granulomatous amebic encephalitis, amebic keratitis
Etiological agent(s)	*Naegleria fowleri*, *Acanthamoeba* spp, *Balamuthia mandrillaris*
Major organ(s) affected	Brain, cornea (only *Acanthamoeba*)
Transmission mode or vector	Contact with amebae in soil or water and contaminated contact lens solutions
Geographical distribution	Worldwide
Morbidity and mortality	Causes encephalitis-like disease with a high level of mortality. The keratitis can lead to blindness if not treated
Diagnosis	Keratitis diagnosed by detecting amebae in corneal scrapings. Diagnosis of encephalitis generally at autopsy
Treatment	Topical formulations of biguanides or diamidines applied as eye drops for the keratitis
Control and prevention	Keratitis prevented by proper cleaning and care of contact lenses

Several different species of free-living amebae are known to cause human disease (Table 17.1). These infections are quite rare and are presumably acquired directly from the environment (e.g., water or soil contact). No human-to-human transmission or vector transmission have been documented. The most serious manifestation involves infection of the brain causing encephalitis-like diseases which generally result in death (case fatality rate >95%). On autopsy the amebae are detected in the brain and are relatively easy to distinguish from *Entamoeba histolytica*, which can also on rare occasions infect the central nervous system (Chapter 3). However, distinguishing the free-living amebae from each other in histological sections can be rather difficult.

These free-living amebae exhibit typical protozoan life cycles consisting of dormant cyst and vegetative trophozoite stages. Trophozoites are active forms that acquire nutrients, undergo replication, and are motile. In general, ameboid trophozoites exhibit no distinctive structures and are somewhat

Table 17.1 Free-living amebae causing human disease

Ameba	Disease manifestations
Naegleria fowleri	Acute primary amebic meningoencephalitis (PAM)
Acanthamoeba spp	Chronic granulomatous amebic encephalitis (GAE)
	Granulomatous skin and lung lesions
	Amebic keratitis
Balamuthia mandrillaris	Subacute to chronic GAE
	Granulomatous skin and lung lesions
Sappinia diploidea	Single case report associated with encephalitis
Paravahlkampfia francinae	Single case report associated with encephalitis

amorphous in their ability to change shape. Motility is based on the classic ameboid movement with the projection of pseudopodia (Chapter 1). When conditions are harsh (e.g., desiccation or low nutrients) the trophozoites secrete a thick cell wall and enter into a dormant period characterized by low metabolic activity. The cyst wall and low metabolic activity allow the organism to survive while food and water are insufficient. When conditions are favorable again, the cysts convert back into trophozoites and expand the population through asexual cellular replication. Analogous cyst and trophozoite stages are also associated with many intestinal protozoa exhibiting a fecal–oral transmission (Chapter 2).

Naegleria fowleri

Naegleria species are found in freshwater habitats and moist soil throughout the world. **Primary amebic meningoencephalitis** (PAM) was first recognized by Fowler in 1965 and *N. fowleri* is the only *Naegleria* species known to be pathogenic to humans. The designation of "primary" distinguishes it from meningoencephalitis caused by the metastasis of *E. histolytica* to the brain from other sites. Documented cases of infections with *Naegleria* are quite rare and to date the total number of cases is approximately 200. Most of the cases are from the United States and Europe. This is probably due in large part to the difficulty in making the diagnosis, which is almost always post-mortem, and the greater resources available for investigating sudden unexplained deaths in developed countries.

The life cycle consists of trophozoite and cyst stages and the trophozoite stage can be either ameboid or flagellated (Figure 17.1). The ameboid trophozoites of *Naegleria* are 10–25 μm and exhibit broad blunt pseudopodia called lobopodia. The ameboid trophozoite feeds on bacteria and other organic matter and undergoes asexual replication. When deprived of nutrients the ameboid form transforms into a flagellated form. The flagellated stage is pear shaped with two flagella emerging from the narrow end and is 10–16 μm long. Flagellated trophozoites are nonfeeding and do not replicate, but will convert back into the ameboid form when nutrients are restored. The ameboid forms can also encyst resulting in a stage resistant to desiccation. *Naegleria* cysts are spherical and double walled. All stages are characterized by a single nucleus.

Naegleria produces a fulminant, and almost always fatal, acute meningoencephalitis in children and young adults who were previously in excellent health (Table 17.2). Almost all reported cases have been associated with swimming or diving in warm or heated waters a few days prior to the onset of symptoms. In addition, *N. fowleri* is thermophilic and can tolerate

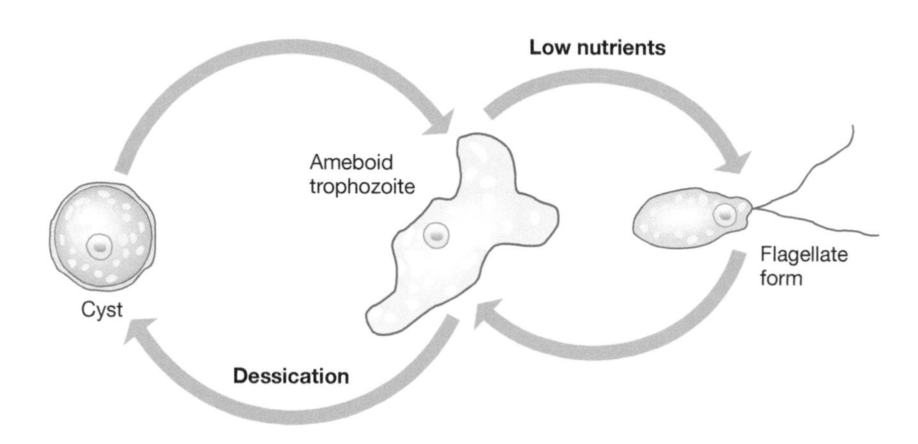

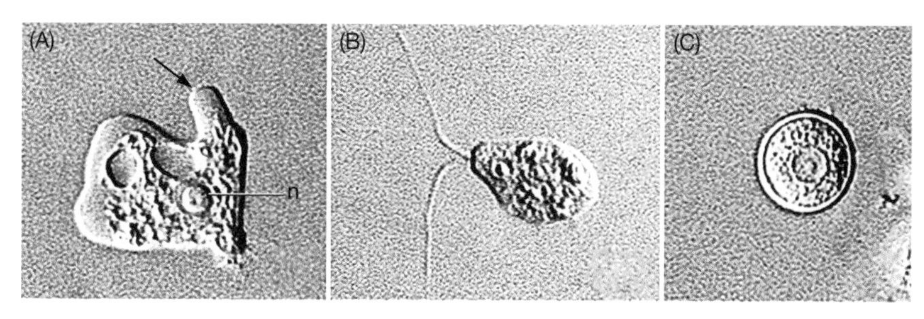

Figure 17.1 Life cycle of *Naegleria*. *Naegleria* can exist in three distinct forms: ameboid trophozoite (A), bi-flagellated trophozoite (B), and double-walled cyst (C). Note the broad blunt lobopodium (arrow), nucleus (n), and uroid (u) defined as the posterior end containing a contractile vacuole and trailing protoplasmic filaments. (Micrographs kindly provided by Govinda Visvesvara, Division of Parasitic Diseases, National Center for Infectious Diseases, Centers for Disease Control and Prevention, Atlanta, Georgia. Originally published in Visvesvara et al., 2007, *FEMS Immunol. Med. Microbiol.* 50: 1–26 and used with permission.)

temperatures up to 45°C, and thus PAM is typically observed in the hot summer months. The portal of entry appears to be the olfactory neuroepithelium in the nasal cavity (Box 17.1). The trophozoites probably migrate along the olfactory nerves into the brain and central nervous system (CNS). Only the ameboid trophozoite form of *Naegleria* is found in brain tissue. The trophozoites exhibit a high rate of replication and ingest erythrocytes and brain tissue leading to a severe hemorrhagic necrosis.

The symptoms of PAM resemble bacterial meningoencephalitis. Rapid deterioration of the patient, failure to find a causative agent, and lack of

Table 17.2 Comparison of PAM and GAE features

PAM	1–14 days incubation period
	Symptoms usually within a few days after swimming in warm still waters
	Infection may be introduced through nasal cavity and olfactory neuroepithelium
	Symptoms include headache, lethargy, disorientation, coma
	Rapid clinical course with death 4–5 days after onset of symptoms
	Trophozoites detected in spinal fluid
	Diagnosis is usually at autopsy
	Eight known survivors treated with amphotericin B + other drugs
GAE	Portal of entry unknown (possibly respiratory tract, eyes, skin)
	Hematogenous or perivascular dissemination to the CNS
	Infection often associated with debility or immunosuppression
	Onset is insidious with headache, personality changes, slight fever
	Progresses to coma and death in weeks to months
	Amebae not detected in spinal fluid
	Cysts and trophozoites detectable in histological samples

Box 17.1 Entry and dissemination of free-living amebae in the human host

Numerous barriers prevent microbes from infecting humans and other animals. Skin, mucus, and other physical and physiological barriers serve as a first line of defense preventing entry into the host. Various phagocytic cells and natural killer cells then eliminate many potential pathogens that manage to break through this first line of defense. Finally, there is an adaptive, or learned, immune response mediated by B- and T-lymphocytes directed at specific pathogens functioning as a third line of defense. As discussed in the previous chapters, parasitic protozoa have evolved specialized means to gain entry into their hosts and then be able to survive and replicate within those hosts. This lack of specialized mechanisms to gain entry into the human host and the highly evolved defensive mechanisms of the host probably explain the relative paucity of infections with free-living amebae in spite of the high levels of exposure from the environment.

Since infections by free-living amebae are rather infrequent it is not completely understood how the amebae infect the host and avoid the immune system. PAM is principally found in healthy children and young adults with a recent history of swimming in warm fresh water. It is believed that the organisms are introduced via the nose when water is inhaled during swimming and diving. The trophozoites cross the nasal epithelium and crawl along the olfactory nerve into the meninges and multiply in the perivascular spaces around the blood vessels. In the case of *Acanthamoeba* and *Balamuthia* lesions in the skin and lung have been noted prior to the neurological symptoms. It is believed that the trophozoites penetrate the epithelium or the skin and reach the CNS via a hemotogenous route. However, Recaverren-Arce et al. [1] have proposed that the amebae disseminate from the primary mucosal lesion via a perivascular route in that the trophozoites have not been observed within the vessels and only in the perivascular space.

It is not clear how the free-living amebae are able to survive the immune system after gaining entry. In the case of *Acanthamoeba* the disease is found primarily in immunocompromised persons or persons with another underlying disease. This is not the case with *Naegleria* and in many cases of *Balamuthia*. Another possible factor is that the perivascular spaces in the CNS may constitute an immunoprivileged site with a limited second line of defense.

1. Recavarren-Arce, S., Velarde, C., Gotuzzo, E. and Cabrera, J. (1999) Amoeba angeitic lesions of the central nervous system in *Balamuthia mandrilaris* amoebiasis. *Hum. Pathol.* 30: 269–273.

response to antimicrobial drugs can be used to rule out a bacterial etiology. PAM is generally characterized by a sudden onset of headache and fever. Nausea, vomiting, and other symptoms related to increased intracranial pressure may also be evident. There is a rapid progression from headache and fever to coma, with occasional seizures, and death. The prognosis is not good and diagnosis is almost always post-mortem. Among the more than 200 confirmed cases of PAM there have only been eight documented survivors and nearly all of them received amphotericin B with or without other drugs.

Acanthamoeba

Acanthamoeba species are also ubiquitous in soil and water throughout the world. They exhibit a typical protozoan life cycle consisting of an ameboid trophozoite stage and a cyst stage. A unique characteristic of *Acanthamoeba* trophozoites are tapering spike-like pseudopodia called acanthopodia (Figure 17.2). Trophozoites range in size from 15 to 50 μm depending on

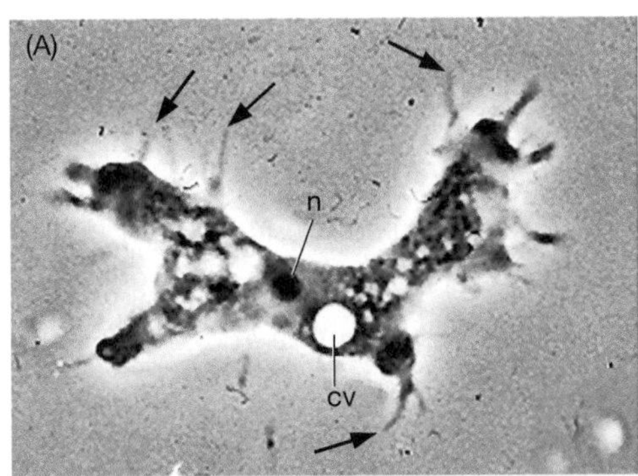

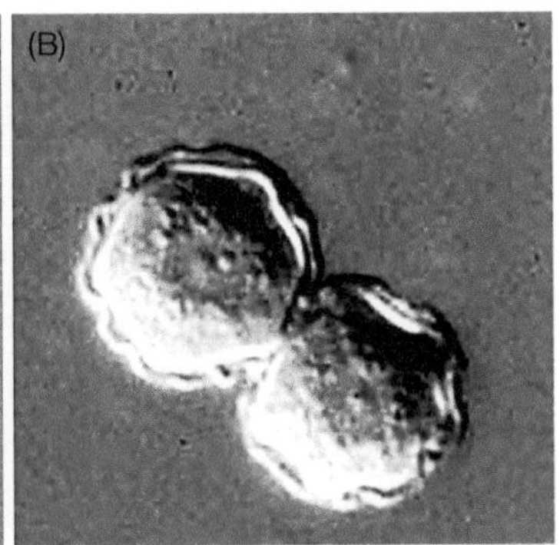

Figure 17.2 *Acanthamoeba.* Differential interference contrast micrographs of (A) trophozoite and (B) cysts. Note the numerous spike-like acanthapodia (arrows); nucleus (n); contractile vacuole (cv). (Micrographs kindly provided by Govinda Visvesvara, Division of Parasitic Diseases, National Center for Infectious Diseases, Centers for Disease Control and Prevention, Atlanta, Georgia. Originally published in Visvesvara et al., 2007, *FEMS Immunol. Med. Microbiol.*50: 1–26 and used with permission.)

species. Cysts of *Acanthamoeba* are double walled and range in size from 10 to 15 μm. The outer cyst wall is wrinkled and the inner cyst wall varies in shape ranging from stellate, polygonal, oval, or spherical.

Their pathogenic potential was first recognized in 1958 by Culbertson who produced encephalitis in mice following inoculation with an *Acanthamoeba*-contaminated cell culture. The first definitive human cases were reported in the early 1970s. In contrast to PAM, *Acanthamoeba* infections are usually associated with chronically ill, immunocompromised, or other debilitated patients. In this regard, *Acanthamoeba* can be considered an opportunistic infection. *Acanthamoeba*, like *Naegleria*, is neurotropic and causes an encephalitis-like disease called **granulomatous amebic encephalitis** (GAE). GAE is more slowly progressing and chronic than PAM (Table 17.2). *Acanthamoeba* infections are also associated with lesions in the cornea (amebic keratitis), lungs, and skin.

The portal of entry for *Acanthamoeba* is not known but believed to be either the respiratory tract via inhalation of cysts or through wounds in the skin that become contaminated by soil. Trophozoites were presumed to be disseminated by a hematogenous route (i.e., via the circulatory system) to the central nervous system (CNS). However, a perivascular route of dissemination has also been suggested (Box 17.1). In other words, the amebae crawl along the outside of the blood vessels to reach the CNS. The onset of symptoms is often insidious and the clinical manifestations include subtle headache, personality changes, and slight fever. GAE has a prolonged clinical course and can take weeks to months to progress to coma and death. Diagnosis is difficult and usually done at autopsy.

Balamuthia mandrillaris

Balamuthia mandrillaris is another free-living ameba capable of causing human disease. It was first reported in a mandrill baboon in 1986 and subsequently shown to be associated with human disease. The trophozoites and cysts of *Balamuthia* are morphologically similar to *Acanthamoeba* and it also causes GAE (Table 17.2). The trophozoites are 12–60 μm and have long thin pseudopodia, but are not as spike-like as *Acanthamoeba* (Figure 17.3). The cysts also appear double walled with a wavy outer wall and a round inner wall and range in size from 12 to 30 μm. Phylogenetic analysis of rRNA sequences indicates that *Balamuthia* is a sister genus of *Acanthamoeba* and thus the two are related. But they are likely to be independent and monophyletic genera.

Many cases of GAE originally ascribed to *Acanthamoeba* have been retrospectively assigned to *Balamuthia*. Since it was first identified, there have been approximately 150 cases of *Balamuthia* infections reported with the majority of the cases being reported in Latin America. The majority of the cases in the United States have been in the southwest (e.g., Texas, California, Arizona). Many of the reported *Balamuthia* cases involve children. In contrast to *Acanthamoeba*, *Balamuthia* can cause a subacute-to-chronic

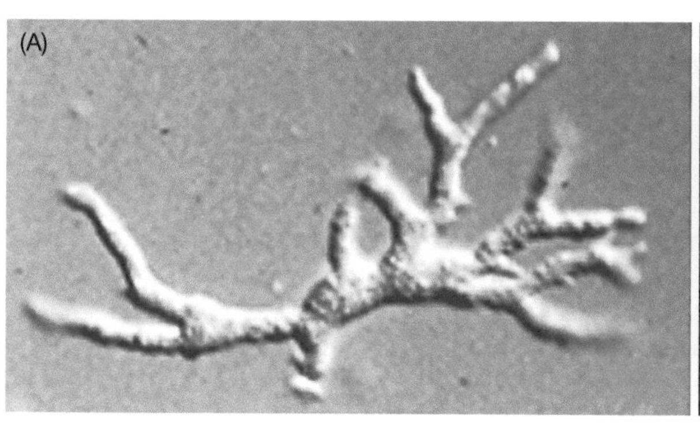

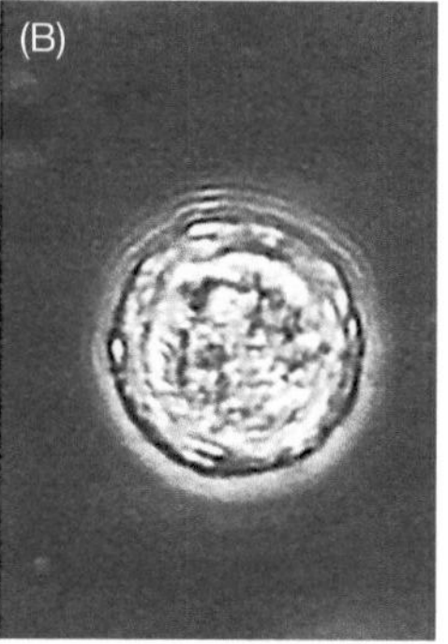

Figure 17.3 *Balamuthia mandrillaris.* Differential interference contrast micrographs of (A) trophozoite and (B) cyst. (Micrographs kindly provided by Govinda Visvesvara, Division of Parasitic Diseases, National Center for Infectious Diseases, Centers for Disease Control and Prevention, Atlanta, Georgia. Originally published in Visvesvara et al., 2007, *FEMS Immunol. Med. Microbiol.* 50: 1–26 and used with permission.)

infection which is more rapidly progressing. In addition, *Balamuthia* appears more capable of infecting healthy individuals. Primary lesions in the respiratory tract or skin have been noted in many of the infections, suggesting that the infection is acquired via inhalation or breaks in the skin. There has been one documented case in which the infection has spread via organ transplantations from a donor who died of neurological problems. However, infection via transplantation does not appear to be likely in the case of *Naegleria* infections in that there have been two cases of transplants from infected donors without infection of the recipients. This may reflect the hematogenous spread of *Balamuthia* versus the direct invasion of the brain exhibited by *Naegleria* (Box 17.1). Thus far, only a few survivors of *Balamuthia* infections have been reported. All were treated for an extensive time period with numerous drugs.

Balamuthia was not identified in the environment until more than a decade after its initial description. The amebae were first successfully grown from soil samples associated with a fatal case in a northern California child and were subsequently cultivated from another soil source. In two other cases involving persons surviving the infection, acquisition of the infection was associated with gardening activities. The amebae are slow growing (generation time of 20–50 hours) and probably feed on other amebae (or cultivated mammalian cells) instead of bacteria. These features, the slow growth and feeding on other amebae, may account for the difficulty in detecting *Balamuthia* in the environment.

Amebic keratitis

Amebic keratitis is a painful and vision-threatening chronic inflammation of the cornea caused by *Acanthamoeba.* It was first reported in 1973 and until the mid-1980s was quite rare and usually associated with ocular trauma. The number of cases has increased since 1985—and quite substantially over the past few years—particularly in the United States, the United Kingdom, and Australia. This increase is attributed to the use of contact lenses and especially incomplete cleaning and disinfection of the lenses. Major risk factors associated with amebic keratitis among contact lens users include the use of extended wear lenses, using homemade saline cleaning solutions (especially those from tap water), disinfecting the lenses less frequently than recommended, and wearing contact lenses while swimming. In non-lens users, amebic keratitis is associated with trauma and exposure to contaminated water or soil. An increased incidence has been observed in the summer and early fall that is probably related to higher levels of *Acanthamoeba* in the environment and increased water-associated recreational activities.

Acanthamoeba presumably gains access to the corneal stroma through small breaks in the corneal epithelium produced as a result of minor trauma or abrasion. A mannose-binding protein expressed on the trophozoite surface is believed to promote adherence to host tissues. Proteases are also expressed by the trophozoites and likely contribute to the pathology by killing host cells and degrading epithelial basement membranes and the underlying stromal matrix. As a result of this damage, the ameba can then penetrate deeper into the cornea and continue to destroy the corneal stroma. A mild to moderate corneal inflammation is also associated with *Acanthamoeba* invasion, as well as secondary bacterial infections. *Acanthamoeba* rarely causes an intraocular infection because neutrophils kill the trophozoites.

Early diagnosis and treatment are essential for a good prognosis. Other common agents of infectious keratitis include bacteria, fungi, microsporidia, and herpes simplex virus and differential diagnosis can be difficult. Furthermore, up to 20% of the amebic keratitis may be due to co-infections. Clinical manifestations of amebic keratitis include severe ocular pain, photophobia, tearing, and corneal lesions refractory to antiviral, antibacterial, and antimycotic drugs. Diagnosis is confirmed by detecting the amebae in corneal scrapings or biopsies through direct observation, following *in vitro* culture, or by PCR.

Topical formulations of biguanides or diamidines applied as eye drops are the two most effective classes of drugs currently available. Among the biguanides, polyhexamethylene biguanide and chlorhexidine exhibit good antiamebic activity and effective diamidines include propamidine isethionate (Brolene®) and hexamidine (Desomedine®). Initially intense therapy in regard to dosage and frequency of application is carried out and then tapered off to avoid toxicity. An average of 6 months of therapy is generally needed to clear the infection. In cases of persistent inflammation a low potency topical steroid (e.g., prednisolone) is recommended. However, the anti-inflammatory treatment should not be started until after most organisms have been killed by the antiamebic agent(s) so as to preclude a recrudescence of the infection. Severe inflammation may require a stronger steroid such as dexamethasone. Surgery may be needed to correct any loss of vision.

Harmful Algal Blooms and Dinoflagellates

Disease(s)	Ciguatera and shellfish poisoning
Etiological agent(s)	Toxins from various planktonic dinoflagellates
Major organ(s) affected	Gastrointestinal tract, central nervous system, respiratory system, muscles
Transmission mode or vector	Ingestion of fish or shellfish contaminated with the toxins
Geographical distribution	Focal endemic areas throughout the world
Morbidity and mortality	A transient toxemia with transient gastrointestinal and neurological symptoms and on occasion longer term effects and rarely death
Diagnosis	History of eating fish or shellfish prior to the symptoms and other possible causes ruled out
Treatment	Intravenous mannitol may ease acute symptoms and prevent chronic effects
Control and prevention	Monitoring shellfish harvesting areas for harmful algal blooms

Algal blooms are the result of large increases in the number of unicellular planktonic organisms. **Dinoflagellates**—claimed by protozoologists as protozoa and by phycologists as algae—are a major component of the microscopic zoo- and phytoplankton. Many of the dinoflagellates contain pigments and therefore will result in the water taking on a distinct colored appearance in the affected areas (usually along the coast), thus are often referred to as red tides. Many of the dinoflagellates have long-lived cyst stages which can survive in the sediment of the ocean floor for months to years (see Box 17.2). When conditions are favorable the cysts will convert to replicating forms and can result in the relatively sudden appearance of a red tide. Increases in the frequency of algal blooms over the last several decades are usually attributed to higher levels of nutrients in estuaries and coastal waters due to pollution and agricultural runoff. The shipping industry has also been blamed for increased worldwide distribution of organisms causing red tides.

These algal blooms can affect human health, as well as the environment, and thus are often referred to as harmful algal blooms. Generally, the harmful effects of algal blooms are indirect. For example, one possible consequence of algal blooms is simply due to an overabundance of metabolically active microorganisms which can cause "dead zones" by depleting the water of oxygen. Fish and other macroscopic organisms will either die or avoid these dead zones, which can therefore have an adverse economic impact on fishermen and the seafood industry. In some cases, though, the fish kills are the result of toxins produced by some dinoflagellates and other planktonic organisms. In addition, many of these toxins also accumulate in fatty tissues or organs (e.g., liver) of finfish and shellfish and pass up the

Box 17.2 Typical dinoflagellate life cycle

Dinoflagellates are found in marine and fresh water habitats. The typical dinoflagellate life cycle consists of trophic forms called dinospores (or zoospores), sexual forms, and cysts (Figure 17A). The cysts are long lived dormant forms found in the sediment of the ocean floor or other aquatic habitats. Increasing temperature or light or other factors will cause the cysts to germinate and motile dinospores will emerge. The dinospore stage is characterized by two unequal flagella. One encircles the cell in a groove and provides propulsive and spinning force. (Dino is derived from a Greek word that means spinning.) The other flagellum is directed posteriorly along a longitudinal groove and presumably acts like a rudder for steering. Another common feature of many dinoflagellates is a cellulose cell wall that is divided into plates called theca. Dinoflagellates with theca are called armored and those without are referred to as naked.

Many dinoflagellates are heterotrophic and feed on diatoms, other dinoflagellates, ciliates, and other planktonic organisms, whereas others are autotrophic (i.e., photosynthetic), or exhibit both heterotrophic and autotrophic lifestyles (i.e., mixotrophic). Many of the heterotrophic dinoflagellates feed upon their prey using a specialized feed tube called the peduncle. The peduncle is a conical shaped structure that is projected from the dinospore and penetrates the membrane of the prey. The contents of the prey cytoplasm are then taken up by phagocytosis and sucked into the dinospore. The dinospores are also a replicating form and undergo mitosis producing more dinospores. High rates of asexual reproduction can lead to massive algal blooms.

The dinospores can convert to gametes instead of undergoing mitosis. Low levels of nutrients or other conditions not favorable for replication may lead to the conversion to gametes. The gametes will fuse to form the diploid planozygote. After a period of motility the flagella are lost from the planozygote, meiosis is initiated, and a multilayer cell wall is secreted to form the cyst which sediments. The cysts can survive for months to years and may be part of a seasonal reproductive cycle. Germination is accompanied by the completion of meiosis resulting in the production of four haploid dinospores.

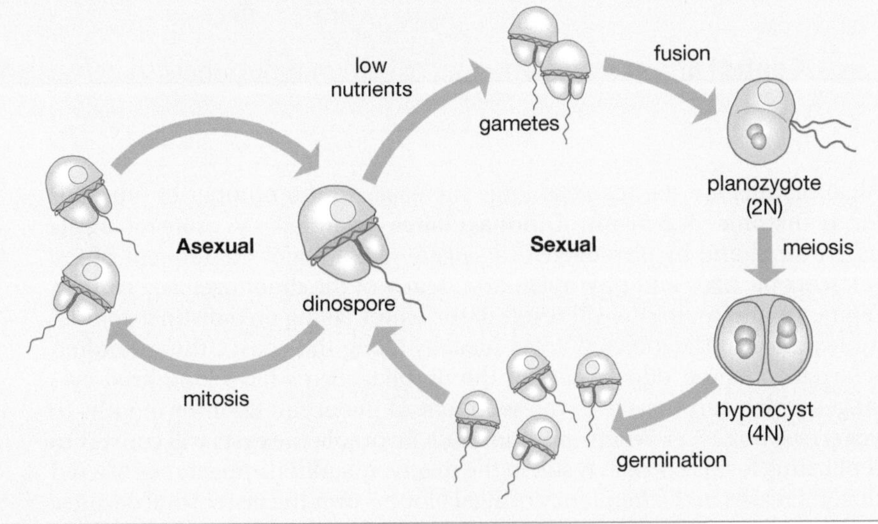

Figure 17A Generalized dinoflagellate life cycle. The typical dinoflagellate life cycle consists of an asexual phase and a sexual phase. The asexual phase usually involves the binary fission of the dinospore stage. Dinospores can also convert to gametes which will fuse and develop into cyst stages. Germination of the cysts involves the completion of meiosis resulting in the production of the haploid dinospores.

Table 17.3 Examples of common toxic dinoflagellates

Organism	Disease	Toxin	Target of toxin
Gambierdiscus toxicus	CFP	Ciguatoxins	Opens sodium channels
Karenia brevis	NSP	Brevetoxin	Opens sodium channels
Alexandrium tamerense	PSP	Saxitoxins	Blocks sodium channels
Prorocentrum lima	DSP	Okadaic acid	Protein phosphatase inhibitor
Pfiesteria piscicida	PEAS	?	

CFP, ciguatera fish poisoning; NSP, neurotoxic shellfish poisoning; PSP, paralytic shellfish poisoning; DSP, diarrheic shellfish poisoning; PEAS, possible estuary-associated syndrome

food chain and ultimately are consumed by humans. In other words, the planktonic organisms do not directly cause disease, but cause a toxemia.

Several toxins associated with planktonic organisms have been identified and in many cases the target of the toxin is known (Table 17.3). The diseases caused by ingesting these toxins are known as **ciguatera** fish poisoning and **shellfish poisoning**. It may also be possible to experience these toxic effects by simply coming in contact with water containing toxic dinoflagellates through recreational or occupational activities. This latter scenario is referred to as possible estuary-associated syndrome. Many of these toxins are potent neurotoxins that interfere with neurotransmission or voltage-sensitive ion channels. This interference with the neurotransmission and channels perturbs the normal function of neuronal and other excitable tissues and leads to clinical symptoms.

Ciguatera fish poisoning

Ciguatera—known in the tropics for centuries and becoming more common elsewhere as tropical fish are more widely marketed—is a type of food poisoning associated with the accumulation of dinoflagellate toxins in fish. As many as 50 000 persons are affected annually and ciguatera is most common in discrete regions of the Pacific Ocean, the western Indian Ocean, and the Caribbean Sea. However, the ciguatera endemic regions appear to be expanding, possibly due to the same factors contributing to the increase in harmful algal blooms.

Herbivorous fish, or perhaps invertebrates in some locations, ingest toxic dinoflagellates, primarily *Gambierdiscus toxicus* (Figure 17.4), while feeding on seaweed. These smaller fish are then eaten by larger carnivorous fish. Thus, the larger and more desirable fish, such as barracuda, grouper, jack, and snapper, accumulate the ciguatoxins in their tissues over time. Some of the fish will accumulate sufficient toxin levels to cause gastrointestinal, neurological, and/or cardiovascular symptoms in humans after a single meal. The occurrence of toxic fish is sporadic and not all fish of a given species or from a ciguatera endemic region will be toxic. The major ciguatoxins have been identified as a family of cyclic polyethers which are heat-stable and lipid-soluble and activate voltage-sensitive sodium channels at nanomolar and picomolar concentrations making them extremely potent neurotoxins.

Initially the symptoms may resemble food poisoning (e.g., nausea, vomiting, diarrhea) occurring within 6 hours of eating toxic fish. Initial symptoms can also include numbness and tingling around the mouth which can spread to extremities. Neurological manifestations include: intensified tingling (paresthesia), headache, pain in muscles and joints, acute sensitivity to temperature extremes, reversal of hot and cold sensations, vertigo, and muscular weakness. Typical cardiovascular manifestations are arrhythmias and reduced blood pressure. Ciguatera poisoning is usually self-limiting

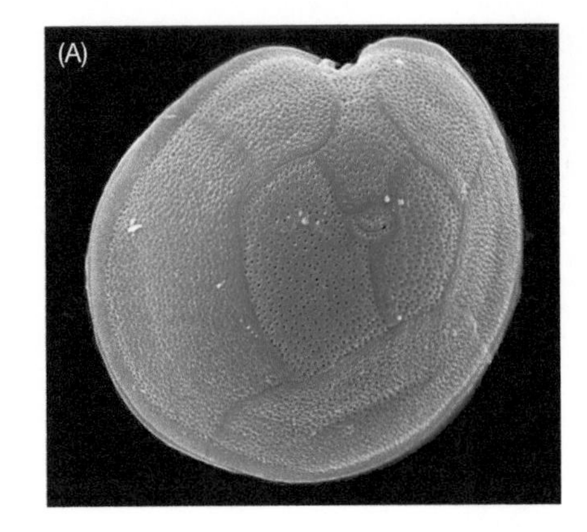

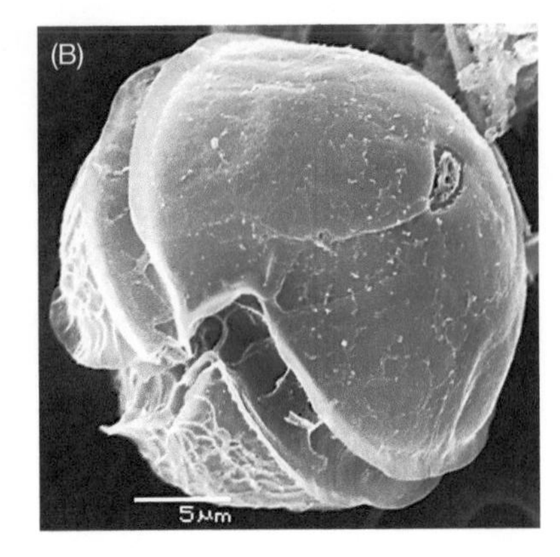

Figure 17.4 Toxic dinoflagellates. (A) Scanning electron micrograph of *Gambierdiscus toxicus*. (B) Scanning electron micrograph of *Alexandrium tamerense*. (Electron micrographs kindly provided by Steve Morton of NOAA/NOS Marine Biotoxins Program.) (C) Chemical structure of ciguatoxin.

and the symptoms typically subside within several days. However, in severe cases the neurological symptoms can persist for weeks to months, and in rare cases, neurological symptoms have persisted for years. There have been a few deaths due to respiratory or cardiovascular failures. The low rate of fatalities is probably due to the limited amount of toxin which can be accumulated by the fish without being toxic to the fish.

There are currently no available biomarkers to confirm exposure to ciguatoxins or other marine toxins. Therefore diagnosis of ciguatera (or shellfish poisoning) relies on the symptoms and history of eating reef fish, as well as excluding other possible causes. The infrequency and irregularity of ciguatera precludes well-designed controlled studies to test treatment. The most studied potential therapy is the intravenous infusion of mannitol. Mannitol appears to provide a rapid and significant improvement in the patient and a reduction in the acute symptoms, as well as preventing the development of chronic symptoms. As in any acute poisoning, patients in shock may require fluid resuscitation with large volume of isotonic fluids and supportive therapies for controlling dehydration and electrolyte balance may be necessary. Support of any depressed vital functions is also critical in severe cases. Prevention is difficult because toxic fish cannot be identified by appearance or behavior and the toxins are heat and cold stable. Avoidance of large reef fish from areas known to be endemic for ciguatera is the most effective preventative method.

Shellfish poisoning

Similarly, shellfish, which are less sensitive to the toxins, will ingest the dinoflagellates or other algae and retain the toxins. The various dinoflagellates produce distinct toxins that have specific targets (Table 17.3), which result in a variety of clinical manifestations (Table 17.4). The exact symptoms will also depend upon the dose in addition to the specific toxin. Shellfish

Table 17.4 Clinical features of shellfish poisoning

Type	Symptoms
DSP	Mild gastrointestinal symptoms (i.e., nausea, vomiting, diarrhea, and abdominal pain)
	Chills, headache, and fever
NSP	Acute gastrointestinal symptoms followed by neurological symptoms
	Tingling and numbness of lips, tongue, and throat
	Muscular aches
	Dizziness
	Reversal of the sensations of hot and cold
PSP	Neurological (i.e., tingling, burning, numbness)
	Drowsiness
	Incoherent speech
	Respiratory paralysis
ASP	Gastrointestinal symptoms
	Confusion, memory loss, and disorientation
	Seizure and coma

poisoning is generally classified as diarrheic (DSP), neurotoxic (NSP), paralytic (PSP), or amnesic (ASP). Diagnosis is based on recent dietary history and the specific symptoms. Developed countries typically monitor shellfish beds and other resources for toxic algae as a means to protect people from the consumption of toxic shellfish. Diarrheic shellfish poisoning is primarily observed as a mild gastrointestinal disorder (i.e., nausea, vomiting, diarrhea, and abdominal pain) accompanied by chills, headache, and fever. Onset of the disease may be as little as 30 minutes and up to 2–3 hours after ingestion of contaminated seafood. Symptoms may last 2–3 days and recovery is usually complete with no long-term effects. The causative organisms are the marine dinoflagellates *Dinophysis* species and *Prorocentrum lima*. Both produce okadaic acid and derivatives, which inhibit protein phosphatase resulting in enhanced sodium excretion from intestinal epithelial cells.

Neurotoxic shellfish poisoning exhibits gastrointestinal and neurological symptoms quite similar to ciguatera. Generally an acute gastrointestinitis (e.g., diarrhea and vomiting) is followed by neurological symptoms (including tingling and numbness of lips, tongue, and throat, muscular aches, dizziness, and reversal of the sensations of hot and cold). Onset occurs within a few minutes to a few hours after ingestion. Duration is from a few hours to several days and recovery is complete with few after effects. No fatalities have been reported. In addition, asthma-like respiratory symptoms resulting from the inhalation of aerosols produced via wave action have been described. The primary causative agent is *Karenia brevis* and the toxins are brevetoxins which exhibit a similar structure and mode of action as the ciguatoxins.

Symptoms of paralytic shellfish poisoning develop within a half to 2 hours after ingestion. They are predominantly neurological (include: tingling, burning, numbness, drowsiness, incoherent speech, and respiratory paralysis). Respiratory paralysis is observed in severe cases and death may occur if respiratory support is not provided. Recovery is usually complete with no lasting side effects if such support is applied within 12 hours of exposure. The dominant species of dinoflagellate associated with PSP are species within the genus *Alexandrium* (Figure 17.4B) which produce a family of toxins known as saxitoxins. The saxitoxins block sodium channels in nerve cells leading to an uncoupling of the communication between nerve and muscle cells, and thus paralysis.

Symptoms of amnesic shellfish poisoning include gastrointestinal disorders (vomiting, diarrhea, abdominal pain) and neurological problems (confusion, memory loss, disorientation, seizure, coma). Gastrointestinal symptoms occur within 24 hours and the neurological symptoms occur within 48 hours of ingestion. It is particularly serious in elderly patients and the symptoms are reminiscent of Alzheimer's disease. All fatalities to date have involved elderly patients. ASP is caused by domoic acid which is produced by the diatom *Pseudo-nitzschia*.

Pfiesteria and possible estuary-associated syndrome

Massive fish kills and reports from fishermen and swimmers complaining of rashes, respiratory problems, and neurological phenomena in North Carolina estuaries and the Chesapeake Bay during the late 1980s and early 1990s led to the identification of a previously undescribed dinoflagellate. This new dinoflagellate is assigned to a new genus *Pfiesteria* and two species, *P. piscicida* and *P. shumwayae*, have been described. Large blooms of *Pfiesteria* were associated with fish kills and human health problems. The Centers for Disease Control (CDC) designated these symptoms as possible **estuary-associated syndrome**. In contrast to shellfish poisoning or ciguatera, the adverse affects on humans and fish are thought to be through contact with the organism or toxins instead of ingestion of the toxins. Originally, *Pfiesteria* was described as having a very complex life cycle involving up to 24 different morphological forms, including several types of ameboid forms. However, more recently it has been suggested that *Pfiesteria* has a more conventional dinoflagellate type of life cycle involving cysts, flagellated forms known as dinospores, and sexual stages (Box 17.2).

The identification of a presumptive toxin associated with the fish killing activity and human toxicity has been elusive. Filtrates from *Pfiesteria* cultures were shown to induce open ulcerative sores, hemorrhaging, and death in finfish and shellfish. Thus far, though, no specific toxin has been isolated from the filtrates. It has also been proposed that *Pfiesteria* can interconvert between nontoxic and toxic forms in the presence of fish as a possible explanation for the inability to isolate the toxin or the lack of correlation between the presence of *Pfiesteria* and the toxic activity. On the other hand, others have shown that the toxicity to fish depends on direct contact between the dinospores and the fish and therefore no excreted toxin is necessary. Furthermore, the dinospores swarm toward the fish and attach directly to them and actively feed on the fish by myzocytosis (Figure 17.5).

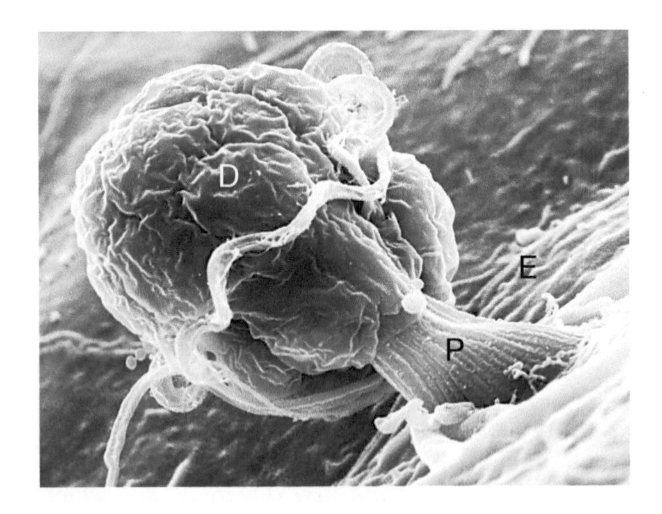

Figure 17.5 *Pfiesteria shumwayae* attached to, and feeding on, fish epidermis. Scanning electron micrograph of a dinospore (D) attached to and feeding on damaged epidermal cell (E). The peduncle (P) is a specialized organelle of some dinoflagellates used in a specialized feeding mechanism called myzocytosis. (Micrograph kindly provided by Vince Lovko, Environmental and Aquatic Animal Health, Virginia Institute of Marine Science.)

Therefore, the killing activity may be due to micropredation, or an indirect result of the micropredation, and not an exotoxin secreted by the *Pfiesteria*. It has been speculated that another dinoflagellate, *Karlodinium veneficum*, may be the organism responsible for the fish kills and possibly the rashes and other symptoms exhibited by the fishermen.

The best evidence for a toxin comes from human exposure to *Pfiesteria* cultures. Several investigators have experienced neurological symptoms after being exposed to aerosols derived from cultures. These have included both dramatic and rapid neurological effects and long-term and chronic effects (Table 17.5). The controversies associated with the toxicity and life cycle of *Pfiesteria* are due in part to working with environmental samples containing a variety of organisms. Under such conditions it is difficult to definitively ascribe specific phenomena to particular organisms. In addition, *Pfiesteria* has not triggered any major fish kills or human illness since 1998.

Table 17.5 Symptoms associated with exposure to *Pfiesteria* aerosols
Narcosis, disorientation
Respiratory distress (asthma-like)
Stomach cramping, nausea, vomiting
Eye irritation, blurred vision
Erratic heartbeat (weeks)
Sudden rages, personality changes
Short-term memory loss

Summary and Key Concepts

- Several normally free-living amebae found in soil and water can, on rare occasions, result in human disease if the organisms gain access to the central nervous system.
- *Naegleria fowleri* produces a fulminant, and almost always fatal, acute meningoencephalitis called primary amebic meningoencephalitis.
- Several *Acanthamoeba* species cause a slow-progressing chronic encephalitis in immunocompromised or otherwise debilitated persons.
- *Balamuthia mandrillaris* causes a subacute-to-chronic granulomatous amebic encephalitis similar to *Acanthamoeba* except that it is more rapidly progressing and is more capable of infecting healthy individuals.
- *Acanthamoeba* also causes an amebic keratitis associated with contaminated contact lens solutions and eye trauma.
- Harmful algal blooms can result in fish kills and impact human health.
- Toxins associated with some dinoflagellates can be passed up the food chain and when ingested by humans can result in ciguatera fish poisoning or shellfish poisoning.
- The symptoms associated with ciguatera and shellfish poisoning can be quite severe and are dependent in part on the specific toxin produced by the dinoflagellate.
- Toxic *Pfiesteria* species are speculated to be involved in human disease through external contact with the organism or a toxin produced by the organism.

Further Reading

Burkholder, J.M. and Glasgow, H.B. Jr (1997) Trophic controls on stage transformations of a toxic ambush-predator dinoflagellate. *J. Euk. Microbiol.* 44: 200–205.

Dart, J.K.G., Saw, V.P.J. and Kilvington, S. (2009) *Acanthamoeba* keratitis: diagnosis and treatment update 2009. *Am. J. Ophthalmol.* 148: 487–499.

Dickey, R.W. and Plakas, S.M. (2009) Ciguatera: a public health perspective. *Toxicon*, in press.

James, K.J., Carey, B., O'Halloran, J., van Pelt, F.N.A.M. and Skrabakova, Z. (2010) Shellfish toxicity: human health implications. *Epidemiol. Infect.* 138: 927–940.

Kaiser, J. (2002) The science of *Pfiesteria*: elusive, subtle and toxic. *Science* 298: 346–349.

Matin, A., Siddiqui, R., Jayasekera, S. and Khan, N.A. (2009) Increasing importance of *Balamuthia mandrillaris*. *Clin. Microbiol. Rev.* 21: 435–448.

Miller, T.R. and Belas, R. (2003) *Pfiesteria piscicida, P. shumwayae,* and other *Pfiesteria*-like dinoflagellates. *Res. Microbiol.* 154: 85–90.

Visvesvara, G.S., Moura, H. and Schuster, F.L. (2007) Pathogenic and opportunistic free-living amoebae: *Acanthamoeba* spp., *Balamuthia mandrillaris, Naegleria fowleri,* and *Sappinia dipoidea. FEMS Immunol. Med. Microbiol.* 50: 1–26.

Glossary

abscess A localized collection of pus in a cavity formed by the disintegration of tissue.

acanthopodium The spike-like pseudopodium observed in *Acanthamoeba* species.

actin A highly conserved protein that forms microfilaments (also known as actin filaments).

acute Any process which has a sudden onset and a relatively brief duration.

adhesive disk The structure on the ventral side of *Giardia* trophozoites composed of cytoskeletal elements and functions in the attachment of the organism to the intestinal epithelium (also known as ventral disk or striated disk).

algal bloom Refers to a sudden expansion in the population of a species planktonic protozoan or algae (also known as red tide). Referred to as harmful algal bloom if the species is toxic.

alveolate A protozoan group consisting of the apicomplexa, ciliates and dinoflagellates.

alveolus Generally refers to the small sac in the vertebrate lung where the exchange of gases takes place between the lungs and blood. Plural = alveoli.

amastigote An intracellular form of a kinetoplastid without a flagellum.

amebapore A protein synthesized by *Enatamoeba histolytica* that is capable of lysing bacterial and host cell membranes. It may be a virulence factor.

amebiasis The disease associate with ameba infections, especially *Entameoba histolytica.*

amebic keratitis A inflammation of the cornea due to *Acanthamoeba* species.

ameboma A tumor like mass in the colon produced by localized inflammation which is occassionally observed during amebiasis.

amylopectin A component of starch consisting of highly branched chains of glucose units.

anemia A lower than normal level of erythrocytes in the blood.

anergy A absence of immune reactions to antigens or allergens.

aneurysm A bulge formed from a localized widening (dilation) of a weakened blood vessel or the heart.

anorexia Lack or loss of appetite for food and the associated wasting syndrome.

anthroponosis A disease that is spread from person to person.

anthropophilic Preferring humans to animals. Generally in reference to blood-feeding insects.

antibody A protein secreted by B-lymphocytes that functions in immunity by binding to antigens resulting in pathogen neutralization or stimulation of other components of the immune system.

antigen Molecule, virus particle, or cell recognized by either antibody or T-cell receptor following processing by antigen presenting cell.

antigenic variation The process by which an infectious organism alters its surface proteins in order to evade a host immune response.

apicomplexa A group of protozoa defined by specialized organelles at their apical end for the invasion of host cells. Members of this group generally exhibit complex life cycles involving asexual and sexual stages and intracellular forms.

apicoplast A plastid like organelle found in apicomplexa.

apoptosis The process of programmed cell death.

arrhythmia Any variation in the normal rhythm of the heart beat.

arthropod Organisms described as having legs with joints and an exoskeleton (e.g., insects).

atrophy A wasting away or decrease in the size or function of a cell, tissue or organ.

autotrophism The ability to build up food substances from simple compounds and external energy such as sunlight (e.g., photosynthesis).

avidity A measure of the strength of the antibody-antigen reaction based on rate at which the complex is formed.

axoneme The central core of the flagellum composed of two central microtubules surrounded by microtuble pairs.

axostyle A complex of microtubules forming a stiff rod rod like structure in certain protozoa (e.g., *Trichomonas*).

basal body Short cylindrical array of microtubules found at the base of cilium and flagellum that serves as the nucleation site for the axoneme. Similar in structure to a centriole.

B-cell A lymphocyte that serves as a precursor for plasma cells that secrete antibodies.

benthic Refers to the flora and fauna at the bottom of oceans.

bradyzoite A slowly growing stage of *Toxoplasma gondii* encapsulate in tissue and responsible for the chronic stage of the disease.

brush border Refers to epithelial surfaces in which the cells have microvilli.

calcification The process by which organic tissue becomes hardened by the deposition of calcium salts.

cardiomegaly Enlargement of the heart.

cardiomyopathy Disease of the heart muscle, often of obscure or unknown etiology.

cell-mediated immunity Immunity involving cellular elements of the immune system, especially T-lymphocytes.

cerebral malaria A clinical manifestation of falciparum malaria characterized by impaired consciousness and other neurological symptoms due to the sequestration of infected erythrocytes in the brain.

Chagas disease A name commonly used to describe the disease caused by *Trypanosoma cruzi* in honor of Carlos Chagas for describing the parasite and the disease.

chagoma A inflammation, especially in the skin, due to *Trypanosoma cruzi* at the site of infection.

chemoprophylactis The use of drugs to prevent a disease or infection.

chitin A polymer of N-acetylglucosamine (i.e., polysaccharide) found in the exoskeletons of arthropods, the cyst wall of some protozoa, and in the cell walls of some fungi.

cholangitis Inflammation of the bile duct.

cholera An acute infectious disease caused by the bacterium *Vibrio cholerae* and characterized by severe diarrhea with extreme fluid and electrolyte depletion.

choroid A layer of pigment cells rich in blood vessels behind the retina of the eye.

chromatin Complex of DNA and associated proteins in eukaryotic cells. The material from which chromosomes are made.

chromatoid bodies Staining structures found in *Entamoeba* cysts and precysts which are considered to be crystalized ribosomes.

chronic Describing a condition or disease that is long lasting.

ciguatera Poisoning caused by ingesting fish contaminated with toxins originally secreted by dinoflagellates and accumulated in the fish (i.e., ciguatoxin) and characterized by gastrointestinal and neurological symptoms.

ciliate A group of protozoa in the alveolates that are characterized by having a vegatative macronucleus and a somatic micronucleus, as well as having rows of cilia during some stage of their life cycle.

cilium A microtubule containing structure found on the surface of cells, usually in rows, and functions in cell motility. *See also* flagellum.

cladogram A branching tree-like diagram used to illustrate phylogenetic relationships between organisms.

colitis Inflammation of the colon.

commensal An organism living within a host organism and deriving benefit from the host but not injuring or benefitting the host.

complement A set of plasma proteins that function to attack and kill extracellular pathogens or promote opsonization of pathogens.

congenital Referring to conditions at birth that are either inherited or acquired during fetal development (e.g., infections).

conjugation A type of sexual reproduction involving exchange of genetic material (i.e., DNA) between two individual organisms that are temporarily paired.

conjunctiva Protective mucus secreting layer lining the cornea and inside layer of the eyelids.

conjunctivitis Inflammation of the conjunctivia.

conoid A hollow cone structure found at the anterior end of invasive stages in some Apicomplexa composed of microtubules.

copro-antigen An antigen found in the feces.

cutaneous Pertaining to the skin.

cyst A stage in the life cycle of some protozoa that represents a dormant or resting stage. During this stage the organism is surrounded by a cyst wall and is more resistant to environmental factors. In parasitic species the cyst stage is often involved in the transmission from host to host.

cytochalasin A fungal metabolite that prevents the formation of actin filaments.

cytokine Proteins made by some cells that affect the behavior of other immune effector cells.

cytoskeleton System of filamentous proteins in eukaryotic cells that give the cell shape and function in cell motility (e.g., actin filaments and microtubules).

cytostome An organelle found in some protozoa that functions in the acquisition of nutrients.

definitive host The host that harbors the adult forms of the parasite. In regards to protozoa sexual reproduction usually takes place in the definitive host.

delayed hypersensitivity A type of cell mediated immunity elicited by injection of an antigen in skin which appears hours or days after the injection.

diarrhea Stools that are loose or liquid that are passed more frequently than normal due to a rapid transit of the intestinal contents.

diffuse cutaneous leishmaniasis A form of cutaneous leishmaniasis characterized by nodules or other skin lesion spread throughout the body and not just localized to the site of infection.

dinoflagellate A group of protozoa found mainly in the oceans that are characterized by the presence of transverse and longitudinal flagella which propel the organisms in a rotating manner through the water.

disaccharidase A enzyme that cleaves disaccharides into two monosaccharides.

duodenum The first or proximal portion of small intestine immediately following the stomach and followed by the jejunum.

dynein A protein associated with microtubules that is responsible for the movement of microtubules.

dysentery A disorder marked by inflammation of the intestines, especially the colon, characterized by abdomenal pain and frequent stools containing blood and mucus.

dyspareunia Difficult or painful sexual intercourse.

dyspnea Difficult or labored breathing.

dysuria Painful or difficult urination.

ectoparasite A parasite that lives on the outside of its host.

edema The presence of excessive amounts of fluids in tissues, especially subcutaneous tissues.

electrocardiogram A graphic tracing of the electric current produced by the heart muscle.

electrolyte A substance that dissociates into ions when put into solution.

ELISA Enzyme-linked immunosorbant assay. An assay used to evalute the amont of antibody (or antigen) in a sample.

embolism The suddden blocking of an artery by a clot or foreign material which has been carried by blood flow.

encystment The process of changing into the cyst stage.

endemic A disease that is presence in a community at relatively constant levels over long periods of time.

endocytosis The internalization of material into a cell by invagination of the plasma membrane and incorportation into a membrane bound vesicle (e.g., phagocytosis).

endodyogeny A specialized type of cell reproduction exhibited by some apicomplexa in which the daughter cells begin to form within the mother cell.

endoplasm The central portion of cytoplasm of a cell, especially in reference to amebic trophozoites.

endopolygeny An asexual replication process in which the progeny are formed by budding within the parent cell preceded by multiple rounds of nuclear division without cytoplasmic division.

endothelium The layer of cells that lines the cavities of the heart, blood and lymphatic vessels, and serous cavities of the body.

enterocolitis Inflammation of both the small intestine and colon.

enterocyte An intestinal epithelial cell.

epidemic A disease characterized by a sudden change in the incidence in a community over a short period of time (e.g., outbreak).

epidemiology The study of the various factors which determine the frequency and distribution of disease. Can also include the analysis of disease etiology and determination of the cause of disease or outbreaks.

epididymis A long coiled tube in the testes of some vertebrates.

epimastigote A morphological form of a kinetoplastid in which the kinetoplast is located near the nucleas and the flagellum emerges from the middle of the parasite and forms and undulating membrane along half the length of the protozoan.

epitope The specific portion of the antigen recognized by the antibody or T-cell receptor (also known as antigenic determinant).

erythema Redness of the skin produced by congestion of the capillaries.

erythropoiesis The production of erythrocytes.

estuary-associated syndrome Skin lesions and other health problems associated with contact with water and possibly due to dinoflagellates or other microorganisms in the water.

etiology The study or theory the factors that cause disease and their introduction to the human host.

extracellular matrix Complex network of polysaccharides and proteins that functions as structural elements in tissues.

facultative pathogen A pathogen that in some situations does not cause a serious disease and exists as a commensal.

fecal–oral Refers to a type of transmission in which the infectious form of a pathogen (e.g., cyst) is found in the feces, and the infection is spread through the ingestion of material contaminated with fecal matter.

feeder organelle An electron dense zone found between the host epithelial cell and the attached *Cryptosporidium* parasite that is believed to function in the uptake of nutrients from the host cell.

fibrosis Formation of fribrous tissue.

fistula An abnormal passage or communication between two internal organs, or leading from an internal organ to the surface of the body.

flagellum A whip-like organelle composed of microtubules that functions in cell motility.

funduscopy The examination of the fundus (base) of the eye with a funduscope.

Gal/GalNAc lectin A protein expressed on the surface of *Entamoeba* trophozoites that probably functions in adherence and may be a virulence factor.

gametocytogenesis The process of forming gametocytes especially in reference to *Plasmodium* species.

gametogenesis The process of forming gametes in apicomplexa (also known as gametogony).

gametogony The process of forming gametes in apicomplexa (also known as gametogenesis).

ganglion A group of nerve cell bodies located outside the central nervous system. Plural = ganglia.

gastroenteritis Inflammation of the stomach and intestines.

genome The complete DNA sequence (i.e., total genetic information) of an organism.

giardin A major protein component of the adhesive disk of *Giardia.*

glycocalyx A polysaccharide (i.e., composed of carbohydrates) covering or surronding a cell (i.e., cell coat).

glycoconjugate Refers to compounds, especially lipids or proteins, that have sugar residues attached (i.e., glycolipid, glycoprotein, etc.).

glycogen Polysaccharide composed of glucose and functions in energy storage.

glycoprotein A protein with covalently attached carbohydrate groups.

glycosome A membrane bound organelle found in the kinetoplastids containing enzymes of the glycolytic pathway. Related to peroxisomes and other microbody-like organelles found in other organisms.

glycosylphosphatidylinositol A glycolipid conjugated to the C-terminus of some membrane proteins and functions as a membrane anchor.

goblet cell Mucus secreting cell in mucous membranes.

granuloma A tumor like mass due to chronic inflammation.

gregarine A group of apicomplexan parasites characterized by specialized attachment and feeding organelles called either a mucron or an epimerite.

HAART Highly active antiretroviral therapy. The use of a combination of antiretroviral drugs to treat HIV infected persons.

hematogenous Produced or derived from the blood or disseminated by the circulatory system.

hematophagous The state of ingesting erythrocytes by phagocytosis.

hematopoiesis The formation and development of red blood cells.

hemocoel The body cavity in arthropods serving for the circulation of blood.

hemozoin Crystalized heme formed by the malarial parasite during the degradation of hemoglobin (also known as malarial pigment).

hepatomegaly Enlargement of the liver.

hepatosplenomegaly Enlargement of the liver and spleen.

heterotrophism Mode of nutrition involving the uptake of organic substances from the environment.

heteroxenous Requiring more than one host to complete the life cycle.

holophyletic A taxon in which all descendants of a singular common ancestor are united into one lineage (i.e., clade). Similar to monophyletic except that not necessarily all descendants need to be included in a monophyletic group.

humoral immunity Immunity based upon antibodies.

hydrocephaly An abnormal amount of fluid in the cranium.

hydrogenosome An organelle in some protozoa containing a metabolic pathway that uses hydrogen as the final electron acceptor instead of oxygen.

hypertrophy Increased size of an organ due to enlargement of the individual cells.

hypnozoite A dormant stage found in the liver of *Plasmodium vivax* and *P. ovale.*

hypoxia A state characterized by low oxygen levels in the tissues.

ileum The last section of the small intestine between the jejunum and colon.

immunoblot A method to evaluate antibodies or antigens used in conjuction with gel electrophoresis.

immunocompromised Having reduced immune responses (i.e., immune deficiency).

immunofluorescence A technique to localize antigens in tissues or cells or to characterized antibodies.

immunogen A substance that elicits an immune response.

incidence (rate) An expression of the rate in which a certain disease occurs. For example, the number of new cases occuring over a period of time.

incubation period The time elapsing between acquiring an infection and the apperance of symptoms.

induration An abnormal hardness of tissue.

inflammation A local immune response resulting from injury or infection characterized by swelling, redness, pain and warmth due to the accumulation of fluid and white blood cells.

intermediate host The host in which larval development occurs. In regards to protozoa this is the host in which only asexual reproduction occurs.

interstitium The area between tissues.

intubation The insertion of a tube through the mouth and larynx.

ischemia Deficiency of blood due to constriction or obstruction of a blood vessel.

isoenzyme A different variant of the same enzyme (i.e., same activity) including enzyme variants that are the products of different genes and enzymes that are the products of different alleles of the same gene (allozymes).

jaundice A condition characterized by the deposition of bile pigment in the skin and mucous membranes giving the patient a yellow color.

jejunum The middle portion of the small intestine between the duodenum and ileum.

karyosome A spherical mass of staining chromatin material within the nucleus corresponding to the nucleolus.

keratitis Inflammation of the cornea.

kinetoplast A Giemsa-staining structure—distinct from the nucleus—found near the base of the flagellum which defines the kinetoplastids as a group. The staining of the kinetoplast is due to mitochondrial DNA found within the single mitochondrion.

lactose The major disaccharide found in milk.

lamina propia A thin layer of loose tissue which lies beneath the epithelium and together with the epithelium constitutes the mucosa. Thus the term mucosa or mucous membrane always refers to the combination of the epithelial layer plus the lamina propria.

latency A state of apparent inactivity.

lectin A protein that has multiple carbohydrate binding sites.

leishmanin skin test A delayed-type hypersensitivity test in which *Leishmania* antigens are injected intradermally used for the diagnosis of leishmaniasis. Called the Montenegro test in Latin American countries.

leucopenia A low white cell count.

leukocyte A white blood cell including polymorphonuclear leukocytes, lymphocytes and monocytes.

lobopodium A broad blunt pseudopodium.

lumen The space enclosed by a duct, vessel, or tubular organ.

lymphadenopathy Disease of the lymph nodes generally manifested as swollen lymph nodes.

lymphocyte A class of white blood cells responsible for the adaptive immune response (e.g., B-cells and T-cells).

lymphocytosis An excess of lymphocytes.

lysosome A membrane bound organelle containing hydrolytic enzymes and an acidic pH.

macrogamete The equivalent of the egg in apicomplexa.

macrogametocyte The cell that produces the macrogametes.

macrogamont Another name for the macrogametocyte.

macronucleus The larger of the two nuclei in ciliates that is required for vegatative reproduction but not for sexual reproduction.

macrophage Large mononuclear phagocytic cells found in most tissues. They function in innate immunity to phagocytose pathogens and in adaptive immunity to present antigens.

malabsorption Impaired intestinal absorption of nutrients.

maxi-circle DNA A component of kinetoplast DNA which is equivalent to the mitochondrial DNA.

median body A staining structure found in *Giardia* that probably directs the formation of the adhesive disk.

megaviscerae A syndrome associated with Chagas' disease in which the hollow viscerae (e.g., esophogus and colon) are enlarged.

meiosis A type of cell division in which the number of chromosomes is halved resulting in a haploid state. Generally associated with sexual reproduction.

meninges The three layers of protective tissue that envelop the brain and spinal cord.

meningoencephalitis Inflammation of the meninges and brain.

merogony A process in apicomplexan parasites resulting in the production of merozoites.

merozoite An invasive stage found in most apicomplexan life cycles.

mesentery Membraneous sheets surrounding various organs and used especially in reference to the small intestine.

metacyclic Refers to the stage in kinetoplastid parasites that is transferred from the arthropod vector to the vertebrate host.

metastasis Process by which disease bearing cells are transferred from one part of the body to another.

MHC Major histocompatibility complex. A group of genes that encode polymorphic glycoproteins that present antigen peptides to T-cells.

microcephaly Abnormal smallness of the head usually associated with mental retardation.

microfilament A cytoskeletal element composed of actin and associated proteins.

microgamete The equivalent of the sperm in apicomplexa.

microgametocyte The cell that produces the microgametes.

microgamont Another name for the microgametocyte.

microneme A membrane bound organelle found at the apical end of invasive stages of the apicomplexa. Contains adhesins that are important in cell invasion and motility.

micronucleus The smaller of the two nuclei in ciliates that is required for sexual reproduction and contains all of the genetic information.

microribbon A cytoskeletal element found in the adhesive disk of *Giardia* trophozoites.

microtubule A cytoskeletal element composed of tubulin and associated proteins.

microvilli Minute projections on the lumenal side of epithelial cells.

mini-circle DNA A component of kinetoplast DNA consisting of a heterogenous mixture of small circular DNA molcules.

mitochondrion A double-membrane bound organelle important in the synthesis of ATP and the production of energy.

mitosis The normal cell division process in which the nucleus splits into two identical daughter nuclei that are passed to the daughter cells.

mitosome A remnant of the mitochondrion found in some eukaryotic microbes that has a limited metabolic function.

mixotrophic Exhibiting both autotrophic and heterotrophic nutritional characteristics.

monocytosis Increased levels of monocytes in the blood.

monophyletic Refers to a group of organisms all derived from a common ancestor.

monoxenous Requiring only a single host species to complete the life cycle.

morbidity The condition of being diseased or sick. The incidence of sickness in a population.

mortality The quality of being mortal. The incidence of death in a population.

mucins The major constituent of mucus composed of polysaccharrides or glycoproteins.

mucocutaneous Pertaining to mucous membranes (i.e., mucosa) and skin.

mucosa The mucus secreting epithelia (i.e., mucous membranes) that line the digestive, respiratory, and urogenital tracts. Other mucosa include the conjuctiva of the eye and the mammary glands.

mucron An attachment organelle of aseptate gregarines similar to an epimerite of septate gregarines.The epimerite is set off from the rest of the gregarine body by a septum.

mucus The viscous liquid secreted by mucous membranes composed of mucins, inorganic salts, shed epithelial cells and leukocytes.

murein The peptidoglyan that forms the meshwork making up the cell wall of bacteria.

mutualism A symbiotic relations in which both organisms benefit.

myocarditis Inflammation of the heart muscle.

myocyte A muscle cell.

myosin A motor protein that uses ATP to generate movement or force along actin filaments.

myositis Inflammation of a voluntary muscle.

myzocytosis A type of feeding in which the predator penetrates the cell surface of the prey and takes up the cytoplasmic contents via a specialized structure.

nagana An often fatal disease of African ungulates and livestock caused by various species of trypanosomes that are transmitted by the tsetse fly.

necrosis Death of a tissue, ususally in a small localized area involving individual cells or small groups of cells.

neurotropic Having a selective affinity for the nervous system or exerting its principal effect on the nervous system.

neutrophil Phagocytic white blood cells that enter tissues in large numbers in response to infection or injury.

nucleus Prominent membrane bound organelle in eukaryotic cells containing DNA organized into chromosomes.

oocyst The stage in apicomplexa following sexual reproduction in which sporozoites are produced.

ookinete A motile stage in *Plasmodium* formed after fertilization of the macrogamete by the microgamete that will lead to the formation of the oocyst.

oomycetes A group of organisms that are related to brown algae and diatoms. Although not related to fungi, they physically resemble fungi and are also known as water molds.

opsonization The coating of a pathogen with a molecule (e.g., antibody or complement) that makes it more readily ingested by phagocytes.

organelle Discrete membrane bound compartment within a eukaryotic cell that carries out a specialized function.

osmotrophy The absorbtion of nutrients by microorganisms from the medium or environment.

parabasal body A structure near the nucleus in some flagellated protozoa that is probably related to the Golgi body.

parabasalid A group of protozoa characterized by having a parabasal body.

paraphyletic A monophyletic group that contains its most recent common ancestor but does not contain all the descendants of that ancestor.

parasitism A relationship between two organims in which one derivies benefit (i.e., parasite) from the other (i.e., host), usually to obtain nutrients or protection.

parasitophorous vacuole The membrane bound vacuole in which an intracellular parasite resolves.

paraxial rod A cytoskeletal element associated with the axoneme in some flagellated protozoa with an undulating membrane.

paresthesia A abnormal sensation such as burning or prickling.

paroxysm A sudden recurrence or intesification of symptons. Commonly used to describe the fever attacks associated with malaria.

pathogenesis The orgin and course of a disease.

pathology Detectable damage to a tissue or the study of disease.

PCR Polymerase chain reaction. A technique to amplify specific DNA fragments.

pellicle A multi-layered membrane found in some protozoa.

peridomiciliary Found around or associated with the house.

peritoneum The serous membrane lining of the abdominal cavity and internal organs (i.e., viscera).

peritonitis Inflammation of the peritoneum or peritoneal cavity.

peritrophic membrane The covering that forms around the blood meal of hematophagous insects.

peroxisome A small membrane bound organelle that functions in the oxidation of organic molecules and often involves the metabolism of hydrogen peroxide.

Peyer's patch A gut associated lymphoid tissue (i.e., GALT) found in the wall of the small intestine.

phagocytosis Engulfment of external solid material or pathogens by a cell.

phagolysosome The membrane bound compartment formed by the fusion of a lysosome with the vacuole containing phagocytosed material in which the material is broken down by hydrolytic lysosomal enzymes.

phenotype The observable character of a cell or organism.

phylogenetic Of or relating to the evolutionary development or history between organisms.

plankton The collection of micro-organisms (e.g., algae and protozoans) that float or drift in fresh or salt water, especially at or near the surface. Can be classified as phytoplankton (plant-like or autotrophic) or zooplankton (animal-like or heterotrophic).

planozygote A stage in the life cycle of dinoflagellates formed after the fusion of gametes.

plastid A double-membrane bound organelle of plants or algae that is often pigmented and in cases functions in photosynthesis (e.g., chloroplast).

pneumonitis Inflammation of the lungs.

polymorphonuclear cells White blood cells with irregulary shaped and multi-lobed nuclei and cytoplasmic granules. Includes neutorphils, basophils and eosinophils.

polyphyletic Unnatural taxa in which members of a group are derived from different ancestors.

predator–prey Refers to a life cycle in which one stage of the parasite is transmitted to the predator by eating the infected prey and another stage of the life cycle is transmitted to the prey by ingesting material contaminated with the feces of the predator.

premunition A state of immunity which depends on the presence of the infectious agent.

prepatent period The period between acquiring infection and detection of parasites.

prevalence The total number of cases of a disease at a certain time in a designated area.

procyclic Refers to a stage in kinetoplastids that replicates within the vector.

prodromal A symptom indicating the onset of illness.

promastigote A morphological form of a kinetoplastid in which the kinetoplast is between the nucleus and anterior end and the flagellum emerges from the anterior end without an undulating membrane.

protease A hydrolytic enzyme that breaks down proteins into either peptides (i.e., smaller proteins) or amino acids.

proteoglycan A class of glycoproteins of high molecular weight that are found especially in the extracellular matrix of connective tissue.

protista A phylogenetic group incompassing all or part of the unicellalar organims while excluding multicellular organisms.

pruritus Various conditions characterized by itching.

pseudopodia A portion of ameba or cells that bulges out from the cell body and is involved in locomotion or to phagocytose particles.

pulmonary edema An abnormal accumulation of fluid in the lung tissues and air spaces characterized clinically by difficulties in breathing.

purulent Consisting of or containing pus.

pus A yellowish-white fluid formed in infected tissue containing of white blood cells, cellular debris, dead microorganisms and necrotic tissue.

PVM Parasitophorous vacuolar membrane, the membrane surround an intracellular parasite making up the parasitophorous vacuole.

pyrethroid Any of several synthetic compounds similar to pyrethrin and used as an insecticide.

pyretic Pertaining to fever.

ray body The gamete stage of *Babesia.*

reactive oxygen intermediate Molecules containing a reactive oxygen group, such as superoxide, hydrogen peroxide, or hydroxyl radical, that can damage cellular macromolecules (e.g., protein, DNA and RNA) and participate in pathogen killing by macrophages (i.e., respiratory burst).

recivida A recurring form of leishmaniasis characterized by a hypersensitivity skin rash.

recrudescence The recurrence of disease symptoms after a tempory abatement, generally on the order of days or weeks. *See also* relapse.

red tide Often used to describe algal blooms in marine coastal areas, particularly associated with dinoflagellate species.

relapse The return of a disease after its apparent cessation. *See also* recrudescence.

reservoir The animal host in which zoonotic infections are maintained.

reticulocyte An immature red blood cell circulating in the blood stream.

reticuloendothelial system Part of the immune system composed of highly phagocytic cells and especially of the monocyte–macrophage lineage.

retinochoroiditis In inflammation of the retina and underlying choroid.

RFLP Restriction fragment length polymorphism. A genetic fingerprinting technique used primarily as a marker to distinguish genotypes and isolates within a species.

rhoptry An organelle of apicomplexa that functions in the invasion of the host cell.

ring form A stage in the *Plasmodium* life cycle describing the intraerythrocytic form shortly after merozoite invasion.

Romaña sign Refers to a swelling of the eye and conjunctiva in response to *Trypanosoma cruzi* infection.

ruminant Any of various hoofed mammals such as cattle, sheep, goats, deer, etc.

sand fly The common name for the vector of *Leishmania.*

sarcocyst The tissue cyst stage of the tissue cyst forming coccidia, especially *Sarcocystis.*

schizont The multinucleated life cycle stage of *Plasmodium* found in the liver or red blood cells.

segmenter The *Plasmodium* life cycle stage in which individual merozoites are visable.

septum A dividing wall or section found in biological systems.

sequela Any lesion or outcome following or caused by a disease. Plural = sequelae.

serosa The outer lining of organs and body cavities of the abdomen and chest, including the stomach.

shellfish poisoning Poisoning caused by ingesting shellfish (e.g., oysters, clams, etc.) contaminated with toxins originating from planktonic organisms (e.g., dinoflagellates) and characterized by gastrointestinal and neurological symptoms.

soil transmission Refers to a type of transmission in which an immature form of the parasite is excreted by the host which must mature in the environment before becoming infectious.

somnolence Sleepiness or unnatural drowsiness.

splenectomize To remove the spleen.

splenomegaly Enlargement of the spleen.

sporoblast An asexual reproductive stage in some apicomplexa.

sporogony An asexual reproductive process in apicomplexan life cycles following the sexual stage and resulting in the production of sporozoites.

sporozoa A term used previously to describe the protozoan group now known as apicomplexa.

sporozoite An invasive stage of apicomplexa produced by sporogony.

steatorrhea Excessive amounts of fats in the feces, especially in malabsorbtion syndromes.

stem cell A cell that has the potential to divide and regenerate tissue.

stercorarian Refers to posterior station, or hindgut, development of infective flagellates in the feces of the vector and tranmission by comtamination.

sterilizing immunity An immunity which completely protects against subsequent infection or disease.

stramenopile A group of protozoa, or alga, characterized by unequal flagella.

stroma The supporting tissue or matrix of an organ.

submucosa The layer of tissue beneath the mucous membrane.

sylvan Pertaining to or located or living in the forest.

symbiosis Two distinct organisms living together or in close association. The relationship can be mutualistic, commensal or parasitic.

tachyzoite A stage in the *Toxoplasma* life cycle characterized by a rapid rate of proliferation and causing an acute infection.

taxonomy The classification of living organisms based on similarities and differences.

T-cell A lymphocyte that responds to antigens presented by the MHC and either directly kills abnormal or infected cells or helps other immune effector cells through the action of cytokines.

telomere The ends of chromosomes characterized by a particular sequence and synthesized in a non-template mediated fashion by telomerase.

thrombocytopenia A decrease in the number of platelets in the blood.

thromboembolism Obstruction of a blood vessel with thrombic material (i.e., clot) that originated at another site and was carried by the blood.

tight junctions A cell-cell juction between adjacent epithelial cells.

transcription Synthesis of RNA from a DNA template.

translation The synthesis of proteins from a messenger RNA transcript.

transovarial Refers to an infectious agent that is passed to the progeny via the eggs or reproctive system (also known as vertical transmission).

triatomine A group of blood sucking insects that transmits *Trypanosoma cruzi.*

trophozoite The motile, feeding, and/or replicating stage of many protozoa.

tropism The specifity of a pathogen for a particular tissue.

trypanosome lytic factor A component of serum for higher primates that kills many of the African trypanosomes found in animals and thereby preventing infection.

trypanothione A glutathione containing molecule found in kinetoplastids that functions in redox metabolism and the detoxification of reactive oxygen intermediates.

trypomastigote A morphological form of a kinetoplastid in which the kinetoplast is at the posterior end of the protozoan. The flagellum also emerges from the posterior end and folds back along the body of the protozoan to form an undulating membrane that runs the length of the cell.

tsetse A fly of the genus *Glossina* that transmits African trypanosomes.

tubulin The primary protein component of microtubules.

undulating membrane Refers to a flagellum attached to and running along the body of a protozoan to form a fin-like structure.

ungulate A large group of mammals all of which have hooves.

urethra The tube used to discharge urine from the bladder.

urethritis Inflammation of the urethra.

vacuole A membrane bound space of cavity within the cytoplasm of an eukaryotic cell.

vaginitis Inflammation of the vagina.

variant surface glycoprotein A protein found on the surface of African trypanosomes that forms a dense protein coat and exhibits antigenic variation.

vector An arthropod or other invertebrate the serves to transmit infectious agents to a vertebrate host.

vertical transmission Transmission of a parasite from one generation to the next through the egg or in utero.

villus A small vascular protrusion, especially in reference to the intestinal epithelium.

virulence The degree of pathogenicity or disease caused by a micro-organism.

visceral leishmaniasis A manifestation of *Leishmania* infection involving the reticuloendothelial system (i.e., liver, spleen and bone marrow).

vomica The sudden expulsion of pus and putrescent matter from the lungs.

xenodiagnosis Diagnosis by allowing the natural vector to feed on the patient and examining the vector for the pathogen. Used almost exclusively for the diagnosis of *Trypanosoma cruzi* (i.e., Chagas' disease).

zoites A generic term used to describe specialized invasive forms of the Apicomplexa.

zoonosis A disease common to both animals and humans.

zoophilic Preferring animals to humans. Generally in reference to blood-feeding insects.

zoospore The trophic stage of dinoflagelletes (also known as dinospores).

zymodeme A pattern of isoenymes as determined by gel electrophoresis which can serve as a marker for a particular population of organisms.

Index

Entries followed by 'f' refer to figures and those followed by 't' and 'b' refer to tables and boxed material respectively.

B